4TH KYOTO SUMMER INSTITUTE

GRAND UNIFIED THEORIES AND RELATED TOPICS

1981

World Scientific

4TH KYOTO SUMMER INSTITUTE

GRAND UNIFIED THEORIES AND RELATED TOPICS

1981

World Scientific

Grand Unified Theories and Related Topics

Proceedings of the 4th Kyoto Summer Institute
Kyoto, Japan, 29 June — 3 July, 1981

Editors:
M. Konuma
T. Maskawa

World Scientific

World Scientific Publishing Co Pte Ltd
P.O. Box 128
Farrer Road
Singapore 9128

ISBN 9971–950–01–4 9971–950–27–8 pbk
Printed by Singapore National Printers (Pte) Ltd

The Fourth Kyoto Summer Institute

Grand Unified Theories and Related Topics

Organized by

Research Institute for Fundamental Physics, Kyoto University

Cooperated by

Institute for Nuclear Study, University of Tokyo,

Research Institute for Mathematical Sciences, Kyoto University,

Institute for Cosmic Ray Research, University of Tokyo,

National Laboratory for High Energy Physics (KEK).

Supported by

The Yukawa Foundation

Yamada Science Foundation

Organizing Committee

Z. Maki (RIFP, Chairman)
K. Fujikawa (INS)
M. Ida (Kobe Univ.)
T. Inami (Univ. of Tokyo-Komaba)
K. Inoue (Kyushu Univ.)
Y. Iwasaki (Univ. of Tsukuba)
M. Konuma (RIFP)
T. Kugo (Kyoto Univ.)
T. Maskawa (RIFP)
T. Suzuki (Kanazawa Univ.)
M. Yoshimura (KEK)

Scientific Advisory Committee

S. Machida (Kyoto Univ.)
Y. Ohnuki (Nagoya Univ.)
S. Otsuki (Kyushu Univ.)
H. Sato (RIFP)
H. Sugawara (KEK)
G. Takeda (Tohoku Univ.)
S. Tanaka (Kyoto Univ.)
H. Terazawa (INS)
Y. Yamaguchi (Univ. of Tokyo)

Executive Committee

M. Konuma (RIFP, Chairman)
M. Bando (Kyoto Univ.)
R. Fukuda (RIFP)
T. Kugo (Kyoto Univ.)
T. Maskawa (RIFP)
T. Muta (RIFP)

This volume is dedicated to the memory of Professor Hideki Yukawa.

FOREWORD

The Fourth Kyoto Summer Institute, called the "KSI '81", devoted to "Grand Unified Theories and Related Topics" was held both at Research Institute for Fundamental Physics and at Research Institute for Mathematical Sciences, Kyoto University, from June 29 to July 3, 1981. This is the fourth of the annual KSI organized by Research Institute for Fundamental Physics.

The number of participants was 141, of which 18 were from abroad.

The KSI provided a series of lectures and seminars delivered by invited lecturers. In addition we organized a discussion session and accepted contributed talks of participants.

These Proceedings include all the lecture notes which have been completed by the lecturers after the KSI. Talks given in the discussion session are summarized by the speakers themselves. We are very grateful to all the authors for their efforts preparing their manuscripts.

The KSI '81 was partially supported by the Yukawa Foundation, the Yamada Science Foundation and other sponsors. To all of these we would like to express our thanks for their financial support. Thanks are also due to Dr. M. Toya, Mrs. C. Hikami, Mrs. K. Honda and Mrs. S. Matsumoto for their secretarial assistance.

Michiji Konuma

Toshihide Maskawa

1981年京都サマーインステイチュート"大統一理論とその周辺"
6月29日～7月3日　京都大学基礎物理学研究所

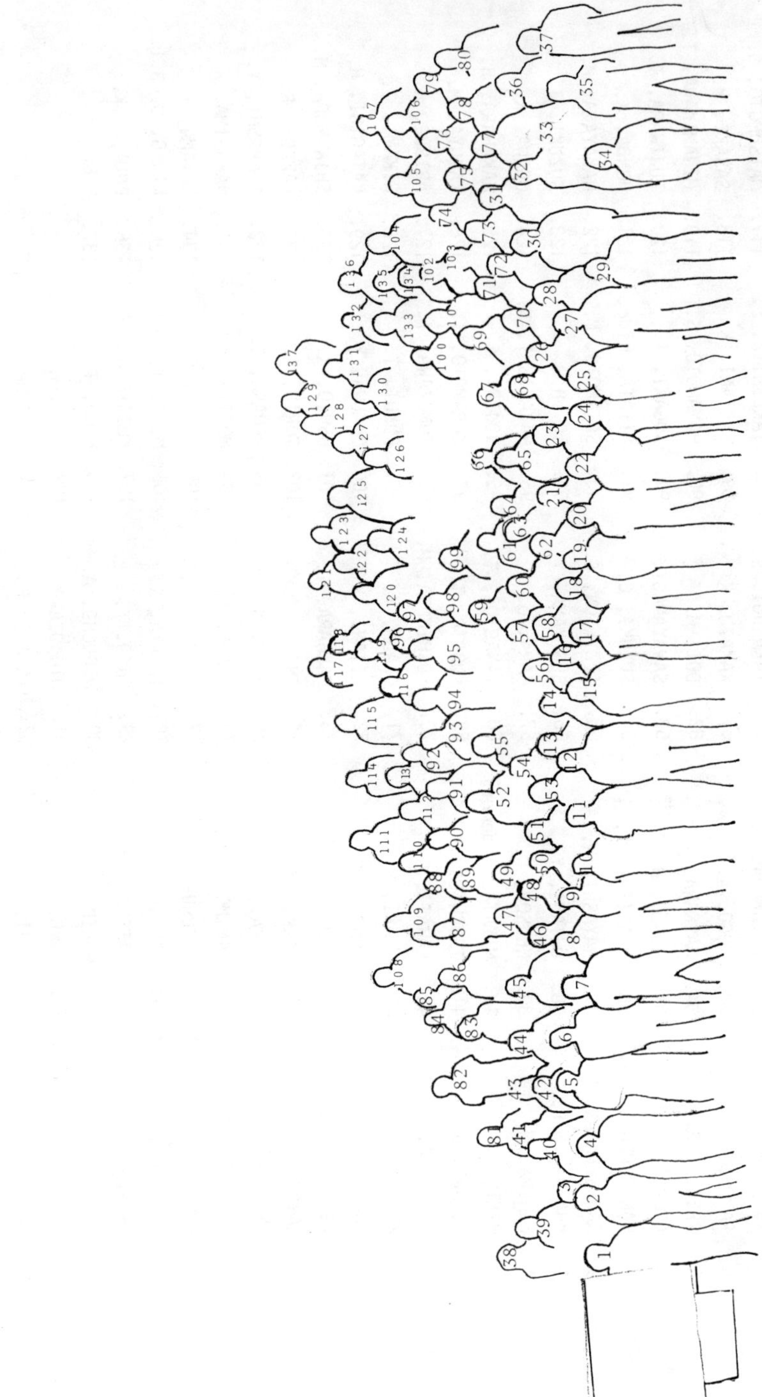

1. KONASHI, H.
2. MIDORIKAWA, S.
3. YAMANAKA, Y.
4. MATSUDA, S.
5. RAMOND, P.
6. PECCEI, R.
7. GEORGI, H.
8. NANOPOULOS, D.V.
9. RAJASEKARAN, G.
10. KHANNA, M.P.
11. KONUMA, M.
12. MAKI, Z.
13. AIZAWA, N.
14. NINOMIYA, M.
15. ZEE, A.
16. IDA, M.
17. IGI, K.
18. HAYASHI, M.
19. SATO. K.
20. SAKAGAMI, M.
21. AOKI, K.
22. OKA, T.
23. HATA, H.
24. KOIDE, Y.
25. HIKASA, K.
26. ICHINOSE, S.
27. KAI, N.
28. KATO, M.
29. OKABAYASHI, M.
30. KASHIBAYASHI, S.
31. IKEDA, M.
32. WATAMURA, S.
33. KOGO, K.
34. MASKAWA, T.
35. KONDO, H.
36. URANISHI. Y.
37. HAYASHI, H.
38. OHTANI, Y.
39. NELSON, C.A.
40. MIYATA, Hidenori
41. TSUCHIDA, T.
42. KAWATI, S.
43. OZAKI, K.
44. SONG. H.S.
45. KIM, J.E.
46. HIOKI, Z.
47. FUKAI, T.
48. TANIMOTO, M.
49. KIZUKURI, Y.
50. YAMAWAKI, K.
51. MATSUDA, M.
52. HORIBE, M.
53. YANG, G.C.
54. FUJII, K.
55. YAMAMOTO, N.
56. HIRAI, S.
57. ISHIKAWA, K.
58. PAKVAKA, S.
59. NISHIURA, H.
60. TAKASUGI, E.
61. KAKAZU, K.
62. DOI, M.
63. SAWAYANAGI, H.
64. SUZUKI, C.
65. ITZYKSON, C.
66. SUGIYAMA, Y.
67. HARA, O.
68. YANAGIDA, T.
69. KUBOTA, T.
70. YOSHIMURA, M.
71. OKADA, Y.
72. KITANI, K
73. MAEHARA, T.
74. INOUE, T.
75. OHTA, N.
76. BANDO, M.
77. OHNUKI, Y.
78. TANAKA, S.
79. SATO, Y.
80. HIGUCHI, A.
81. OTOKOZAWA, J.
82. TAKAGI, F.
83. EBATA, T.
84. KOMATSU, H.
85. TOYA, M.
86. MATUMOTO, K.
87. MURAYAMA, A.
88. TAKESHITA, S.
89. IKETANI, K.
90. HAMAMOTO, S.
91. SOGAMI, I.
92. MIYATA, Hideo
93. SAKAMOTO, M.
94. SO. H.
95. NAKANISHI, N.
96. NOJIRI, S.
97. KURAMOTO, T.
98. TOMOZAWA, Y.
99. FUKUI, T.
100. OHKUWA, Y.
101. USHIO, K.
102. TORIU, T.
103. UEHARA, S.
104. MORITA, K.
105. KOH, I:G.
106. MUTA, T.
107. KOBAYASHI, A.
108. KUGO, T.
109. OHYA, K.
111. KOIKAWA, T.
112. KAWAI, E.
113. TAKEUCHI, Y.
114. INOUE, K.
115. FUKUDA, R.
116. ITO, H.
117. MATSUMOTO, S.
118. SASAKI, M.
119. MINAMIKAWA, T.
120. FUJIWARA, T.
121. HONDA, K.
122. NAKAYAMA, R.
123. SUZUKI, T.
124. OGAWA, K.
125. NAKAHARA, H.
126. KATUYA, M.
127. ARISUE, H.
128. HIKAMI, C.
129. KATAGIRI, H.
130. SHIN-MURA, M.
131. KAWABE, R.
132. IGARASHI, Y.
133. NAKAJIMA, H.
134. KAYAMA, Y.
135. KIANG, D. B.I.
136. PHUA, K.K.
137. KUBO, N.

CONTENTS

CONTENTS

OPENING ADDRESS

Michiji Konuma
Research Institute for Fundamental Physics
Kyoto University, Kyoto 606, Japan

Dear Friends and Colleagues,

I am very glad to open the 4th Kyoto Summer Institute on Grand Unified Theories and Related Topics.

As the secretary of the Organizing Committee, I welcome domestic and foreign participants from various areas in the world.

The series of Kyoto Summer Institutes, called the "KSI", has been organized by Research Institute for Fundamental Physics, Kyoto University. It has been held annually. The KSI started in 1978 just after the 19th International Conference on High Energy Physics, Tokyo. The topic was particle physics and accelerator projects. The second and the third KSI's were respectively on the physics of low dimensional systems and on the fundamental physics of amorphous semiconductors. The fourth KSI is again in particle physics. It is also in connection with cosmology and astrophysics.

Now, the unified theory of the interactions of fundamental particles is a very important subject in particle physics.

I wish to call your attention to Yukawa's contribution in the direction to unify strong and weak interactions. He is the director emeritus of our institute, RIFP. We regret that he can not be present at this KSI due to his health condition. He arrived at an idea in his paper on meson theory in 1935 that the β decay is mediated by a bose particle and wanted to unify the strong nuclear interaction with the weak β decay interaction. This was the first attempt to unify these two interactions.

The object of the KSI is to provide a series of invited lectures and several seminars on recent active fields in fundamental physics. Such summer schools or institutes are held frequently in American and European countries. These meetings have played an important role in developing later research activities. East Asia, including Japan, is located far from American and European countries. So the chance for direct contact by young East Asian and Japanese physicists and opportunities to present and to exchange their achievements and ideas are rather limited. In this sense it is very important to hear of recent developments and of their present status directly from active researchers at the KSI. Young people are able to learn of important recent results and of future prospects from these physicists.

Now, nearly half of participants are young graduate students and research fellows. No participants are expected to be passive members of the audience. Instead, they are encouraged to ask questions and to make comments to the lectures. This is easier in such a small scale meeting than in large scale conferences. We hope that all of you will actively contribute at this KSI.

This Summer Institute was supported by Research Institute for Mathematical Sciences, Kyoto University and by Institute for Nuclear Study, University of Tokyo. Financial support is due to the Yamada Science Foundation, the Yukawa Foundation and various companies.

Finally I express our gratitude to lecturers and invited seminar speakers for their participation and contribution.

We encourage all participants to make the KSI '81 very successful.

Thank you.

TALES OF THE GUT AGE

Demetrios V. Nanopoulos
CERN, CH-1211 Geneve 23, Switzerland

in

Proceedings of the Fourth Kyoto Summer Institute on Grand Unified Theories and Related Topics, Kyoto, Japan, June 29 - July 3, 1981, ed. by M. Konuma and T. Maskawa (World Science Publishing Co., Singapore, 1981).

Contents

TALES OF THE GUT AGE

D.V. NANOPOULOS

CERN - Geneva

Prologue

These lecture notes contain a tour through some recent developments in Grand Unified Theories. After some general remarks on the structure of Grand Unified Theories we turn to the possibility of understanding the hierarchical fermion mass spectrum inside Grand Unified Theories. A new mechanism for cosmic baryon production through the decays of superheavy fermions is discussed in some detail and a new connection between the still unobserved neutron electric dipole moment and the observed baryon asymmetry is studied. Finally, the possibility of solving the strong CP-violation problem inside Grand Unified Theories is discussed in some length and the consequences of such a possibility are analysed. The table of contents that follows, hopefully, makes clear the scope of these lectures.

Table of Contents

I. The Guts of Grand Unified Theories (GUTs)

1. In this section we present some general characteristics of grand unified theories (GUTs). It seems that there is a lot that can be said for grand unification without going to a particular model, it is merely enough to go to the guts of GUTs.

2. "The Fermi-Syndrome"

Let us suppose that we are living in a world where only strong (QCD) and electromagnetic (EM) interactions are present acting on the "standard" three generations of fermions

$$(u,d,e,\nu_e)\ ;\ (c,s,\mu,\nu_\mu)\ ;\ (t,b,\tau,\nu_\tau)\ . \tag{1}$$

As usual, strong and electromagnetic interactions are described by a renormalizable gauge theory based on the unbroken groups $SU(3)_{COLOUR}$ and $U(1)_{EM}$, respectively. So the relevant group is $G_0 \equiv SU(3)_C \times U(1)_{EM}$. Then, one can prove[1)] in perturbation theory that, under such circumstances, there is a whole class of quantum numbers that are conserved, like the baryon number (B), lepton number (L), strangeness, charm, etc., corresponding to some unbroken global symmetries. It should be stressed right away that the restriction of renormalizability, i.e., allowing only couplings of dimension equal to or less than four plays a rather essential role. Suppose that we find experimentally that the neutron is unstable or that strangeness is violated, what will we do then? Well, we have to put by hand terms in our Lagrangian that violate these quantum numbers, i.e., we are bound to introduce new couplings of dimension five or more between our "known" fields, i.e., non-renormalizable interactions.

For example, one may consider terms like (generically written)

$$\frac{g^2}{M_W^2}\,(\bar{s}u)(\bar{\mu}\nu_\mu) \tag{2}$$

where g is some gauge coupling constant O(e) and M_W is some scale, determined experimentally, but definitely much bigger than the energy scale at our disposal.

Actually, we have just introduced a new kind of interactions, which are characterized for energies $E_{CM} \ll M_W$ by an effective constant $(E_{CM}^2/M_W^2)g^2$ much smaller than g^2 and thus they are properly called "weak interactions".

For the bulk of the phenomena in our "low energy" world, weak interactions play essentially a very minor role, though Eq.(2) cannot be the whole story. As we increase our energy capacities and specifically as E_{CM} exceeds M_W, terms like the one of Eq.(2) cannot be neglected anymore $(E_{CM}^2/M_W)g^2 \sim 1$

which is unfortunate because they lead to violation of unitarity and other horrible and unpleasant things. We have to do a bit better. If gauge theories work for strong and electromagnetic interactions why not use them also for weak interactions? The basic form of weak interactions is then

$$g(\bar{\psi}_1\gamma_\mu(1 - \gamma_5)\psi_2)W_\mu \ , \tag{3}$$

where Eq.(2) is now translated as an "effective" interaction emerging from (3) at energies much smaller than M_W, the mass of the gauge bosons (W) mediating weak interactions.

An important point to stress here is the fact that, while looking for a renormalizable theory of weak interactions, it transpires that the unification of weak and electromagnetic interactions is a necessity. Such an outcome is difficult to overestimate. We have quite strong experimental evidence[2] that the group that unites weak and electromagnetic interactions is SU(2)×U(1).

So we reach the point where strong, electromagnetic and weak interactions are described by a renormalizable gauge theory based on the group $G_1 \equiv SU(3) \times SU(2) \times U(1)$. The main difference between strong/electromagnetic and weak interactions is that SU(2)×U(1) is spontaneously broken [$SU(2)\times U(1) \to U(1)_{EM}$ at $M_W \simeq 100$ GeV] while $SU(3)_C$ and $U(1)_{EM}$ remain unbroken (exact symmetries). Again, we can prove[1),3)] that under such circumstances [SU(3)×SU(2)×U(1) gauge theory, "standard" fermions, one Higgs doublet and only renormalizable couplings] the only unbroken global symmetries are those that correspond to B and L conservation. The fact that we restrict ourselves to renormalizable couplings again plays an essential role in the proof.

Suppose now that we find experimentally that protons decay, and/or that neutrinos have a Majorana mass ($m_\nu \nu\nu$), what then? Well, we have now learned the trick. Mutatis mutandis, replace (2) by

$$\frac{g^2}{M_X^2}\,(uu)(\bar{e}^+ d) \tag{2'}$$

and (3) by

$$g(\psi_1\ \gamma_\mu\ (1 \pm \gamma_5)\psi_2)X_\mu \tag{3'}$$

where Eq.(2') should now be considered as an effective interaction emerging from (3') at energies much smaller than M_X, the mass of the gauge bosons (X) mediating these "exotic" new interactions, sometimes called grand unified interactions. No surprises, dull repetition of history[4),5)]! B or L number violation would signal the existence of new interactions beyond the SU(3)

$\times SU(2)\times U(1)$ and by analogy with the previous case, i.e.,

strangeness violation → weak-electromagnetic unification .

We would like to argue that

B and/or L number violation → strong-electromagnetic-weak unification ,

in other words, there is some big group G such that $G\supset SU(3)\times SU(2)\times U(1)$, i.e., G, is the grand unification group. Lower bounds[6] on the proton lifetime ($\tau_p \gtrsim 10^{30}$ years) indicate that $M_X \gtrsim O(10^{14}$ GeV), indeed a superhigh value, at least twelve orders of magnitude bigger than M_W and at most five orders of magnitude smaller than the Planck mass, $M_{P\ell} \simeq 1.2\cdot 10^{19}$ GeV. The effects of such a superlarge mass scale will be discussed later on.

The examples used up to now show us the general rules of discovering new interactions. Since Fermi taught us all, I call them the "Fermi-Silver rules":

1) Find some conserved quantum number under the "established" (known) interactions.
2) Find some "reason" as to why this "conserved" quantum number is "slightly" violated.
3) Find some experimentalists to look for such "tiny" violations. If they find it, we have made it.

Nowadays, there is a frenzied activity of looking for proton decay deep in the mines and justifying it by employing theoretical arguments from the very very early hot Universe. Grand unified theories have indeed changed experimentalists into mine engineers and theorists into cosmologists. Our present approach to physics is certainly dominated by the "Fermi-Syndrome", the introduction of more and more new "mass scales" which signal the existence of more and more new interactions.

3. Grand Unification

Grand unified theories (GUTs), theories that unify[4),5)] weak, electromagnetic and strong interactions beyond their aesthetical appeal, answer a lot of questions and resolve many of the mysteries contained in the "standard" ($SU(3) \times SU(2)\times U(1)$) model. Furthermore, by making some spectacular and dramatic predictions like the instability of the proton, or the massiveness of the neutrino, it will, hopefully soon be possible, to test such grand ideas. Since many reviews[7)] exist by now on the "classical" applications of grand unified theories (GUTs), let me briefly mention the highlights of these grand applications. In grand unified theories, there is a big "simple" group G that contains $G_1 (\equiv SU(3)\times SU(2)\times U(1))$, characterized by a unique gauge coupling constant g, and where quarks and leptons are sitting in the same representation(s) of the group G.

Then, because the group G is "simple", the following facts are immediate consequences.

1) The electric charge is quantized. No more mystery as to why $Q_{electron} + Q_{proton} = 0(<10^{-20})$. Incidentally, grand unification implies that α, the EM fine structure satisfies the inequality[8]

$$\frac{1}{170} < \alpha(0) < \frac{1}{120} \qquad (4)$$

if the proton decays! Not a trivial bound.

2) At superhigh energies, there are "simple relations" (involving at most Clebsch-Gordan coefficients) between the different coupling constants (g_1, g_2, g_3) because they are all trivially related to g.

For example, at superhigh energies, we get[5]

$$\sin^2\theta_W \equiv \frac{e^2}{g_2^2} = \frac{3}{8} . \qquad (5)$$

Then at "lower" energies each gauge coupling constant is renormalized in a different way[9], and so eventually at our "low energy regime" we find the big differences in "strength" between strong, electromagnetic and weak interactions.

For example, at "low energies", we get[9),10]

$$\sin^2\theta_W(M_W) = \frac{1}{6} + \frac{5}{9}\frac{\alpha\ (M_W)}{\alpha_3(M_W)} \simeq 0.210 - 0.217 \qquad (6)$$

to be compared with the "experimental" value of $\sin^2\theta_W \simeq 0.215 \pm 0.012$ - a rather successful prediction of GUTs. The fact that quarks and leptons are in the same multiplet(s) has some very important consequences.

1) There are relations between quark and lepton quantum numbers. For example, one finds that $Q_d = (1/3)Q_{e^-}$, where the factor of three comes in because there are three colours. Also, if for example the u and d quarks are thrown in a left-handed doublet of $SU(2)_{weak}$ $((u,d)_L)$, then it necessarily follows that ν_e and e have also to be in a left-handed doublet $((\nu_e, e)_L)$, etc. An extensive analysis of these phenomena is given in Ref.11). In other words, we get a satisfactory explanation of the startlingly similar behaviour of quarks and leptons under weak and electromagnetic interactions.

2) At superhigh energies there are "simple" relations (involving at most Clebsch-Gordan coefficients) between quark and lepton masses.

For example, a "natural" relation of the form[5),10]

$$m_b = m_\tau \qquad (7)$$

at superhigh energies is often found in many grand unified models (GUMs). Again at "lower energies" quark masses and lepton masses are renormalized in a different way[10] (basically quarks feel the strong interactions, while the leptons are insensitive to them), with a result that, at "present energies"[10]

$$\frac{m_b}{m_\tau} = \left[\frac{\alpha_3(2m_b)}{\alpha_3(M_X)}\right]^{\frac{12}{33-2f}} \qquad (8)$$

where Eq.(7) has been used and f indicates the number of quark flavours. Using f=6 and presently accepted values for the strong coupling constant α_3 $(\equiv\frac{g_3^2}{4\pi})$ we find[10]

$$\frac{m_b}{m_\tau} \simeq (2.7 - 3) \qquad (9)$$

just what we see experimentally! Another rather successful prediction of GUTs. A simple quark-lepton mass relation, like the one of Eq.(7), seems also to work (?) for the second generation ($m_s=m_\mu$), but it is definitely wrong for the first generation ($m_d \neq m_e$). Some modification is needed. We will come to this point later. The fact that we used f=6 in computing the renormalization is very crucial. If we start increasing the number of flavours, as it is apparent from (9), the m_b/m_τ ratio will increase unchecked[12]. Thus we claim that in GUTs the number of flavours is more or less determined to be[10),12)]

$$f = 6 \,. \qquad (10)$$

This is a rather drastic prediction and not difficult to test experimentally.

3) Since quarks and leptons coexist in the same multiplet(s), there are gauge interactions (mediated by the X boson) that transform one to another and thus B and L numbers are violated. Protons are not forever. The predicted proton lifetime[13]

$$\tau_p \simeq 10^{31\pm2} \text{ years} \qquad (11)$$

is not far above the experimental lower bound[6] $\tau_p > 10^{30}$ years. Thanks to the presently intense experimental search[14] for proton decay, hopefully we will know soon.

With the form of L violation generally found in GUTs, we find that neutrinos do indeed acquire a <u>naturally</u> small Majorana mass[15] ($m_\nu \nu\nu$) of the type discussed earlier, where

$$m_\nu \simeq (\frac{M_W}{M}) m \simeq (10^{-5} \text{ eV} \rightarrow \text{few eV}) \tag{12}$$

and, as usual, m represents a "low energy" charged fermion mass (most likely the corresponding q=2/3 quark mass), and M represents a "superheavy" mass scale. Again experiments will tell, if we do not already know[16).

4. Consequences of the Existence of Superheavy Mass Scale $M(\sim 10^{15}$ GeV)

It is important to notice that the proton quasistability, the fact that the strong interactions are strong ($\alpha_3 \simeq 1$) around 1 GeV, the observed value of $\sin^2\theta_W$ and the observed ratio $\frac{m_b}{m_\tau}$ all indicate independently and uniquely that the energy scale (M) where the grand unification occurs is extremely (exotically) large[5),9),10),13)]

$$M \simeq 10^{15} \text{ GeV} . \tag{13}$$

Actually it is a triumph of grand unification that, given Eq.(13), all the above mentioned experimental facts, completely unrelated at first sight, are subject to a common explanation in the guts of which lies the mere fact of the existence of a new superheavy mass scale, M. Here are some more implications of the existence of a new superheavy mass scale.

4.1 The Survival Hypothesis[17),18)]

Up to now we have accepted the existence of at least two energy scales related with new phenomena according to the following line of symmetry breaking:

$$G \xrightarrow[M \sim 10^{15} \text{ GeV}]{} G_1 \; (\equiv SU(3)\times SU(2)\times U(1)) \xrightarrow[M_W \sim 100 \text{ GeV}]{} G_0 \; (\equiv SU(3)\times U(1)) . \tag{14}$$

It is only natural to assume that anybody who "feels" the mechanism that triggers the $G \rightarrow G_1$ breaking will also get a superheavy mass of order M. So we better be careful that the observed "low energy" particles are "insensitive" to the superheavy symmetry breaking. This is not difficult to arrange for the "low energy" gauge bosons. Choose particular directions of the spontaneous symmetry breaking and then gauge invariance guarantees the masslessness, at this stage of breaking, of the "observed" gauge bosons. There is also no difficulty in arranging things for the fermions. Put the "observed" fermions in such representations so that it is impossible to get invariant mass terms under G_1. This is the so-called[18)] "Survival Hypothesis": "Low mass fermions are those that cannot receive G_1 invariant masses".

In other words, the mere fact of the existence of "low mass" fermions puts constraints on their representation content. Actually, since nature chooses the exact (unbroken) groups of strong (SU(3)) and electromagnetic ($U(1)_{EM}$) interactions to be vectorlike (non-chiral), i.e., L and R fermionic components transform identically under these groups, our only hope left is that SU(2) is a chiral group, i.e., the weak interactions violate P and C both in the charged and neutral currents. Indeed, this is the case experimentally. What a strange connection, the existence of "low mass" fermions implies parity violation in weak interactions. If Fermi only knew! A detailed analysis of the implications of the survival hypothesis on the "low mass" fermion spectrum may be found in Ref. 11).

4.2. The Gauge Hierarchy Problem

We have seen that some gauge bosons remain massless because a certain gauge symmetry remains unbroken, while the existence of some unbroken chiral symmetries keeps some fermions massless. Then we say that the masslessness of these (gauge and fermion) particles is a natural effect as emerging from some symmetry. Unfortunately, there is no symmetry able to keep the Higgs bosons needed to trigger the $G_1 \to G_0$ spontaneous symmetry breaking, massless after the superheavy symmetry breaking.*) Scale invariance may come to mind but, as is well known, it is explicitly broken in higher orders. Here lies the heart of the gauge hierarchy problem: how is it possible to understand naturally the huge gap between M_W and M_X

$$\frac{M_W}{M_X} \simeq 10^{-12} . \tag{15}$$

This is, for the time being, a big puzzle. Maybe it is related to our ignorance of how to handle quantum gravity, or maybe it is a rather technically complicated problem and it needs more hard thought and calculations before it is resolved.[19)] Whatever the answer to the gauge hierarchy problem may be, I believe that the spectacular successes of GUTs make us feel that we are on the right track. Just remember that all the successes of the "old-fashioned" Fermi-type (non-renormalizable) weak interactions remained intact after the invasion of the real thing, i.e., gauge theories.

4.3 Renormalizability

One may wonder why a renormalizable gauge theory, such as "standard" SU(3)

---

*) Supersymmetry seems to be an exception to this rule. For a recent discussion of the relevence of supersymmetry to the solution of the gauge hierarchy problem see Ref. 19).

$\times SU(2)\times U(1)$, gives such a successful description of the world, at least inside the presently explored energy range? That is to say, what is so deep about renormalizability? The argument that perturbation theory makes no sense for non-renormalizable interactions is not totally convincing, since one may imagine some other approach to field theory which bypasses perturbation theory. What then? The answer lies in the so-called "zero mass decoupling theorem"[20). This theorem states that if a large mass (M in our case) occurs in field theory, and if a subset of fields or states remains massless as $M\to\infty$, then these states decouple from the rest in the limit, and their effective interactions at low energy are renormalizable (with additional non-renormalizable interactions inversely proportional to some power of M).

The renormalizability then of the "low energy" interactions of the "low mass" particles is due to the mere existence of a superheavy mass scale M. Many of the mysteries of the "low energy" world are suddenly dissolved. The set of "elementary fields" of low energies should not contain fields with spin higher than one, since their interaction will be necessarily non-renormalizable. Similarly, spin 1 fields must be[21) gauge fields, and spin 1/2 fields have to form an anomaly free set[22). Unfortunately, no constraints are imposed on the scalar fields. Finally, one may replace M by $M_{P\ell}$, the scale which sets in by gravity and ditto justifies the renormalizability of grand unified theories.

If the tacit assumption that nothing happens between a few hundred GeV and 10^{15} GeV (the "desert" or "grand plateau" hypothesis) is correct, as experiments may soon tell, then physics certainly takes a completely new turn. Building larger and larger accelerators may not provide the solution to our problems. Much harder thinking, presumably quite unconventional, experimentally and theoretically, will be needed if we would like to make real progress. It is not accidental that during the last few years quite an activity has developed in studying the interconnections between GUTs and cosmology. The very early and thus extremely hot Universe is a first-class "laboratory" to test a lot of "grand ideas". The GUTs-cosmology amalgamation[23) has produced quite exciting and interesting results, so that I foresee more and more interaction between GUTs and cosmology if, after all, we would like to understand the finest details of the world around us. For a recent review on the GUTs cosmology front see Ref. 23).

Next I would like to discuss some new ideas concerning the fermion mass spectrum.

5. The Problem of Mass

We have already seen that, in the framework of GUTs, we get a satisfactory explanation[10) of the quark lepton mass differences [see Eqs.(8) and (9)] and we understand[15] in a natural way why the neutrino has a very tiny mass, if it is not massless [see Eq.(12)]. Our next task is then to try to understand the so-called generation-(mass)-gap. It is an experimental fact that the masses of the first generation (u,d,e,ν_e) are $O(\alpha^2 M_W)$, while for the second generation (c,s,μ,ν_μ) are $O(\alpha\, M_W)$ and finally some particles of the third one, (t,b,τ,ν_τ) is $O(M_W)$. Apparently, nature tries to tell us something. Let us try to listen! The first thought that immediately comes to mind is that third generation gets a "tree-level" mass by directly "feeling" the mechanism that breaks $SU(2)_{weak}$, while the second generation gets a radiative "one-loop" mass, with the first generation getting a radiative "two-loop" mass, both generations being "insensitive" to the $SU(2)_{weak}$ symmetry breaking mechanism.

Here we have essentially two problems to solve. First, find the mechanism that allows the third generation to get a "tree-level" or "direct" mass, while at the same time prevents the other two generations from getting "direct" mass. In this case, the mass spectrum for the three generations should look like $(0,0,M_W)$ or for short (0,0,1) in units of M_W. I call it the <u>(0,0,1) problem</u>.

Secondly, find the suitable higher order diagrams that create the appropriate masses for the first two generations. Here, the mass spectrum of the three generations should look like $(\alpha^2 M_W, \alpha M_W, M_W)$ or for short $(\alpha^2,\alpha,1)$ in units of M_W. I call it the <u>$(\alpha^2,\alpha,1)$ problem</u>. For the sake of clarity, let us discuss each problem separately.

5.1. <u>The (0,0,1) Problem</u>

Clearly, at the level of SU(3)×SU(2)×U(1) or even at the level of conventional GUTs like SU(5), O(10), or E_6, the appearance of more than one generation is redundant. Since all three generations have identical properties under such interactions, there is no hope to finding a solution to our problem inside the framework of conventional GUTs. Certainly, we should involve generation changing interactions in the game, and let us call G' the "simple" group that unifies all interactions (except gravity) including generation changing interactions. Then Eq. (14) generalizes to

$$G' \xrightarrow[M_{P\ell}>M'>M]{} G \xrightarrow[M]{} G \xrightarrow[M_W]{} G_0 \,. \qquad (16)$$

Now let us proceed by analogy. The basic reason for the presence of superheavy and low mass fermions in the conventional GUTs (G) is that these two classes of fermions transform differently under G; that is the heart of the survival hypothesis. This piece of information is embodied in the structure of the Yukawa interactions invariant under G. For example, in the SU(5) model[5] since $\bar{5}\times 10$ $(=5+\overline{45})$ and 10×10 $(=\bar{5}+45+50)$ do not contain the <u>24</u>, the Higgs representation that causes the superheavy symmetry breaking, fermions belonging to the $\bar{5}+10$ representation remain massless at the $G \to G_1$ breaking. SU(5) invariance prevents the appearance of any $\bar{5}\times 10\times 24$ or $10\times 10\times 24$ coupling. On the other hand, suppose that some fermions belong to the $5+\bar{5}$ representations of SU(5). Then clearly since $\bar{5}\times 5$ (=1+24) contain the 1 and 24, these fermions will become superheavy. In this case, SU(5) invariance allows the appearance of a $\bar{5}\times 5\times 24$ coupling. It should be stressed that the L and R components of the $(5+\bar{5})$ superheavy fermions belong to equivalent representations of $SU(2)_{weak}$, while those of the $(\bar{5}+10)$ "low mass" fermions belong to non-equivalent representations of $SU(2)_{weak}$. This is one way of creating naturally superlarge mass differences between fermions. "Low mass" and superheavy fermions belong to inequivalent representations of G and then simply G invariance forbids some of the fermions to be involve in Yukawa couplings containing superheavy Higgs, and thus G invariance protects some of the fermions from getting superheavy masses. This is not the whole story. It is possible to create naturally superlarge mass differences even for fermions belonging to the same representation. For example, in E_6, a promising group for grand unification, we may put fermions in the fundamental 27 representation. But now, 27×27 $(=\overline{27}+351+351')$ contains the 351, the Higgs representation that causes the superheavy symmetry breaking, and thus some of the fermions contained in the 27 will become superheavy[18]. Why do some and not all of the fermions become superheavy? It is the form of the 351 vacuum expectation value (VEV) that leaves some fermions massless. In other words, while E_6 invariance allows a $27\times 27\times \overline{351}$ coupling, the form of <351> is such that $27\times 27\times \langle\overline{351}\rangle$ contains some zero spots.

Let us now put the whole act together. "Low mass fermions are those that either G invariance prevents from participating in Yukawa interactions involving Higgs bosons (Φ) that cause the superheavy symmetry breaking, i.e., there is no $\psi\psi\Phi$ terms, or the particular form of the vacuum expectation values of these superheavy Higgs fields (ϕ) causes the vanishing of the $\psi\psi\langle\Phi\rangle$ terms".

This is a more technical form of the survival hypothesis. It is exactly this form of the survival hypothesis that, suitably twisted (translated), I want to use[11] in order to solve the (0,0,1) problem. Just re-read the aforementioned, deliberately detailed familiar analysis of natural ways to create superlarge fermion mass differences, using the following substitution rules:

Superheavy fermions	⟶	Third generation fermions
Low mass fermions	⟶	First and second generation fermions
G invariance and/or particular form of VEVs	⟶	G' invariance and/or particular form of appropriate VEVs .

This is my proposed solution to the (0,0,1) problem. It is G' invariance and/or some particular forms of the VEVs of appropriate Higgs fields that are responsible for the (0,0,1) zeroth order form of the fermion mass matrix. This is what I will call the "sidewise" or "horizontal" survival hypothesis.

It is the "standard" or "vertical" survival hypothesis that prevents all three generations from getting a superheavy mass, while it is the "sidewise" or "horizontal" survival hypothesis that protects the first and second generation from getting a "tree level" or "direct" low mass. A happy state of affairs? Almost. We have to find a way of producing naturally very small masses for the first two generations, as we try to do next.

5.2. The $(\alpha^2,\alpha,1)$ Problem

Let us proceed again by analogy. After the lengthy discussion in Section 2, it is apparent that if there are no post $-G_1$ ($\equiv SU(3)\times SU(2)\times U(1)$) interactions and given the observed particle content of the theory, baryon (B) and lepton number (L) conservation follows automatically. It is the existence of post G_1 interactions, i.e., grand unified interactions contained in G, that are responsible for the naturally "tiny" breaking of these global symmetries (B and L). We all know that this is an immediate consequence of the fact that quarks and leptons belong to the same representation(s) of G. There are gauge (or Yukawa) interactions involving post G gauge or Higgs bosons that transform quarks to leptons and thus the very existence of these couplings breaks the global symmetries that were responsible for B and L conservation. The next step is clear, just make the substitution:

B, L quantum numbers	⟶	First and second generation masses
Global symmetry responsible for B and L conservation	⟶	Chiral symmetry responsible for $m_{1,2} = 0$
G interactions break B and L	⟶	G' (or sometimes even G) interactions, after G' breaking, are responsible for $m_{1,2} \neq 0$ by appropriately breaking the relevant chiral symmetries

and the way to solve the $(\alpha^2,\alpha,1)$ problem becomes apparent. Presumably the fermions of the first two generations (call them "light") are sitting in the same representation(s) of G' together with other fermions of the superheavy and/or third generation type (call them "heavy"). The mere existence of gauge or Yukawa couplings involving a "light", a "heavy" and a "superheavy" gauge or

Higgs boson that, after the G' breaking (some "heavy" fermions develop mass), breaks our <u>naturally</u> imposed zeroth order chiral symmetries. Then, the road leading to the creation of "light" fermion masses, <u>necessarily</u> in higher orders (involving G' interactions), is widely open. What a world! G' invariance prevents some of the fermions from getting "tree level" masses, while after G' breaking, G' interactions are responsible for the "tiny" masses of the first two generations. My claim is, that the "light" fermions would have been strictly massless if they were not sitting in the same G' representation(s) with "heavy" fermions. That is, in a strict SU(3)×SU(2)×U(1) world not only would we have B and L conservation, but also masslessness of the first two generations*)! It is through the post-SU(3)×SU(2)×U(1) interactions that the first two generations are getting naturally "tiny" masses. It is amusing to notice that we can carry this analogy even further. Just notice that B-L violation in GUTs occurs in higher order compared to B,L violation. In amplitude[24], $(B-L)_{violat} = (M_W/M)\,(B \text{ or } L)_{violat}$, which is exactly the same relation between the neutrino and charged fermion masses, Eq.(12).

According to the "Fermi-Syndrome" philosophy exposed in Section 2, at the SU(3)×SU(2)×U(1) level the mass terms for the first two generations will have an "effective" non-renormalizable form, for example

$$\left(\frac{\alpha}{\pi}\right)^n \frac{1}{(M)^k} (\bar{\psi}\psi)(\phi_1)^k\, \phi_2 \qquad (17)$$

where ϕ_1 and ϕ_2 are Higgs fields (the subscript indicates their $SU(2)_{weak}$ multiplicity) able to get VEV of order $\langle\phi_1\rangle \sim M$ and $\langle\phi_2\rangle \sim M_W$, and n indicates the number of loops of the diagram from where this term emerges. Then (17) implies

$$m_{\text{"light"}} \simeq \left(\frac{\alpha}{\pi}\right)^n M_W \qquad (18)$$

a highly desirable formula, since n=1 corresponds to the second generation mass and n=2 to the first generation mass. Then, applying the philosophy of Section 2, we should look for the corresponding diagrams that produce the "effective" mass terms (17) or (18).

---

*) Conversely, the observed small masses of the first two generations may be taken as a signal of the existence of post SU(3)×SU(2)×U(1) or even post-G interactions.

i) n = 1

One may consider diagrams of the form[18),25)]

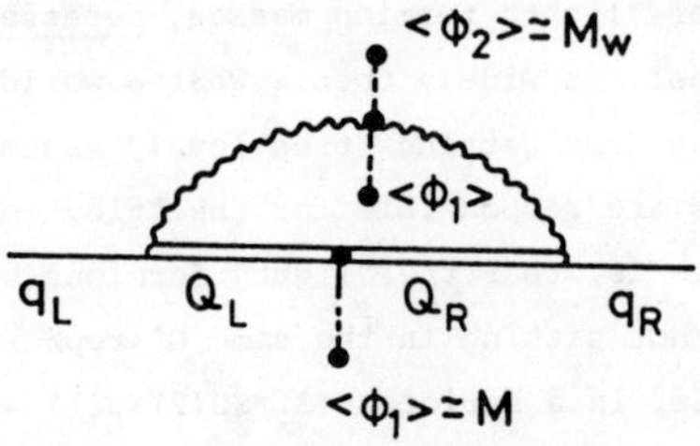

FIG.1

where everything in the loop is superheavy. This diagram produces an "effective" mass term of the form (17) with n=1 and k=2. Thus, in this case, we get $m_2 \simeq (\alpha/\pi) M_W$. The physics here is very interesting because the crucial mixing of the two gauge bosons is a direct consequence of the fact that Q and q have necessarily different $SU(2)_{weak}$ transformation properties as a consequence of the survival hypothesis.

ii) n = 2

An example of a diagram belonging this category looks like[26)]

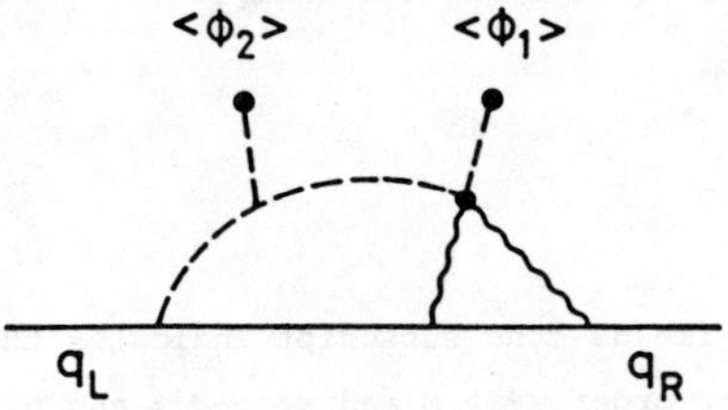

FIG.2

in an obvious notation.

This diagram produces an effective mass term of the form (17) with n=2, k=1 and thus $m_1 \simeq (\alpha/\pi)^2 M_W$. Of course, it is tacitly assumed here that, again because of the "internal" differences (under G') of the first and second generation, it is impossible to give mass to the first generation through one-loop diagrams. In a few words, the third generation gets its mass from the same source that breaks $SU(2)_{weak}$ (i.e., gives the W bosons their masses) while the second and first generation cannot "feel" this SU(2) breaking source. It takes the involvement of post-SU(3)×SU(2)×U(1) interactions to create masses for the first two generations, which then necessarily have to appear as "radiative" corrections and thus naturally small. In a strict SU(3)×SU(2)×U(1) world the first two generations would be massless. Actually, since the first two generations have "radiative" masses, we expect deviations from "direct" or "tree level" quark-lepton mass relations like Eq.(7). Presumably such deviations will be much more drastic for the first than for the second generation and thus we expect $m_d \neq m_e$ and $m_s \simeq m_\mu$, as seen experimentally.

Finally, as a by-product of this way of thinking about the fermion masses, one gets in many GUTs a firm prediction[18] of the top quark mass

$$m_t = m_c \cdot \frac{m_\tau}{m_\mu} \tag{19}$$

which, appropriately renormalized at low energies [like Eq.(8)], gives[18]

$$m_t \simeq (18 - 22) \text{ GeV} \tag{20}$$

not outside of the energy range capacities of PETRA or PEP.

We come next to some explicit realizations of the above ideas concerning the hierarchical fermion mass spectrum.

II. GUTs and Hierarchical Fermion Masses

I present here two models based on the grand unified groups SU(5) and E_6 respectively, as examples of an explicit realization of the mechanism for generation of hierarchical fermion masses discussed in section I. In the SU(5) example[27], done in collaboration with R. Barbieri and D. Wyler, the "driving force" responsible for the generation of a hierarchical fermion mass spectrum are the superheavy Higgs bosons while in the case of the E_6 example[28], done in collaboration with R. Barbieri and A. Masierò, superheavy fermions, naturally present[18] in E_6, play the major role.

A. SU(5) and Fermion Masses[27]

1. Let us go back to the problem of the fermion masses in an SU(5) grand unified model. As is well known, the so-called "minimal" version of it — only one $\underline{5}$ of Higgs with Yukawa couplings to all fermion families — leads to a successful prediction: $m_\tau/m_b=1$ at the grand unification mass $M \simeq 10^{14}$ GeV, and to massless neutrinos. In contrast, one gets, for the lighter families, $m_\mu/m_s=m_e/m_d$ which is not compatible with the current algebra relation $m_d/m_s=m_\pi^2/2m_K^2$, and also $m_\mu/m_s=1$ at M which leads, after renormalization, to a value of m_s about 3÷5 times larger than the presumably observed one[29].

It is tempting to observe that these diseases might simply reflect the very fact that the relative smallness of the lighter families (hierarchy) is in itself not understood within the minimal model. The purpose here is to reconsider these problems by focusing precisely on the hierarchical structure of the fermion masses. We will, in fact, show that such a basic feature of the fermion spectrum can be "naturally" produced in SU(5), exploiting the idea[25)] that the smallness of some fermion masses is due to an exact global symmetry of the "low energy" SU(3)×SU(2)×U(1) approximation of SU(5), broken only by including the effects of superheavy particles. More precisely, we demand that the third family (t,b,τ) gets a direct mass through a $\underline{5}$ of Higgs ($\underline{5}_H$) and that the Yukawa Lagrangian possesses a global symmetry such that the fermions of the second family (c,s,μ) have only a radiatively generated one-loop mass and those of the first family (u,d,e) only a two-loop mass, with loops involving superheavy particles.

Insisting on the mentioned successes of the minimal SU(5) model, we shall not allow any $\underline{45}$'s and/or $\underline{15}$'s of Higgs to couple to the fermions since their vacuum expectation values (VEV) might respectively undo the relation $m_b=m_\tau$ and give direct masses to neutrinos.

Irrespective then of our specific Lagrangian, we shall show that the one-loop generated mass for the second family leads <u>necessarily</u> to $m_\mu=3m_s$ at the unification mass, whereas the two-loop generated mass for the first family gives rise to a ratio m_d/m_e non-simply related to an SU(5) Clebsch-Gordan coefficient. Furthermore, the only possible one-loop mass for the charge Q=2/3 sector (a candidate for the charm quark mass) being necessarily off-diagonal in family space, leads to an eigenvalue of the same size as that of an effective two-loop mass. As such, <u>we are forced</u> to also give the charm quark a direct mass, something which of course may not be unrelated to its actual value.

Finally, again as a general consequence of our basic philosophy and in contrast with the minimal SU(5) model, we are bound to have non-vanishing neutrino masses of size $\simeq(\alpha/\pi)M_W^2/M$.

2. The idea that naturally small fermion masses are generated through loops with virtual superheavy particles is an appealing one, especially since it may lead to masses of order $(\alpha/\pi)M_W$ or $(\alpha/\pi)^2 M_W$ (one-loop and two-loop, respectively) independent of the value of the exchanged large masses M[25)]. In unified models based on a group G where standard (f) and superheavy (F) fermions sit together in an irreducible representation of G, the necessary breaking of chirality for the standard fermions may be provided by gauge couplings of the type $\bar{f}\gamma_\mu F A_\mu$ where A_μ is a superheavy boson. These kinds of couplings are, however, absent in the SU(5) model with standard fermions assigned to the representation $\underline{\bar{5}}+\underline{10}$; possibly relevant in this case are Yukawa couplings $f^T C^{-1} f\phi$ to superheavy scalars ϕ. The general way of guaranteeing the "calculability" (the finiteness) of a radiatively generated fermion mass is to demand that the Yukawa Lagrangian possesses some global symmetry, broken only by soft terms (quadratic/cubic interactions) in the Higgs potential. The global symmetry prevents a direct mass at the tree level, whereas the soft breaking makes

it occur in higher orders through finite contributions.

Within this context, the general one-loop diagram for a radiative mass is given in Fig.3 with tadpoles denoting Higgs vacuum expectation values. The

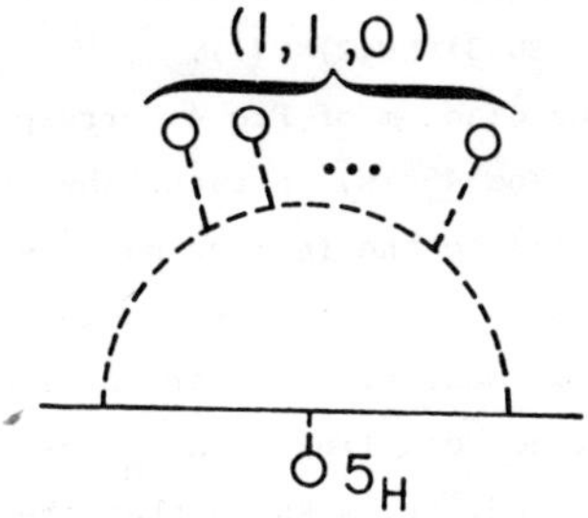

Fig. 3. A general one-loop mass term. The solid lines are fermions and the dashed lines are Higgs fields.

tadpole connected with the fermion line is necessarily a $\underline{5}$ of Higgs. The other Higgs lines attached to the fermion can be a $\underline{5}_H$ (coupled to $\underline{\bar{5}}\times\underline{10}$ and $\underline{10}\times\underline{10}$), a $\underline{10}_H$ (coupled to $\underline{5}\times\underline{5}$) and/or a $\underline{50}_H$ (coupled to $\underline{10}\times\underline{10}$). Of these Higgs fields only the $\underline{5}_H$ can get a VEV*) consistent with unbroken $SU(3)_C\times U(1)_{EM}$, since the others have no neutral, colourless component. The tadpoles connected with the internal superheavy Higgs line, which are necessarily there to provide the breaking of the global symmetry, must all be singlets under $SU(3)\times SU(2)\times U(1)$ in order not to give a totally negligible mass term $\lesssim(\alpha/\pi)\ M_W^2/M$ (see later).

Without making any restriction on the Higgs content of the model, other than excluding $\underline{45}_H$ and $\underline{15}_H$ coupled to fermions for the above-mentioned reasons, it follows that there exists only one specific diagram of the type in Fig.3. The diagram is given in Fig.4. It necessitates the introduction of a $\underline{50}$ and

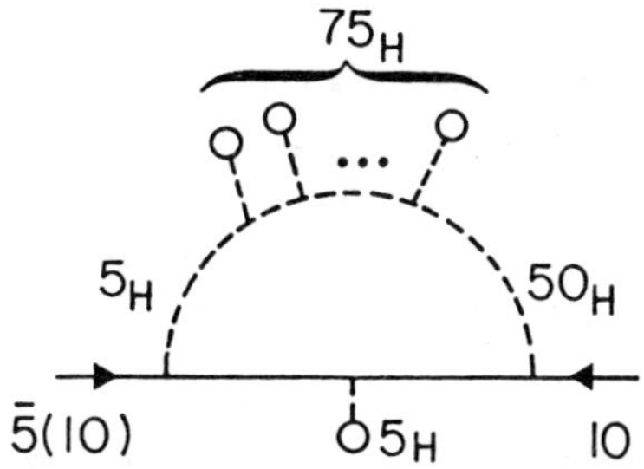

Fig. 4. Actual one-loop mass diagram.

---

*) We actually make the natural assumption that all neutral colourless Higgs components get non-zero VEV.

a $\underline{75}$ of Higgs. In this diagram all tadpoles connected with the internal Higgs line must form altogether a $\underline{75}$. Its uniqueness follows by the explicit construction, asking which products of two Higgs fields possibly coupled to fermions contain, in their SU(5) reduction, a self-conjugate representation, the only one with singlet components under SU(3)×SU(2)×U(1).

Interestingly enough, the diagram of Fig.4 corresponds to an effective $\underline{45}_H$ coupled to $\underline{\bar{5}}\times\underline{10}$ or to $\underline{10}\times\underline{10}$. The $\underline{45}$ is, in fact, the only representation which is contained in $\underline{75}\times\underline{5}$ and couples to the fermion mass terms ($\underline{\bar{5}}\times\underline{10}$ and $\underline{10}\times\underline{10}$). Now, since the coupling $\underline{45}_H\times\underline{10}\times\underline{10}$ is antisymmetric in the two $\underline{10}$'s, the diagram of Fig.4 cannot give a diagonal mass term. Therefore, a sizeable contribution to m_c can only come from a direct coupling to a $\underline{5}_H$, as for the fermions of the heavy family. On ther other hand, it is known that the coupling $\underline{45}_H\times\underline{\bar{5}}\times\underline{10}$ gives, at the grand unification mass M, a ratio of 3 between the lepton and the Q=-1/3 quark masses[30]. For the second family, we are thus led to $m_\mu=3m_s$ at the unification mass scale and to $m_s\simeq(3\div 5/3)m_\mu$ at low energy, which is in spectacular agreement with the usually quoted value[29] of $m_s\simeq 100$-150 MeV.

To estimate the over-all size of the diagram, we associate a unique mass scale to the Higgs system, most likely (but not necessarily) the unification mass M and one dimensionless parameter h perhaps not too different from the gauge coupling itself. More specifically, we think of a situation where, after the breaking, all Higgs fields will have masses of order M except for one "light" Glashow-Weinberg-Salam doublet*), Yukawa couplings of order h**) and trilinear couplings of order hM. Then we have, for the diagram of Fig.4.

$$\delta m^{(1)} \simeq \frac{h^2}{4\pi^2}\, m \qquad (21)$$

where m is a direct mass, typical of the third family.

3. We now ask if a "natural" two-loop mass can be given to the first family (e,d,u). The answer is yes, provided one breaks the chirality of the corresponding $\underline{\bar{5}}_1+\underline{10}_1$ SU(5) representation (the subscript 1 stands for the first family) only through Yukawa couplings to Higgs fields $\underline{10}_H$ and $\underline{50}_H$ which cannot get any VEV. Typical terms are $\bar{5}_1\bar{5}_3 10_H$ and $10_1 10_3\overline{50}_H$. The corresponding min-

---

*) This minimizes the unnaturalness of the gauge hierarchy problem.

**) Of course, $h\simeq e(m/M_W)$ where m is an average value of the masses for the third generation. Based on this we need not impose the artificial condition h<<e.

imal two-loop diagrams are given in Fig.5, where it is understood that any num-

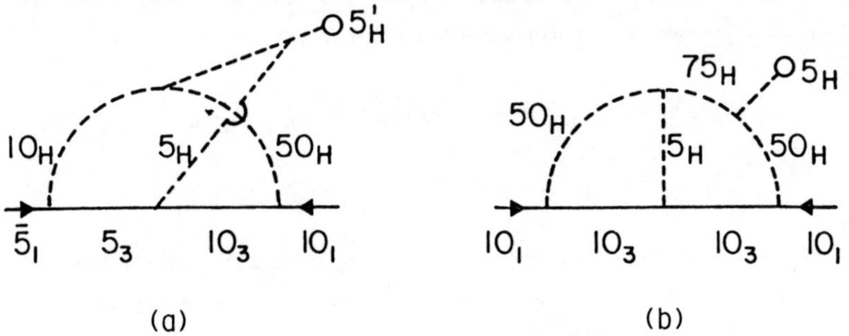

Fig. 5. Possible two-loop mass diagrams.

ber of superlarge tadpoles (e.g., $\underline{24}_H$) can stick out of any internal Higgs line and go into the vacuum without extra-suppressing the diagram. Both diagrams give finite masses of order

$$\delta m^{(2)} \simeq \left(\frac{h^2}{4\pi^2}\right)^2 m \ . \tag{22}$$

In particular, the diagram 5a gives masses to d and e in a ratio which is, however, not simply related to an SU(5) Clebsch-Gordan coefficient, (as in the case of Fig.4), because for different large tadpole insertions the SU(5) transformation properties are different*). Actually, by allowing any superlarge tadpole, [Higgs with SU(3)×SU(2)×U(1) singlet components], other two-loop diagrams can be constructed. They are always finite and no one-loop mass term can be found for the first generation.

4. The features of the generation of radiative masses that we have discussed so far are general. The fact that they can be "naturally" implemented in a specific model is shown by the following Yukawa Lagrangian

$$L_Y = a\ \bar{5}_3 10_3 \bar{5}_H + b\ 10_3 10_3 5_H + c\ \bar{5}_2 10_3 \bar{5}'_H + d\ 10_2 10_2 5'_H$$
$$+ e\ 10_3 10_2 \overline{50}_H + f\ 10_3 10_1 \overline{50}'_H + g\ 5_1 5_3 \overline{10}_H + \text{h.c.} \tag{23}$$

L_Y, together with the standard gauge Lagrangian, is readily seen to possess a global $U(1)^4$ which forbids any other coupling of the fermions to the same Higgs

*) We have calculated a few diagrams which, in fact, show that the mass term is indeed a mixture of a $\underline{5}$ and a $\underline{45}$, as it is required to reproduce the correct ratio m_d/m_e.

fields. Through the Higgs self-interactions, the neutral components of 5_H, $5'_H$ will get a VEV, giving direct masses to t,b, and c. The global $U(1)^4$ is broken by cubic terms of the Higgs potential, e.g.,

$$L_H^{(3)} = h_1 10_H 5_H 5'_H + (h_2 5_H + h_3 5'_H + h_4 \overline{50}_H + h_5 \overline{50'}_H)\, 10_H 10_H$$
$$+ (h_6 5_H 50_H + h_7 5_H 50'_H + h_8 5'_H 50_H + h_9 5'_H 50'_H)\, 75_H + \text{h.c.} \qquad (24)$$

where a $\underline{75}_H$ has also been introduced. L_Y, along with $L_H^{(3)}$ and the remainder of the Higgs potential (respecting the $U(1)^4$ symmetry), is easily seen to give radiative masses as indicated*).

As in the minimal SU(5) model, the global symmetry of L_Y leads, even after spontaneous symmetry breaking, to the conservation of B-L, baryon minus lepton number. Here, however, the necessary breaking of the global symmetries by $L_H^{(3)}$, in order to produce radiative masses, also spoils the conservation of B-L itself. "Calculable" neutrino masses are therefore radiatively generated. At one-loop, the only contributing diagram is given in Fig.6 with mass terms of

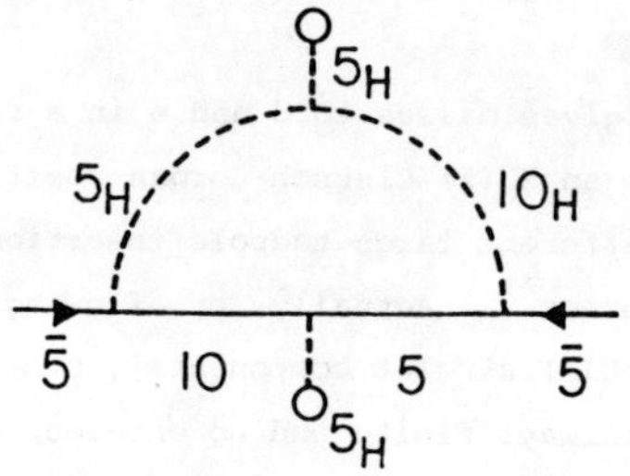

Fig. 6. Neutrino mass diagrams.

typical order

$$\delta m_\nu^{(1)} \simeq \frac{h^2}{4\pi^2} \left(\frac{h}{e}\right)^2 \frac{M_W^2}{M} \, . \qquad (25)$$

For $h \simeq e$ and $M \simeq 10^{14}$ GeV, $\delta m_\nu^{(1)} \simeq 10^{-3}$ eV. The possible relevance of these masses

*) The uniqueness of the one-loop mass generation mechanism presented here and contained in this Lagrangian is tied to the natural assumption that all colourless neutral Higgs components get a non-zero VEV. By relaxing it, and having say a $\underline{5}_H$ without a VEV, we might construct a model in which the mass generation is stepwise, that is in which the mass of a family comes as a one-loop effect of the mass of the preceding family. This mechanism would exhibit an interesting feature: while m_u is proportional to m_s, m_d is proportional to m_c, thus explaining the puzzling inversion of the fermion spectrum $m_c > m_s$, $m_u < m_d$.

in astrophysics has already been emphasized[31)].

The particular Lagrangian (L_Y) gives, in leading order, only two neutrino masses as well as a definite mixing pattern. We cannot, however, really say how much these special features have a general character or if they rather depend on the Lagrangian chosen. The postulates stated at the beginning strongly restrict its form, but not completely. To understand its physical naturalness, the incorporation of SU(5) in a more complete physical theory (with gravity in it?) seems required.

We will show later (section V3), how the SU(5) model considered here solves naturally the so-called strong CP problem.

B. E_6 and Fermion Masses[18),28)]

1. A particular mechanism to produce naturally small masses, making use of superheavy fermions, has been suggested in Ref.25). Here, a realization[18),28)] of it is given in the context of a grand unified E_6 model[18)] where superheavy fermions naturally occur[18)]. On the basis of what we have just said, this model is not complete in so far as it does not provide any clue to the appearance of the mentioned global symmetries, some of which are, in fact, assumed at the beginning. It presents, however, two results: i) giving a driving mass term $O(M_W)$ to the third family (t,b,τ), the second (c,s,μ) and first families (u,d,e) will get masses $O((\alpha/\pi)M_W)$ and $O((\alpha/\pi)^2M_W)$ respectively, as a result of radiative corrections; ii) some plausible reasons for the hierarchy $m_c > m_\mu \sim m_s$ appear naturally in our scheme.

2. The fermions of each family are assigned to the fundamental $\underline{27}$ representation of E_6. Its decomposition under SU(5):

$$\underline{27} = \underline{10} + \underline{\bar{5}} + \underline{1} + \underline{1} + \underline{\bar{5}} + \underline{5} \tag{26}$$

reveals that the $\underline{27}$ contains the usual 15-plet of the known fermion (f) plus a self-conjugate 12-plet of fermions (F) getting a superheavy mass according to the "chirality" argument known as "survival hypothesis".

The classification of the F's according to their colour, weak isospin and electric charge is given by:

$\underline{5}$: N(1,1/2,0) , E(1,-1/2,-1) , D(3,0,-1/3)

$\underline{\bar{5}}$: N^c(1,-1/2,0) , E^c(1,1/2,1) , D^c($\bar{3}$,0,1/3)

$\underline{1}$: L(1,0,0)

$$\underline{1} : \nu^c(1,0,0) \tag{27}$$

where ν^c is the member of the 16-plet of SO(10), singlet under SU(5). The Higgs bosons giving mass to the fermions and, at the same time, providing the mechanism of spontaneous symmetry breaking of E_6 down to $SU(3)_C \times U(1)_{EM}$ *) are contained in the product:

$$\underline{27} \times \underline{27} = (\underline{\overline{27}} + \underline{351})_{\text{symm.}} + \underline{351'}_{\text{antisymm.}} \tag{28}$$

We shall only make use of the symmetric part of Eq.(28) (our notation will be: $\underset{\sim}{\overline{27}} \to H$ and $\underset{\sim}{351} \to \Phi$). We assume that H develops small ($h \sim M_W$) and superlarge (H) vacuum expectation values (VEV's), whereas $\underset{\sim}{\Phi}$ is allowed to develop only superlarge (Φ) VEV's. This assumption is as "unnatural" as the hierarchy itself: we can only hope some explanation will come from an understanding of the "hierarchy problem". The mass terms then come from the following Yukawa couplings:

$$\begin{aligned}\psi\phi\psi \to\ & \phi(1)LL + \phi(16)\ {}^cL + \phi(126)\nu^c\nu^c \\ & + \phi(54,24)(D^cD - 3/2\ E^cE - 3/2\ N^cN) \\ & + \phi(144,24)(d^cD - 3/2\ E^ce - 3/2\ N^c\nu)\end{aligned} \tag{29}$$

$$\begin{aligned}\psi H\psi \to\ & H(1,1)(D^cD + E^cE + N^cN) \\ & + H(16,1)(d^cD + E^ce + N^c\nu) \\ & + h(16,\bar{5})(D^cd + e^cE + N^c\nu^c) \\ & + h(10,\bar{5})(d^cd + e^ce + N^cL) \\ & + h(10,5)(u^cu + \nu^c\nu + LN)\end{aligned} \tag{30}$$

where the number in parentheses gives the transformation properties under SO(10) and SU(5) respectively, and u, d, e and ν indicate a generical $Q=2/3$,

---

*) E_6 is the only Lie group with the following features : i) it is automatically anomaly free : ii) it possesses useful complex representations : iii) the fermions of each generation belong to an irreducible representation : iv) all the symmetry breaking can be realized by using only those Higgs representations which give mass to fermions.

$Q=-1/3$, $Q=-1$ and $Q=0$ fermion, respectively.

We notice that from (29) and (30), a substantial difference emerges between the charged lepton and $Q=-1/3$ quark sectors on one side, and the $Q=2/3$ quark sector on the other. Indeed, in the latter, the current eigenstate u coincides with the mass eigenstate u_0, while in the former the mass eigenstates are given by the mixings:

$$(d_0, d_0^c) = (d, d^c\cos\theta + D^c\sin\theta) \tag{31a}$$

$$(e_0, e_0^c) = (e\cos\theta + E\sin\theta, e^c) \ , \tag{31b}$$

where θ is a function of $\phi(54,24)$ and $\phi(144,24)$. As we shall see, the mixings (31a) and (31b) could provide a possible key for the interpretation of the hierarchy $m_c > m_\mu \sim m_s$.

We restrict the Yukawa Lagrangian to the following couplings:

$$\psi_3 H\psi_3 + \psi_2\phi\psi_2 + (\psi_1\psi_3 + \psi_3\psi_1)\phi \tag{32}$$

where the indices 1, 2 and 3 refer to the u, c and t family, respectively.

The expression (32) can be naturally guaranteed by imposing the global $U(1)\times U(1)$ symmetry:

$$\psi_1 \to e^{-i\theta}\psi_1 \ , \quad \psi_2 \to \psi_2 \ , \quad \psi_3 \to e^{i\theta}\psi_3 \ , \quad H \to e^{-2i\theta}H \ , \quad \phi \to \phi;$$

$$\psi_1 \to e^{2i\chi}\psi_1 \ , \quad \psi_2 \to e^{i\chi}\psi_2 \ , \quad \psi_3 \to \psi_3 \ , \quad H \to H \ , \quad \phi \to e^{-2i\chi}\phi \ . \tag{33}$$

The symmetry (33), broken by the VEV's, must also be broken explicitly by soft terms in the Higgs potential, e.g., $HH\phi$.

If the components of $\underset{\sim}{\phi}$, which are non-singlet under $SU(2)_W\times U(1)_Y$, do not develop a VEV getting a large positive mass squared, then the only "low energy" Yukawa coupling is $\sim\psi_3\psi_3\underset{\sim}{H}$, which respects the global symmetry (33) preventing ψ_1 and ψ_2 from getting direct "low" mass terms. Notice that, in fact, the diagrams providing a "low" radiative mass to ψ_1 and ψ_2 respect the original $U(1)\times U(1)$ in (33) before symmetry breaking (Fig.7). Clearly, it is only h that can ultimately give rise to a "low energy" mass term $\psi_2\psi_2$ (or $\psi_1\psi_3$). However, $\psi_2\psi_2 H$ (or $\psi_1\psi_3 H$) are forbidden by (33). It is actually in this way that the finiteness of the radiative mass diagrams of Fig.7 is guaranteed, since $\psi_1\psi_3\underset{\sim}{H}\underset{\sim}{H}^+\underset{\sim}{\Phi}$ and $\psi_2\psi_2\underset{\sim}{H}\underset{\sim}{H}^+\Phi$ are non-renormalizable couplings.

We would like to stress why the gauge boson mixing through $\underset{\sim}{H}$ bosons is unavoidable in Fig.7: the survival hypothesis compels the internal fermion

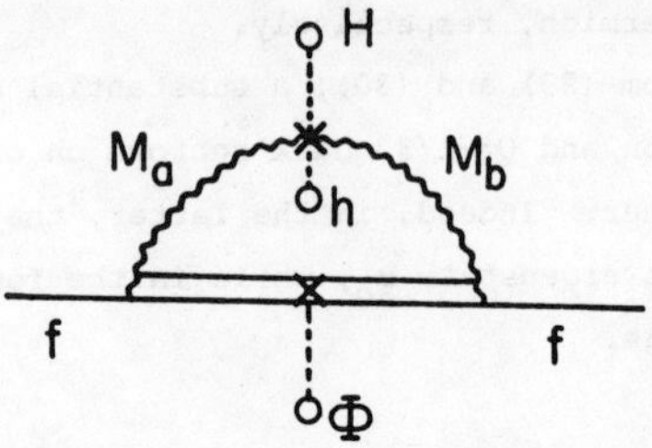

Fig. 7. The basic diagram assuring a "radiative" mass for the fermions of the second and first generations.

mass term to be a SU(3)×SU(2)×U(1) singlet, whereas the external leg fermions must exhibit different SU(2) transformation properties. Thus, the two exchanged gauge bosons carry different weak isospin numbers, so that their mixing can be allowed only through the VEV's h of the $\underset{\sim}{H}$.

Two ingredients are required to construct explicitly all the possible diagrams of the kind represented in Fig.7: the gauge currents of E_6 and the relevant quantum numbers of the neutral $\underset{\sim}{H}$ components.

The 78 gauge currents of E_6 transform under SO(10) as:

$$\underline{78} = \underline{45} + \underline{16} + \underline{\overline{16}} + \underline{1} \ . \tag{34}$$

The 45 currents of SO(10) are well known[32], while it can be useful to give the currents in the 16:

$$(1,2)\begin{cases} Q = 1 \quad (J_\mu)_1 = \bar{D}\gamma_\mu u + \bar{N}^c\gamma_\mu e^c + \bar{E}\gamma_\mu \nu^c + \bar{e}\gamma_\mu L \\ Q = 0 \quad (J_\mu)_2 = \bar{D}\gamma_\mu d + \bar{E}^c\gamma_\mu e^c + \bar{N}\gamma_\mu \nu^c + \bar{\nu}\gamma_\mu L \end{cases}$$

$$(1,1)\begin{cases} Q = 0 \quad (J_\mu)_3 = -(\bar{D}^c\gamma_\mu d^c + \bar{E}\gamma_\mu e + \bar{N}\gamma_\mu \nu + \bar{\nu}^c\gamma_\mu L) \\ Q = 1 \quad (J_\mu)_4 = -(\bar{D}^c\gamma_\mu u^c + \bar{N}^c\gamma_\mu e + \bar{E}^c\gamma_\mu e + \bar{e}^c\gamma_\mu L) \end{cases}$$

$$(3,2)\begin{cases} Q = 1/3 \quad (J^i_\mu)_5 = \varepsilon^{ijk}\,\bar{D}^c_J\gamma_\mu u_k - \bar{d}_i\gamma_\mu L + \bar{D}_i\gamma_\mu \nu + \bar{E}\gamma_\mu u^c_i - \bar{N}^c\gamma_\mu d^c_i \\ Q = -2/3 \quad (J^i_\mu)_6 = \varepsilon^{ijk}\,\bar{D}^c_j\gamma_\mu d_k + \bar{u}_i\gamma_\mu L - \bar{D}_i\gamma_\mu e + \bar{N}\gamma_\mu u^c_i - \bar{E}^c\gamma_\mu d^c_i \end{cases}$$

$$(3,1) \quad Q = -1/3 \quad (J^i_\mu)_7 = \varepsilon^{ijk}\,\bar{D}_j\gamma_\mu u^c_k + \bar{d}^c_i\gamma_\mu L - \bar{D}^c_i\gamma_\mu \nu^c + \bar{N}^c\gamma_\mu d_i - \bar{E}^c\gamma_\mu u_i$$

(35)

where, with (a,b), we have indicated the colour and $SU(2)_W$ numbers of each current.

The $\underset{\sim}{H}$ neutral components can be usefully classified according to their $SU(3)_C$, T_{3L}, T_{3R}, Y_L and Y_R quantum numbers:

$$H(1,1) = (1,0,0,1/3,-1/3)$$

$$H(16,1) = (1,0,-1/2,1/3,1/6)$$

$$h(10,5) = (1,-1/2,1/2,-1/6,1/6)$$

$$h(10,\bar{5}) = (1,1/2,-1/2,1/6,-1/6)$$

$$h(16,5) = (1,1/2,0,-1/6,-1/3) \ . \tag{36}$$

The contribution of the diagrams of Fig.7 is:

$$\delta m \simeq 3\,\frac{\alpha}{\pi}\,m_F\,\frac{M_W}{M_a-M_b}\left[\frac{M_a^2}{m_F^2-M_a^2}\,\lg\left(\frac{m_F}{M_a}\right)^2 - \frac{M_b^2}{m_F^2-M_b^2}\,\lg\left(\frac{m_F}{M_b}\right)^2\right] \tag{37}$$

where M_a and M_b indicate the masses of the exchanged gauge bosons.

We shall show in a moment that the leading contribution from (37) turns out to be $\sim O((\alpha/\pi)M_W)$. Therefore, the following mass matrix arises for any fermion of definite charge, once the one-loop radiative corrections (37) are taken into account:

$$m = \begin{pmatrix} 0 & 0 & c(\frac{\alpha}{\pi})M_W \\ 0 & b(\frac{\alpha}{\pi})M_W & 0 \\ c(\frac{\alpha}{\pi})M_W & 0 & aM_W \end{pmatrix} \tag{38}$$

with a, b and c numerical factors O(1), dependent upon the charged sector under consideration.

The three eigenvalues of the matrix m are of order M_W, $(\alpha/\pi)M_W$ and $(\alpha/\pi)^2M_W$, thus accounting correctly for the order of magnitude of the masses of the three families.

3. We now discuss in more detail the contributions (37) to the second families in order to get the distinct values m_c, δm_μ and δm_s. The rich gauge structure of E_6 allows for such a quantity of diagrams of the kind represented in Fig.7, that we need some criterion to decide where the leading contributions

come from. We assume SO(10) to be a good intermediate symmetry, so that the symmetry breaking scheme is:

$$E_6 \xrightarrow[M_{E_6}]{} SO(10) \xrightarrow[M_{SO(10)}]{} SU(3)_C \times SU(2)_W \times U(1)_Y \xrightarrow[M_W]{} SU(3)_C \times U(1)_{EM}. \quad (39)$$

From (39) and (29), the following splitting in the F masses holds*):

$$m_L \sim M_{E_6} \;;\; m_{\nu^c} \sim m_D \sim m_E \sim m_N \sim M_{SO(10)} \,. \quad (40)$$

The other crucial quantity in the evaluation (37) is the mass of the exchanged gauge bosons. A major difference between diagrams whose internal fermion line is a ν^c or a D, E and N comes about: as ν^c belongs to the 16-plet of SO(10), the exchanged bosons belong to the $\underset{\sim}{45}$ of SO(10) with a mass $\sim O(M_{SO(10)})$, whereas in all the other cases (i.e., with D, E, N and L in the internal line) at least one gauge boson belongs to the $\underset{\sim}{16}$ (or $\overline{\underset{\sim}{16}}$) of SO(10) in Eq.(34) thus being $\sim O(M_{E_6})$. Therefore, the diagrams exchanging D, N or E are suppressed with respect to those exchanging ν^c, as their gauge bosons are heavier, and to those exchanging L as $m_D \sim m_N \sim m_E < m_L$. Indeed, it can easily be seen that the diagrams with L and ν^c are of the same order of magnitude, so that one would be led to conclude that all (and only) the diagrams of Fig.8 give the leading contribution in (37). Things are not so: the graphs 8c and 8e of Fig.8 are absent as they involve gauge bosons not present in E_6 (35). For example, the boson coming out of the vertex $s^c \nu^c$ should be a triplet of colour and $SU(2)_R$ and a singlet of $SU(2)_L$.

From the remaining graphs of Fig.8, we can read the following contributions [we make use here of a simplified version of Eq.(37), which is valid when the mass of the internal F is smaller than the gauge boson masses]:

$$\delta m_c \sim \frac{\alpha}{\pi} [\alpha m_{\nu_c} \frac{H(16,1)^* h(16,\bar{5})}{M_a^2 - M_b^2} \lg \frac{M_a^2}{M_b^2} + \alpha m_L \frac{h(10,\bar{5})^* H(1,1)}{M_6^2 - M_8^2} \lg \frac{M_6^2}{M_8^2}] \quad (41a)$$

$$\delta m_s \sim \frac{\alpha}{\pi} [\alpha m_L \frac{h(10,5)^* H(1,1)}{M_6^2 - M_7^2} \lg \frac{M_6^2}{M_7^2}] \cos\theta \quad (41b)$$

*) As $\underset{\sim}{H}$ couples only to the third generation, the VEV H(1,1) gives mass $\sim O(M_{E_6})$ only to the D, E and N belonging to this generation, whereas the superheavy fermions taking part in the radiative contributions to the charm and up-generation belong to the second and first generations. This justifies why we must simply consider the Yukawa coupling of ϕ_{351}, thus getting the splitting (40).

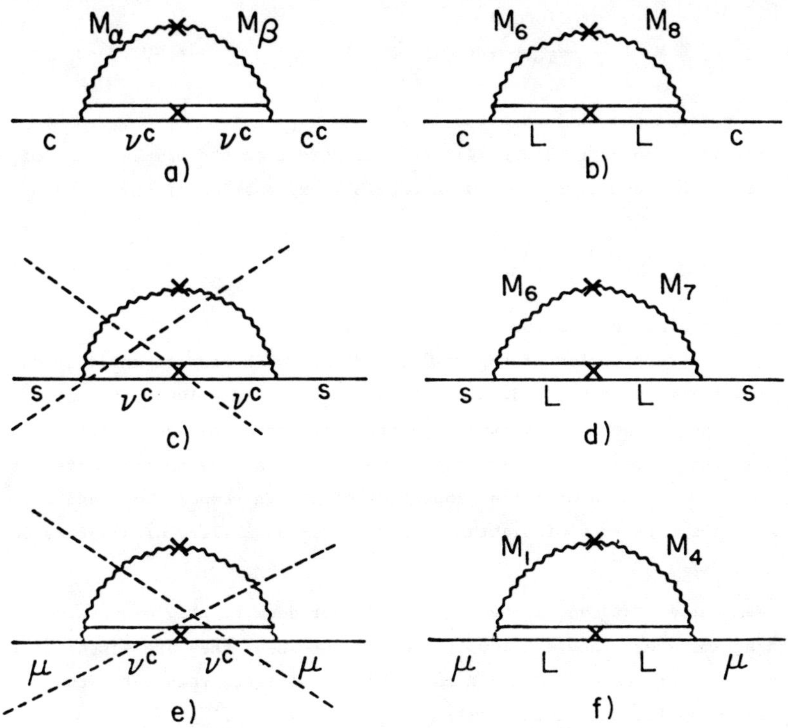

Fig. 8. Leading diagrams according to the symmetry breaking scheme (39). Notice that the diagrams c) and e) cannot exist as they involve gauge bosons which do not belong to E_6.

$$\delta m_\mu \sim \frac{\alpha}{\pi} [\alpha m_L \frac{h(10,5)^* H(1,1)}{M_1^2 - M_4^2} \lg \frac{M_1^2}{M_4^2}] \cos\theta \ . \tag{41c}$$

The numbers put as indices of the masses refer to the classification given in (35), while a and b in (41a) refer to gauge bosons belonging to the $\underset{\sim\sim}{45}$ of SO(10) and thus are not contained in (35).

Let us consider (41b), for example, in which:

$$M_6^2 - M_7^2 \sim O(M_{E_6} \cdot M_W) \qquad \text{and} \qquad \lg \frac{M_6^2}{M_7^2} \sim O(\frac{M_W}{M_{E_6}})$$

so that we get:

$$m_s \sim \frac{\alpha}{\pi} O(M_{E_6}) \frac{O(M_W) O(M_{E_6})}{O(M_{E_6} \cdot M_W)} \cdot O(\frac{M_W}{M_{E_6}}) \sim \frac{\alpha}{\pi} O(M_W) \qquad \text{(at GUT mass)}$$

as we had preannounced just after (37).

However, the expressions (41) tell us more than the simple order of magnitude of the masses, suggesting a possible explanation of the hierarchy:

$$m_c > m_s \sim m_\mu . \qquad (42)$$

Indeed, two distinct reasons are available:

i) the suppression of the cosθ in (41b~c) with respect to (41a) due to the fact that s and μ must "share" their radiative contributions with their superheavy partners (differently from the quark c);

ii) the presence of an additional term in (41a) due to the different T_{3L} and T_{3R} tranformation properties of c with respect to s and μ (causing the existence of diagram 8a and the impossibility of diagrams 8c and 8e).

We notice that even though D and E do not seem to play a decisive role as the diagrams where they are involved are suppressed, they are essential in our scheme: the presence of D and E could be the ultimate reason for the hierarchy (42) allowing for the mixings (31).

For the first generation, we must limit ourselves to the correct order of magnitude as provided by the mass matrix (38).

We have shown that the mass mechanism of Ref.25), based on the presence of superheavy fermions, can be successfully implemented in an E_6 model. This adds more interest to the study of the superheavy fermion presence in grand unified theories, which has also been advocated[33)] in connection with the production of a net baryon asymmetry in the early Universe. We next trun to this subject.

III. GUTs and a New Mechanism for Cosmological Baryon Production: A Case for Superheavy Fermions

1. One of the biggest successes of Grand Unified Theories, is the quantitative understanding of the observed baryon asymmetry in the Universe. Since, by now, this is a rather well-known subject,[23)] I will just review quickly here the basic ingredients, in order to make better the connection with the new proposed mechanism. Grand unified interactions contain normally and naturally baryon number, CP and C violating component. Also, the expression of the Universe makes it easy for the aforementioned interactions, to go out of equilibrium. But then we have all the necessary and sufficient conditions[23)] to

create a net baryon number. What we believe happened is something like the following: Some superheavy particles (the Myrtons (A) of Ref.23)) have been kept at a large density (of order n_γ) at temperatures below their masses

$$\left.\begin{array}{l} n_A = n_{\bar{A}} \sim n_\gamma \\ T < M_A \end{array}\right\} . \tag{43}$$

Clearly, at some temperature T_A their decay rate will become equal or bigger than the expansion rate of the Universe. At that moment, they will decay like mad, but the decay products (because of the Boltzmann suppression factor $e^{-M_A/T_A} << 1$, (remember $T_A << M_A$), will be unable to remake their parents. So, we have the almost free decays of an equal number of Myrtons (A) and anti-Myrtons ($\bar{A}$) (and actually $n_A \sim n_\gamma$), which as we all know from our basic particle physics course, may create a baryon asymmetry given all the needed "violating" interactions. Numerous calculations[23)] have shown, that the most likely candidate to be identified with the Myrtons (A-particles) are the superheavy Higgs bosons (H). In this case, it is easy to see that Eq.(43) is satisfied, if

$$\left.\begin{array}{l} M_H \geqq O(\alpha^2 M_{P\ell}) \\ T_A \geqq O(10^4 \text{ GeV}) \end{array}\right\} \tag{44}$$

and in the "standard" 2 $\underline{5}$ of Higgs SU(5) model one gets[34)]

$$\frac{n_B}{n_\gamma} \sim 10^{-2} (\Delta B)_H \simeq 10^{-2} (G_F \cdot m_b^2)\varepsilon , \tag{45}$$

where $(\Delta B)_H$ is the CP-violating asymmetry in the decays of H and $\bar{H}$, $m_b \simeq 2$ GeV (the mass of the b quark at 10^{15} GeV), $G_F \cdot m_p^2 \simeq 10^{-5}$ is the usual Fermi constant and ε is a CP-violation parameter. We immediately see that Eq.(45) is compatible and pretty close to the observed experimental number, as given by

$$\frac{n_B}{n_\gamma} \sim 10^{-9\pm1} . \tag{46}$$

2. All this is fine, though, some people may have a hard time swallowing the idea that all our existence depends on (superheavy) Higgs bosons, while even the low energy Higgsons seem to be trouble-makers (of the gauge hierarchy problem) and indeed they have very elusive properties. Some other people may have difficulties in believing, that a Baryon asymmetry created so early has not been erased one way or another, before the temperature come down to 1 GeV.

For example, the existence of intermediate interaction which violate baryon number (and any global symmetry involving baryon number), would impede the conservation of the already created baryon asymmetry. Maybe all such thoughts are too pessimistic, and everything at the end will turn out to be O.K, i.e., there are no problems in having Higgsons around, and there are no intermediate scale, B-violating interactions around, but let us keep an open mind and look for alternatives. The most obvious candidate is of course the superheavy gauge bosons. However, there are rather severe problems[23)] related to the creation of baryon asymmetry; through superheavy gauge boson decays. In this case the contribution of superheavy gauge boson decays to $\frac{n_B}{n_\gamma}$ is

$$\frac{n_B}{n_\gamma} \simeq (1.5\cdot 10^{-1})\ (\Delta B)_G \cdot [160\ (\frac{10^{15}\ \text{GeV}}{M_G})]^{-1.3} \tag{47}$$

where the factor is square parentheses in Eq.(47) represents the suppression effects of $2\leftrightarrow 2$ interactions which tend to dilute the asymmetry generated by gauge boson decay and inverse decays. The factor $(\Delta B)_G$ is the CP-violating asymmetry in the decays of gauge particles and antiparticles. The parenthesised suppression factor in Eq.(47) is quite fierce if $M_G \simeq 6\cdot 10^{14}$ GeV as expected[23)] in minimal SU(5). Also in simple models, $(\Delta B)_G$ of Eq.(47) turns out to be one order higher in α_G than $(\Delta B)_H$ of Eq.(45). Moreover Eq.(45) does not involve any suppression factor, at least for a physically interesting range of mass and coupling parameters. Simply, superheavy gauge bosons seem to be a disfavoured candidate for the creation of baryons. What is left? Very natural candidates are superheavy fermions. I would like now to discuss some work[33)] that R. Barbieri, A. Masiero and myself have been doing at CERN, concerning a possible mechanism to create baryons through superheavy fermion decays. Grand unified theories, except the minimal SU(5) model, contain naturally superheavy fermions. Even the minimal SU(5) model may be trivially extended to include superheavy fermions. After all, if there are low and superheavy mass gauge bosons, low and superheavy mass Higgs bosons, why not low and superheavy mass fermions?

3. Let us concentrate on a special category of superheavy fermions, those that decay mainly to three light fermions through the exchange of a superheavy Higgs boson. The reader may envisage, this decays as a kind of $\mu\to e\bar{\nu}_e\nu_\mu$ decay with the only difference that the W-boson is replaced by a Higgs boson. In this case the width is given by

$$\Gamma_F = (\frac{\lambda^2}{4\pi})^2 \frac{m_F^5}{12\pi\ M_H^4} \tag{48}$$

in an obvious notation, where λ indicates an average universal Higgs coupling, m_F is the mass of the superheavy fermion and $M_H \simeq 10^{15}$ GeV is the mass of the exchanged superheavy Higgs boson. Then we distinguish the following stages in the history of the superheavy fermions, during the cooling of the early Universe:

i Superheavy Fermions in Equilibrium

In general, the two body interactions mediated by superheavy bosons with rates $\simeq \alpha_G^2 \frac{T^5}{M^4}$, will remain in equilibrium down to a temperature

$$T_0' \simeq M\cdot\left(\frac{1.6\cdot N^{1/2}}{\alpha_G^2}\frac{M}{M_{P\ell}}\right)^{1/3} \simeq 10^{15}\text{ GeV} \tag{49}$$

($\alpha_G \sim \frac{1}{40}$ is a typical GUT fine structure constant)

where the expansion rate of the Universe

$$H = \frac{(1.6)T^2 N^{1/2}}{M_{P\ell}} \tag{50}$$

becomes dominant. Here, N=10-100 is the suitably weighted number of all particles with masses $<T_0'$, and $M\simeq 10^{15}$ GeV is a typical grand unified mass scale. Since some of these interactions will in general be B-violating, one expects no net baryon (or antibaryon) survival below T_0' *). If the superheavy fermions are not SU(3)×SU(2)×U(1) singlets, they will still be kept in equilibrium below T_0', or even below the mass, m_F, by annihilations and Compton scattering with "light" gauge bosons. In this case Eq.(49) should be replaced by (assuming $m_F<10^{14}$ GeV)

$$\left.\begin{aligned} T_0 &\simeq m_F\ [\ln Q + \tfrac{1}{2}\ln\ln Q]^{-1} \\ Q &\equiv \alpha_G^2\cdot\left(\frac{\pi}{8N}\right)^{1/2}\frac{1}{\pi^2}\frac{M_{P\ell}}{m_F} \end{aligned}\right\} \tag{51}$$

while the relative number density is given by

$$\left.\frac{n_F}{n_\gamma}\right|_{T_0} \simeq Q^{-1}\ln Q\ . \tag{52}$$

In the following we will use Eqs.(51) and (52), as being realistic and general.

*) It is a remarkable property of superheavy fermion interactions that they violate, in general, any global symmetry involving the B-number. With superheavy fermions around, it is difficult to keep alive any primordial asymmetry.

ii Superheavy Fermions out of Equilibrium

At later times ($T<T_0$), until they decay, the relative number density of the F's (Eq.(52) remains constant, with their energy density being dominated by their rest mass, $\rho_F \simeq n_F \cdot m_F$. Actually, at $T \simeq \frac{m_F}{3N}$, this same energy density (ρ_F) will start dominating the full energy density of the Universe, down to the temperature T_F of their decays. So, if this scenario has something to do with reality, it introduces a new epoch in the early Universe for $T_F < T < \frac{m_F}{3N}$, where the Universe is matter dominated! This new era may have some consequences for the subsequent evolution of the Universe.

iii Superheavy Fermions Decay

When the time comes, i.e. when

$$H \simeq \left(\frac{8\pi\rho_F}{3M_{P\ell}^2}\right)^{1/2} = \left(\frac{8\pi}{3M_{P\ell}^2} \frac{m_F \cdot T^3}{\pi^2} \left.\frac{n_F}{n_\gamma}\right|_{T_0}\right)^{1/2} \tag{53}$$

falls below the decay rate given by Eq.(48) superheavy fermions will decay. This happens at a temperature

$$T_F \simeq \frac{1}{10} \left(\frac{\lambda^2}{4\pi}\right)^{4/3} \left(\frac{m_F}{M}\right)^3 (MM_{P\ell}^2)^{1/3} \left(\left.\frac{n_F}{n_\gamma}\right|_{T_0}\right)^{-1/3} . \tag{54}$$

In these decays some generation of a baryon asymmetry is expected. In fact, the general scenario analyzed in the beginning, is applicable. F's can be identified as Myrtons and indeed one finds:

$$\frac{\tilde{n}_B}{n_\gamma} \simeq (10^{-3} - 10^{-9}) \left.\frac{n_F}{n_\gamma}\right|_{T_0} \tag{55}$$

depending on the parameters of the particular model. However, one should keep in mind that, since $\rho(T_F)$ is dominated by the rest mass of the F's, their decay will also reheat the Universe up to a temperature T_F', so that

$$\rho(T_F) \simeq \left(\frac{m_F T_F^3}{\pi^2}\right) \left.\frac{n_F}{n_\gamma}\right|_{T_0} \simeq \frac{3N\, T_F'^4}{\pi^2} \tag{56}$$

and in turn dilute the baryon asymmetry by a factor

$$d \simeq \left(\frac{T_F}{T_F'}\right)^3 = \left(\frac{3N\, T_F}{m_F}\right)^{3/4} \left(\left.\frac{n_F}{n_\gamma}\right|_{T_0}\right)^{-3/4} . \tag{57}$$

The final observed asymmetry will then be given by

$$\frac{n_B}{n_\gamma} = \frac{\tilde{n}_B}{n_\gamma} \cdot d \simeq (10^{-3} - 10^{-9}) \left(\frac{3N\, T_F}{m_F}\right)^{3/4} \left(\left.\frac{n_F}{n_\gamma}\right|_{T_0}\right)^{1/4} \equiv (10^{-3} - 10^{-9})d' . \tag{58}$$

Demanding that, the factor d' should not be less than 10^{-5}, and using Eqs.(52), (54) and (58), we put a lower limit on m_F, as well as on T_F:

$$\left.\begin{aligned} m_F &\gtrsim 10^{11} - 10^{12}\ \text{GeV} \\ T_f &\geqq 10^{3} - 10^{4}\ \text{GeV} \end{aligned}\right\} \qquad (59)$$

where a typical value $\frac{\lambda^2}{4\pi} \sim 10^{-4} - 10^{-5}$ has been taken. We notice that, we may create a baryon asymmetry through F's at temperatures as low as $\underline{T_F \simeq 1\ \text{TeV}}$. Even if Higgsons do not exist, or even if there are dangerous, "wipe-out" intermediate scale interactions still there is some hope inside GUTs to create a respectable baryon asymmetry at a very low energy scale (~1 TeV), through superheavy fermions. I believe, we do not have to go to extremes. The following scenario looks to me highly probable. As the Universe cools down, F's and H's eventually get out of equilibrium (in the case of F's at least their B-violating interactions get out of equilibrium). Then, the H's decay and may (if we have a complicated enough Higgs structure) create a substantial baryon asymmetry. As the Universe cools down further intermediate scale B-violating interactions (if they exist) may partially or completely erase any preexisting baryon asymmetry. Eventually, the Universe cools down, to a temperature at which the relevant F's decay and thereby provide some baryon asymmetry. This baryon asymmetry should be added up (hopefully constructively) to any preexisting baryon asymmetry, to make up what we observe today. GUTs have enough complexity to cover any exceptional case, and provide enough baryons to build up our Universe.

Finally, I will like to close this section with a few comments on the specific new mechanism proposed here.

1 The key point is that we should have around very long-lived superheavy particles, which make the important condition imposed by Eq.(43) trivial to satisfy, and at the same time create baryons of comparatively "low temperatures" (~1 TeV), so that subsequent "accidents" are easier to avoid.

2. Very long-lived particles are not strangers. Remember the protons! Actually, one may push the analogy further. Maybe, at some point there was a microscopic copy of our Universe. Galaxies and even "people" made up from "Superheavy protons" were around, etc., etc. The very existence of a "superheavy matter"-dominated era, as an intermezzo in a radiation dominated Universe, may have some important consequences for the subsequent evolution of the Universe.

3 I would also like to stress the fact, that the occurrence of the scale of 1 TeV, not far above the electro-weak phase transition, may not be accidental.

Since, our estimates are rather naive, one may think that what happens is that superheavy fermions may live up to the moment where the electroweak phase transition takes place. F's may even help a first order phase transition to occur faster. There are some amusing questions that deserve further thought.

4. If, indeed, superheavy fermions are the main source of baryon production, then the recent connection (see next section) between the electric dipole of the neutron and the observed baryon asymmetry does not hold at least in it's present form.

On the other hand, if this connection will be found to be experimentally unacceptable, and the other tacit assumptions, in its derivation do hold, then we will have A CASE FOR THE SUPERHEAVY FERMIONS.

The above discussions bring us naturally to our next topic.

IV. GUTs, the Neutron Electric Dipole Moment, Baryon Asymmetry and Galaxy Formation

1. As we saw, one of the most striking successes of Grand Unified Theories (GUTs) is the possibility of understanding, for the first time in physics, qualitatively and quantitatively the observed baryon asymmetry in the Universe[23]. GUTs contain all the basic ingredients for creating the baryon asymmetry, i.e. B, C and CP violation, and the expanding Universe provides for free an excellent way to get out of equilibrium at the appropriate times. A very interesting question which is often asked is whether there is some connection between the observed low energy CP-violation in the K-system and the CP-violation operating at superhigh energies and thus responsible for the observed baryon asymmetry. The usual answer is <u>no!</u> At superhigh energies there are more phases[35] than just the usual Kobayashi-Maskawa (K-M) phase and so any hope to find some connection between "low energy" and "superhigh energy" CP-violation seems to be extremely difficult[35], though, may be there is still some hope. Recently, J. Ellis, M.K. Gaillard, S. Rudaz and I have done some work[36] towards a connection that may exist between the magnitude of the (still unobserved) electric dipole moment of the neutron (d_n) and the magnitude of the observed baryon asymmetry ($\frac{n_B}{n_\gamma}$). This section is a review of that work, but with strong emphasis on the physical consequences of such a possible relation between d_n and $\frac{n_B}{n_\gamma}$ and less discussion of the technical details that lead to such a connection. The interested reader should consult Ref.36) to fill his/her self up with the details of the calculation.

2. The idea is very simple[36]. We all know that there is a possible contribution to the neutron electric dipole moment (d_n) through non-perturbative QCD effects. A quantitative expression[37] of this fact is given by

$$\Delta d_n \simeq 3\cdot 10^{-16}\, \theta \text{ e.cm.} \tag{60}$$

where θ is the famous "instanton angle" which shows up, as the coefficient multiplying a term meaningless in perturbation theory, but meaningful when non-perturbative effects are taken into account,

$$\theta \frac{g_3^2}{32\pi^2} \varepsilon_{\mu\nu\rho\sigma} G^{\alpha}_{\rho\sigma} G^{\alpha}_{\mu\nu} \,. \tag{61}$$

As usual, g_3 is the strong coupling constant and G stands for the gluon field. Notice that the term given by Eq.(61) violates P and CP invariance and thus legitimately contributes to d_n (see Eq.(60)). On the other hand, GUT interactions that contribute to the build-up of $\frac{n_B}{n_\gamma}$, do also contribute to θ and thus to d_n. Since GUT interactions are totally responsible for $\frac{n_B}{n_\gamma}$ and partially responsible for θ, at least in some cases, we should not be surprised if we succeeded in setting a lower bound on θ, and according to Eq.(60) on d_n, by using the presently observed lower bound on the baryon asymmetry[38)]

$$\left.\frac{n_B}{n_\gamma}\right|_0 \geq 2\cdot 10^{-10} \,, \tag{62}$$

where the subscript 0 indicates quantities measured at the present time.

3. Actually the present experimental upper bound on d_n is rather stringent[39)]

$$|d_n| \leq 6\cdot 10^{-25} \text{ e.cm} \tag{63}$$

which, through Eq.(60) implies

$$\theta \leq 2\cdot 10^{-9} \tag{64}$$

a rather absurdly small angle.

But why is θ so small? An honest response would be that we do not have yet a clear-cut answer. Notice that one of the most remarkable successes of perturbative QCD, i.e., automatic conservation of P, CP,··· invariance[1)] is lost, swamped under the non-perturbative effects. Let us see how the problem emerges. Suppose that we have the right to impose at some stage, the condition $\theta=0$. Then, in the absence of non-strong interactions, θ would remain zero in any order of the strong interactions, but the existence of non-strong interactions can mess things up! The quark mass matrix (M_q) receives in general, complex and off-diagonal contributions in each order of perturbation theory in the non-strong interactions, and thus the necessity emerges at imposing chiral rotations on the quark fields so that the redefined mass matrix is again real and diagonal. BUT, because of the colour anomaly any innocent, in perturbation theory, quark chiral rotation, becomes nasty because of the non-perturbative effects, i.e. we get a shift in the value of θ. So, even if we started with $\theta=0$, we are, generally speaking, going to end up with $\theta\neq 0$. The hope is that $\delta\theta$ is not going to violate the limit given by Eq.(64). The QCD θ-parameter appearing in calculations of the neutron electric dipole moment is presumably the renormalized value at some typical hadronic momentum scale of order 1 GeV. It can be written in the form!

$$\theta_{QCD}(1 \text{ GeV}) \simeq \delta\theta_{K-M} + \delta\theta_{GUT} + \theta_{TOE} , \qquad (65)$$

where $\delta\theta_{K-M}$ denotes the renormalization of θ at scales $\mu << M_X$ [$\sim 10^{15}$ GeV, the grand unification scale (GUT)] due to the weak interactions above, $\delta\theta_{GUT}$ denotes the renormalization of θ at scales $\mu = O(M_X)$ and θ_{TOE} (theory of everything) denotes the value of θ at some scale $\mu >> M_X$. It is attractive to believe that some ultimate theory will set $\theta_{TOE}=0$ at some "relaxation" scale M_X, perhaps at the Planck mass. In other words, we need to take into account all the "kicks" that θ suffers all the way from 10^{19} GeV down to 1 GeV. Ellis and Gaillard[40], in a heroic effort a few years ago, have calculated for us the Kobayashi-Maskawa "kick" to θ

$$\delta\theta_{KM} \simeq O(10^{-16}) \qquad (66)$$

a rather painless "kick". We do not know what the original θ_{TOE} is, but it seems reasonable to suppose that in order of magnitude

$$\theta_{QCD}(1 \text{ GeV}) \gtrsim \delta\theta_{GUT} \qquad (67)$$

and so we are out to calculate $\delta\theta_{GUT}$.

4. It is no secret by now that the most promising mechanism for baryon creation involves the decays of superheavy Higgs bosons[23]. As a first approximation the baryon-to-photon ratio is given by[34]

$$\left(\frac{n_B}{n_\gamma}\right)_{H\text{-decay}} \simeq \frac{N_H}{N}(\Delta B)_H \simeq (10^{-1} - 10^{-2})(\Delta B_H) \qquad (68)$$

where N_H and N are essentially the numbers of helicity states of scalars and of all particles in the theory, and $(\Delta B)_H$ is the baryon asymmetry intrinsic to Higgs decay. Since in the quark basis where gauge couplings are diagonal, CP-violation is confined to the Yukawa couplings, the leading order contribution to $(\Delta B)_H$ can be obtained by considering only Higgs exchange radiative corrections to the $H \to f_1 f_2$ vertex. But here is the HEART OF THE MATTER: the same type of corrections are contributing also to $\delta\theta_{GUT}$, which in general is given by the imaginary part of scalar-exchange radiative corrections to a mass insertion on a fermion loop divided by the uncorrected mass insertion. Thus, since there may be additional contributions to $\delta\theta_{GUT}$,

$$\delta\theta_{GUT} \geq (\Delta B)_H \, \mathrm{Tr}(H_d H_d^+) \tag{69}$$

where H_d is the lowest order coupling matrix of the decaying superheavy Higgs boson, and we guess that as an order of magnitude

$$\mathrm{Tr}(H_d H_d^+) \simeq \sqrt{2}\, G_F\, m_t^2 \geq 6\cdot 10^{-4} \tag{70}$$

by taking $m_t \geq 18$ GeV at present energies and rescaling to an effective $m_t \geq 6$ GeV at grand unification. Then combining Eqs. (60), (68), (69) and (70) we get

$$d_n \geq 2.5\cdot 10^{-18} \left(\frac{n_B}{n_\gamma}\right)_{H\text{-decay}} . \tag{71}$$

Assuming that all the GUT created baryon asymmetry is due to Higgs decays and allowing a subsequent dilution factor E (≥ 1 by definition), i.e.

$$\left.\frac{n_B}{n_\gamma}\right|_{GUT} \equiv \left.\frac{n_B}{n_\gamma}\right|_{H\text{-decay}} = E \left.\frac{n_B}{n_\gamma}\right|_0 . \tag{72}$$

Eq. (71) boils down to

$$d_n \geq 2.5\cdot 10^{-18}\, E \left.\frac{n_B}{n_\gamma}\right|_0 . \tag{73}$$

Actually, one finds in practice that, because the Higgs structure has to be rather "involved" in order to produce an acceptable amount of baryon asymmetry $\delta\theta_{GUT}$ gets contributions at least one power of α_G ($\sim \frac{1}{40}$, the GUT-fine structure constant) less than the ones coming from the baryon-asymmetry diagrams. Thus we are tempted to simplify Eq. (73) and take as our master formula

$$d_n \geq 2.5\cdot 10^{-18} \left(\frac{\alpha_G}{\pi}\right)^{-1} E \left.\frac{n_B}{n_\gamma}\right|_0 . \tag{74}$$

Before discussing in some detail the assumptions and possible loopholes in the derivation of Eq. (73) or (74), let me first exploit the physics implications of our bound (Eq. (74)).

5. There are some rather amusing consequences of the bound (74).

A. Particle Physics Phenomenology

i) Using the observed $\left.\frac{n_B}{n_\gamma}\right|_0$ (as given by Eq.(62)) and knowing that $E \geq 1$, Eq.(74) says that

$$d_n \geq O(6 \cdot 10^{-26}) \text{ e.cm} . \tag{75}$$

Just an order of magnitude below the present upper bound (see Eq.(63)). Hopefully we will know soon! Here a clarifying remark is needed. In the standard K-M model one gets at most[40),41)]

$$\left. d_n \right|_{K-M} \simeq 10^{-30\pm 1} \text{ e.cm} \tag{76}$$

while realistic multi-Higgs models of CP-violation[42)], may predict a d_n close to the bound given by Eq.(75), but they have a hard time explaining the observed CP-violation in the K-system ($\left|\frac{\varepsilon'}{\varepsilon}\right|$ too big, etc.)[43)]. So if these multi-Higgs models are out and since other ways of exorcising θ predict d_n to be at most as large as given by Eq.(76) (see below), the discovery of d_n with a value say between 10^{-25} and 10^{-27} e.cm. will be a great service to the community!

ii) We may also put a limit on the mass (m) of the heaviest possible quark coupled to H_d in a very simple way. Remember that in deriving Eq.(71) (and thus Eq.(74)) we have used Eq.(71) with $m_t \geq 6$ GeV, at M_X. If we let m as a free parameter, we get

$$d_n \geq 6 \cdot 10^{-26} \; E \left(\frac{m}{6 \text{ GeV}}\right)^2 \tag{77}$$

which combined with the experimental bound (Eq.(63)) entails that

$$m \leq 60 \text{ GeV} \tag{78}$$

at "low energies", a rather restrictive limit. Please notice that Eq.(78) does not apply only to the top-quark mass but to any possible heavy quark that may couple to H_d. We come now to the:

B. Cosmic Phenomenology

i) Clearly Eq.(74) combined with the present experimental upper bound on d_n (Eq.(63)) may put a limit on

$$\left.\frac{n_B}{n_\gamma}\right|_{GUT} \equiv E \left.\frac{n_B}{n_\gamma}\right|_0 \leq O(2 \cdot 10^{-9}) \tag{79}$$

or using the definition of $E(E\geq 1)$ and the observed lower bound on $\left.\frac{n_B}{n_\gamma}\right|_0$ (Eq.(62)) we put a bound on

$$1 \leq E \leq O(10) . \tag{80}$$

The bound of Eq.(80) is of fundamental importance. It says that the Universe is not allowed to "hiccup" too much, subsequent to baryon generation. In other words, during the whole history of the evolution of the Universe from say $T\sim 10^{15}$ GeV till now ($T_0\sim 3°K$) not too much entropy is allowed to be produced. Especially the bound (80) puts severe constraints on the electroweak (~100 GeV) phase transition. It is well-known that because of the supercooling effect, a first order phase transition creates a lot of entropy, which in turn dilutes[44] the baryon asymmetry much more than is allowed by Eq.(80). With our present knowledge, a first-order electroweak phase transition and our bound on E (Eq. (80)) do not look to be compatible. So, if we should eventually be convinced that indeed a large entropy factor E must have been generated after the GUT-baryon generation, e.g. during the electroweak phase transition, then our bound (80) would be violated and our picture would be wrong!

ii) The bound on $\left.\frac{n_B}{n_\gamma}\right|_{GUT}$ as given by Eq.(79), may be very important in restricting the amount of any possible primordial shear[36),45)]. For illustrative reasons we use a Bianchi-I type of cosmology

$$H^2 = \frac{8\pi G_N\rho}{3} + \frac{\sigma^2}{2} = \frac{8\pi G_N\rho}{3}(1 + \Sigma(T)) \tag{81}$$

where as usual, ρ is the energy density and $\Sigma(T)$ denotes the ratio of the shear term (σ) to the energy density (ρ)

$$\Sigma(T) \equiv \frac{\sigma^2}{8\pi G_N\rho} \propto T^2 . \tag{82}$$

The success of the standard Robertson-Walker-Friedman (R-W-F) cosmology during nucleosynthesis implies[45)] $\Sigma(1\text{ MeV})<1$ or via Eq.(82)

$$\Sigma(M_X) \simeq \Sigma(1\text{ MeV})\left(\frac{M_X}{1\text{ MeV}}\right)^2 \leq 10^{36} . \tag{83}$$

But we can do much better! Too much primordial shear would create through drastic dissipative processes (involving grand unified viscosity[46)]) too much entropy, i.e. too big dilution factor that would violate our bound (see Eq.(80)). Thus an upper bound may be imposed on $\Sigma(M_X)$. After some non-trivial gymnastics we find[36)]

$$\Sigma(M_X) = 10^{10}\, E^{8/3} \le 10^{10} \left(\frac{d_n}{6\cdot 10^{-26}\ \text{e.cm}}\right)^{8/3} \tag{84}$$

or using Eq.(80) or equivalently Eq.(63) we get[36)]

$$\Sigma(M_X) \le 10^{12} . \tag{85}$$

Just <u>24 orders of magnitude</u> below the previously existing bound on $\Sigma(M_X)$ (see Eq.(83))! Notice that Eq.(85) is comfortably far above the lower limit on $\Sigma(M_X)$ ($\delta\Sigma(M_X) \sim 10^{-3}$) that may be needed for Galaxy Formation[45)]. Actually, if recent data[47)] indicating the existence of a quadrupole anisotropy in the cosmic background radiation are confined beyond doubts, then some primordial shear may be indeed needed[45)] if we want to take seriously the cosmological implications of GUTs. The point being that in the GUT scenario, since $\frac{n_B}{n_\gamma}$ is constant in space, only adiabatic density fluctuations are allowed in the standard R-W-F cosmology. Then one shows[48)] that adiabatic fluctuations alone cannot consistently explain the observed dipole and quadrupole anisotropy in the background radiation. One may need[48)] isothermal fluctuations arising e.g. by deviating from a strict R-W-F cosmology by adding a shear term à la Eq.(81). So our new bound on $\Sigma(M_X)$ ($\le 10^{12}$) may not be of only academic interest.

6. Finally let us discuss the assumptions and possible "traps" that go into the derivation of our Master Formula (74).

We have assumed that there is <u>no symmetry</u> that sets $\theta=0$, except possibly that some ultimate theory may set $\theta_{TOE}=0$ at some "relaxation" scale >> M_X, say at the Planck mass. Still we have found that the renormalization of θ from all kinds of non-strong interactions lies comfortably below the present experimental upper bound given by Eq.(64). We have assumed that the GUT-Higgs decays are the dominant contributors to the observed baryon asymmetry. Furthermore, we have assumed that the Higgs responsible for the fermion masses and the Higgs responsible for the baryon asymmetry are not orthogonal. For example in the SU(5) model the triplet of the $\underline{5}_d$ ($\equiv H_d$) that creates the baryon asymmetry and the doublet of the $\underline{5}_m$ ($\equiv H_m$) that is responsible for fermion masses, should at least partially lie in the same multiplet, e.g. $\underline{5}_d \perp \underline{5}_m$ should not happen.

Suppose now that d_n is found <u>not</u> to satisfy our bound, as given by Eq.(74): what then? Trivial ways out would be that $H_m \perp H_d$ or that GUT-Higgs decays were not the dominant contributors to the observed baryon asymmetry. Since gauge bosons have difficulties in producing large amounts of baryons, the only candidate left would be superheavy fermions, which as we discussed in section III may provide viable contributions to cosmic baryon production. A more advanced, less trivial way out would involve the existence of <u>a symmetry</u> that keeps $\theta=0$ naturally to high accuracy. An obvious candidate would be a Peccei-Quinn symmetry[49], as we will discuss in detail in the next section.

V. GUTs and a Solution to the Strong CP-Problem

1. One of the most severe problems facing unified gauge theories today is the so-called strong CP violation problem. Why is the famous, non-perturbatively created θ parameter at least smaller than 10^{-9}? The most natural solution is to make θ the "unobservable" phase of an U(1) global chiral symmetry. Since there are no massless quarks around, Peccei and Quinn suggested[49] that the theory should necessarily be invariant under a built-in U(1) symmetry, called here $U(1)_{PQ}$. This $U(1)_{PQ}$ symmetry will eventually be spontaneously broken, say at some scale M and since $U(1)_{PQ}$ is also explicitly broken by the Adler-Bell-Jackiw anomaly, a pseudo-Goldstone boson called axion[50] will be formed with mass

$$m_A \simeq \frac{f_\pi m_\pi}{M} \tag{86}$$

and a pseudoscalar coupling

$$f_A \simeq \frac{f_\pi}{M} \quad . \tag{87}$$

Peccei and Quinn identified the scale M with the scale M_W of unification of weak and electromagnetic interactions. Negative experimental searches[51] and astrophysics[52] entail that

$$M > 10^9 \text{ GeV} \; . \tag{88}$$

Recently Dine, Fischler and Srednicki (DFS)[53] made the observation that a natural candidate for M is the grand unification scale $M_X \sim 10^{15}$ GeV. An explicit realization of this idea in SU(5) has been given in Ref.54). Here we present[55] an SU(5) model where the DFS trick is naturally realized since the highly desired $U(1)_{PQ}$ symmetry is part of the global symmetry that is needed in order to produce a hierarchical fermion mass spectrum.

2. Before getting to the specific example[55], let us discuss in general terms how we believe that the $U(1)_{PQ}$ appears and why it is broken at superhigh energies. As is well known, grand unified theories do not solve the family problem. We need something more, so we should not be surprised if, at some energies above or equal to M_X ($\sim 10^{15}$ GeV), something like $G_{world} \to G_{GUT} \times G_{family} \to G_{GUT} \times U(1)^n$ is happening. To us, the family problem and the hierarchical fermion mass spec-

trum are closely interrelated, if not two views of the same coin. So by buying this bigger symmetry, we assume that we produce three families and hierarchical fermion masses. Then, it is easy to see that some combination of the n U(1)s may be identified with the $U(1)_{PQ}$ which will be spontaneously broken at 10^{15} GeV thus creating a PHANTOM AXION and exorcising sizable strong CP violation.

3. We next move to a specific example materializing the above scenario[55]. Essentially we use the SU(5) model of section IIA, where the Yukawa coupling is given by

$$L_Y = a\,\bar{5}_3 10_3 \bar{5}_H + b\,10_3 10_3 5_H + c\,\bar{5}_2 10_3 \bar{5}'_H + d\,10_2 10_2 5'_H$$

$$+ e\,10_3 10_2 \overline{50}_H + f\,10_3 10_1 \overline{50}'_H + g\,5_1 5_3 \overline{10}_H + \text{h.c.} \tag{23}$$

Here, 5_i, 10_i denote the usual three fermion generations and the fields with a subscript H are scalars. The above couplings a, b, c, d <u>follow uniquely</u> if one requires that the first and second generations get only two-loop and one-loop masses, respectively, and charm is not massless at the one-loop level, while the third generation has a direct mass. The couplings to the other Higgs fields (notice that $\langle 10\rangle=\langle 50\rangle=0$ because of colour - and charge conservation) are the minimal ones in order to obtain the above masses, e.g., m (first generation): m (second generation): m (third generation)$=\alpha^2:\alpha:1$.

L_Y has, in addition to the local SU(5) symmetry, a global $U(1)^4$ symmetry, given by

$$U(1)_A:\quad 5_2 \to e^{i\alpha}5_2 \quad 5_3 \to e^{i3\alpha}5_3 \quad 10_3 \to e^{i\alpha}10_3$$

$$10_H \to e^{i3\alpha}10_H \quad 5_H \to e^{-2i\alpha}5_H \quad 50_H \to e^{i\alpha}50_H \quad 50'_H \to e^{i\alpha}50'_H \tag{89}$$

$$U(1)_B:\quad 10_2 \to e^{i\beta}10_2 \quad 5_2 \to e^{i2\beta}5_2$$

$$5'_H \to e^{-2i\beta}5'_H \quad 50_H \to e^{i\beta}50_H \tag{90}$$

$$U(1)_C:\quad 5_1 \to e^{i\gamma}5_1 \quad 10_H \to e^{i\gamma}10_H \tag{91}$$

$$U(1)_D:\quad 10_1 \to e^{i\delta}10_1 \quad 50'_H \to e^{i\delta}50'_H\,, \tag{92}$$

fields not occurring behave neutrally under the respective U(1)s. The four U(1)s protect the lighter fermions from getting a direct mass and ensure B-L conservation. Notice that all U(1)s are anomalous*).

In order that the light quark masses are non-zero, the U(1)s must be broken. In section IIA, we considered only breaking by soft terms in the Higgs potential; in this section we will modify this. The minimal terms needed to break the chiral symmetries and to obtain satisfactory Cabibbo mixings are

$$L_H^3 = A\,(\overline{10}_H\, 5_H 5'_H) + B\,(10_H 10_H \overline{50}_H) + C\,(5_H 50'_H 75_H) + D\,(5'_H 50_H 75^+_H) + \text{h.c.} \qquad (93)$$

where we have introduced a (complex) 75-dimensional Higgs multiplet. The four terms then transform under (89)~(92) as follows:

$$\begin{array}{lllll} U(1)_A : & A \to e^{-i5\alpha}A & B \to e^{i5\alpha}B & C \to e^{i(-1+\bar a)\alpha}C & D \to e^{i(1-\bar a)\alpha}D \\ U(1)_B : & -2 & -1 & \bar b & -1-\bar b \\ U(1)_C : & -1 & 2 & \bar c & -\bar c \\ U(1)_D : & 0 & 0 & 1+\bar d & -\bar d \end{array} \qquad (94)$$

where $\bar a$, $\bar b$, $\bar c$, $\bar d$ are the transformation properties of the 75 under the U(1)s [e.g.; $U(1)_A$: $75 \to e^{i\bar a\alpha}75$].

One might ask whether (93) must not be amended by additional terms, transforming like A, B, C, D under the U(1)s acting as renormalization counterterms. For example, take a term $\bar 5_H \overline{50}_H 75^+_H$, which transforms like D. It can be constructed from a one loop-diagram involving D and the invariant term $50_H 50_H 5_H 5'_H$. But since it is linear in D its renormalization is multiplicative and need therefore not be added to L^3 [56] as an extra counterterm. We have verified that L^3 is indeed complete.

In section IIA the 75_H was taken to be real ($\bar a=\bar b=\bar c=\bar d=0$). Then all U(1)s are explicitly broken by L^3. But if 75_H is complex we may ask if there exists an unbroken $U(1)=xU_A+yU_B+zU_C+sU_D$ with appropriate x, y, z, s. Using rules (24) we find**) this to be the case if

$$x = -1 \quad y = 1 \quad z = 3 \quad s = 1 \qquad (95a)$$

$$-\bar a + \bar b + 3\bar c + \bar d = -2\ . \qquad (95b)$$

---

*) This is so although the U(1)s are not "chiral" in the usual sense.

**) Of course, x, y, z, s are given only up to a common factor.

Equation (95b) indicates that the 75 transforms <u>non-trivially</u> under U(1), in fact $75_H \to e^{-2i\phi} 75_H$. For further use we notice that U(1) is also anomalous.

In the following we take the 75 to satisfy (95b). Before continuing we determine if there exist terms such as $(5'_H 5'_H) 75_H$, $(5_H \bar{5}'_H) 75^2$ *) in the Higgs potential (see below). The first term is not allowed because of (95b). Depending on the choice of $\bar{a}$, $\bar{b}$, $\bar{c}$, $\bar{d}$, the second may be allowed ($\bar{a}=1$, $\bar{b}=-1$, $\bar{c}=\bar{d}=0$) or not. We will consider both cases.

We now see that U(1) can be taken as PQ symmetry. (For a different realization see below). That is, it can be used to rotate the vacuum angle way in the tree Lagrangian.

When the 75_H and then the 5_H, $5'_H$ get a non-vanishing vacuum expectation value we expect a massless (Goldstone) boson associated with the breaking of the U(1), which is a mixture of neutral components of the 75_H, 5_H and $5'_H$. Since the vacuum expectation value of the 75_H is much bigger than the ones of the 5_H, $5'_H$, we expect it to lie predominantly in the 75_H with a small coupling to the 5_H and $5'_H$ of order $M_W/M_{GUT} \sim 10^{-13}$, as pointed out by DFS[53]. At this stage the terms

$$k(\bar{5}'5)_1 (75^2)_1 \quad \text{and} \quad k'(\bar{5}'5)_{24}(75^2)_{24} \tag{96}$$

are important. It is easy to see that the following combinations of neutral scalar**) field form mass eigenstates (in an obvious notation):

$$① \quad \frac{U_5 f_5 + U_{5'} f_{5'}}{\sqrt{U_5^2 + U_{5'}^2}} \to \text{eaten by Z-boson} \tag{97a}$$

$$② \quad \frac{1}{N}\left(\frac{-U_{5'} f_5 + U_5 f_{5'}}{\sqrt{U_5^2+U_{5'}^2}} + \frac{2U_5 U_{5'}}{\sqrt{U_5^2+U_{5'}^2}\cdot V_{75}}\cdot f_{75}\right) \quad \text{massive state} \tag{97b}$$

$$\text{with mass} = \frac{\sqrt{2|\bar{k}|}\cdot V_{75}\cdot\sqrt{U_5^2+U_{5'}^2}}{\sqrt{U_5 U_{5'}}}$$

where $\bar{k}$ is given by that linear combination of k and k' in which the doublet

---

*) There are, of course, two terms of this form in SU(5) : $(\bar{5}'5)_1(75^2)_1$ and $(\bar{5}'5)_{24}(75^2)_{24}$ in an obvious notation.

**) We write the neutral component of a Higgs multiplet as $\phi = e^{i\beta}(U+\eta+if)$ where U is a real number, β the phase of the vacuum expectation value and η and f real fields.

pieces of 5_H , $5'_H$ enter and

$$N^2 = 1 + \frac{4U_5^2 U_{5'}^2}{V_{75}^2 (U_5^2 + U_{5'}^2)}$$

$$③ \quad \frac{1}{N} \left(\frac{-U_{5'} f_5 + U_5 f_{5'}}{(U_5^2 + U_{5'}^2)} \cdot \frac{2U_5 U_{5'}}{V_{75}} - f_{75} \right) . \qquad (97c)$$

This last state will be identified with the axion, for when the instanton effects are taken into account, it is not strictly massless but has a mass $m \sim \Lambda_{QCD}^2/M_X \sim 10^{-15}$ GeV. This bears out the above expectations and ensures that the axion's coupling to matter is negligible.

As was pointed out before, the terms (96) are only possible if $\bar{a}=1$, $\bar{b}=-1$, $\bar{c}=\bar{d}=0$. Then both terms occur and in general one phase is allowed since only one of the k, or a suitable linear combination can be made real by appropriately choosing the 75_H. If the above assignment of $\bar{a}$, $\bar{b}$, $\bar{c}$, $\bar{d}$ is not met, the terms (96) vanish at tree level; but it is easy to see that they are generated in higher order. (See, for example, Fig. 9.) Notice that in this case there

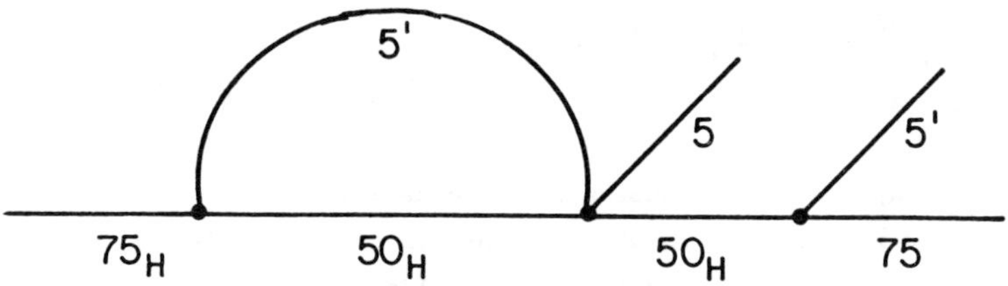

Fig. 9. Effective $5_H^+ 5'_H\ 75_H 75_H$ coupling.

is only one term of the form $(75_H^2) 5_H^+ 5'_H$, as seen from a group theoretical analysis of Fig. 9. Of course, in even higher order, both terms (96) may be generated, which can have interesting consequences for the phenomenology of CP violation. While the size of the coupling corresponding to the figure is reduced, say by a factor $\sim 10^{-2}$, the above results about the mass and coupling of the axion are altered only slightly.

In the above analysis we have taken B of the four U(1)s to be broken softly. In the light of the above discussion it is easy to see how to construct a situation in which all U(1)s are broken spontaneously without observable (Goldstone bosons). For example, we can modify L^3 as follows

$$L^3_H \rightarrow L^{3'}_H = A\, 5_H 5'_H \overline{10}_H \phi^+_1 + \cdots \qquad (98)$$

where the ϕ_i are complex singlets and the 75 might be taken real. If the ϕ_i transform under the U(1)s just opposite to the four couplings in Eqs.(94), then the U(1)s are good symmetries. (Any of them could be taken as PQ symmetry.) Spontaneous symmetry breaking, due to $<\phi_i>\approx M$, gives four essentially uncoupled massless scalars, thus preserving all nice results of section IIA and of this section.

4. Next, we wish to comment on the cosmological baryon generation. Unlike the model presented in Ref.54), the model described here leads, in general, to baryon generation in accord with observations. It will be instructive to see[57),58)] why the SU(5) hypoaxion model of Wise, Georgi and Glashow (WGG)[54)] yields a too-small[57),58)] cosmological baryon to photon ratio, so that future model builders know what they must avoid. This model contains two $\underset{\sim}{5}$ of Higgs $\underset{\sim}{H}_1$ and $\underset{\sim}{H}_2$ and a complex $\underset{\sim}{24}$ of Higgs Σ. The hypoaxion is the anti-Hermitian SU(3)×SU(2) singlet from Σ, while there are two colour triplet mass eigenstates $\underline{H}$ and $\underline{\tilde{H}}$ which are arbitrary orthonormal complex superpositions of the colour triplet fields $\underline{H}_i$ in $\underset{\sim}{H}_i$ (i=1,2):

$$\left.\begin{aligned} \underline{H} &= a\underline{H}_1 + b\underline{H}_2 \\ \underline{\tilde{H}} &= b^*\underline{H}_1 - a^*\underline{H}_2 \end{aligned}\right\} \quad |a|^2 + |b|^2 = 1\ . \qquad (99a)$$

Similarly there are two SU(2) doublet mass eigenstates H (which is hierarchically light) and $\tilde{H}$ (which is heavy). They are in general different orthonormal complex superpositions of the SU(2) doublet fields H_i in $\underset{\sim}{H}_i$ (i=1,2):

$$\left.\begin{aligned} H &= \alpha H_1 + \beta H_2 \\ \tilde{H} &= \beta^* H_1 - \alpha^* H_2 \end{aligned}\right\} \quad |\alpha|^2 + |\beta|^2 = 1\ . \qquad (99b)$$

To realize the U(1) Peccei-Quinn[49)] symmetry, only $\underset{\sim}{H}_1$ couples $\underset{\sim}{\bar{5}}$ to $\underset{\sim}{10}$ fermions with a coupling matrix A, while only $\underset{\sim}{H}_2$ couples $\underset{\sim}{10}$ fermions to $\underset{\sim}{10}$ fermions with a symmetric coupling matrix K. At this point we begin to smell trouble because we recall that the Georgi-Glashow[5)] (GG) minimal version of the SU(5) model with only one $\underset{\sim}{5}$ of Higgs fields also had only two coupling matrices to fermions. This paucity prevented the GG model from obtaining a non-vanishing (n_B/n_γ) through diagrams of lower than 8th order[59)], which gave an unacceptably small number. Our only hope of getting $(n_B/n_\gamma)\neq 0$ from a lower order fermion trace in the WGG model[54] is to pick up extra phase factors from the complex superposition parameters (a,b,α,β) in Eqs.(99). It is immediately obvious that no such factor is forthcoming from the 4th order trace of Fig.10. It is easy

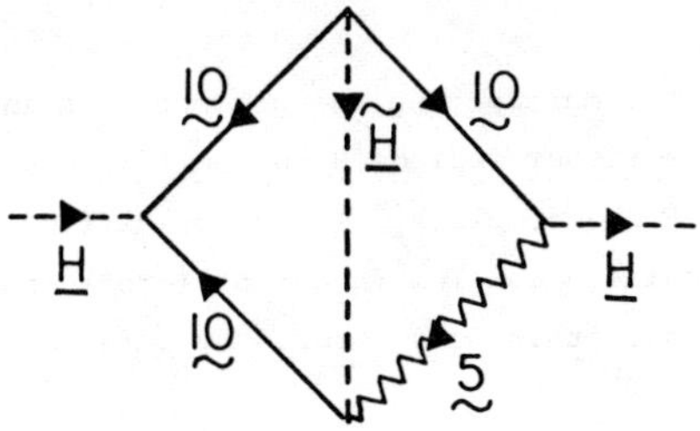

Fig. 10. Fourth order fermion trace which is a candidate for generating a baryon asymmetry

to prove that the a's and b's do not provide a complex factor for any diagram of the generic type of Fig.11, where there is a single fermion trace with the

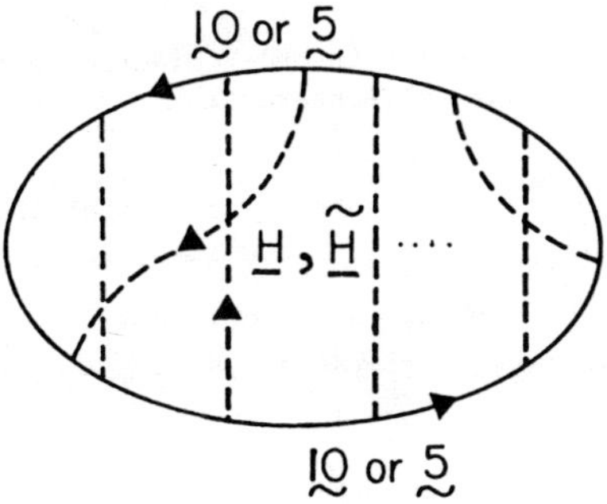

Fig. 11. Generic class of diagrams with one fermion trace and no Higgs self-interactions.

colour triplet Higgses not interacting between their emission and absorption. If one emits n $\underline{H}$ and m $\tilde{\underline{H}}$ Higgs from $\underset{\sim}{5}$-$\overline{\underset{\sim}{10}}$-Higgs vertices, and absorbs n' H and m' $\tilde{H}$ Higgs on $\bar{\underset{\sim}{5}}$-$\underset{\sim}{10}$-Higgs vertices, while emitting k $\underline{H}$ and ℓ $\underline{H}$ from $\underset{\sim}{10}$-$\underset{\sim}{10}$ Higgs vertices and absorbing k' $\underline{H}$ and ℓ' $\tilde{\underline{H}}$ on $\overline{\underset{\sim}{10}}$-$\overline{\underset{\sim}{10}}$-Higgs vertices, then one has the obvious sum rules

$$k' = k + n - n' , \quad \ell' = \ell + m - n' , \quad m' = n + m - n' . \tag{100}$$

The resulting over-all prefactor from the complex mixing parameters (99a) is just

$$(b)^{n+\ell+m-m'}\ (b^*)^{n'+\ell}\ (a)^{m+k}\ (a^*)^{k+n-n'+m} \qquad (101)$$

which is clearly real.

To get a complex factor multiplying the product of A and K coupling matrices we should therefore either include Higgs self-interactions, and/or include exchanges of SU(2) Higgs doublets. It is not difficult to verify that while the a's and b's provide a complex factor of (a*b*) coming from the fourth order fermion trace in Fig.12 this is countered by a factor of (ab) coming from

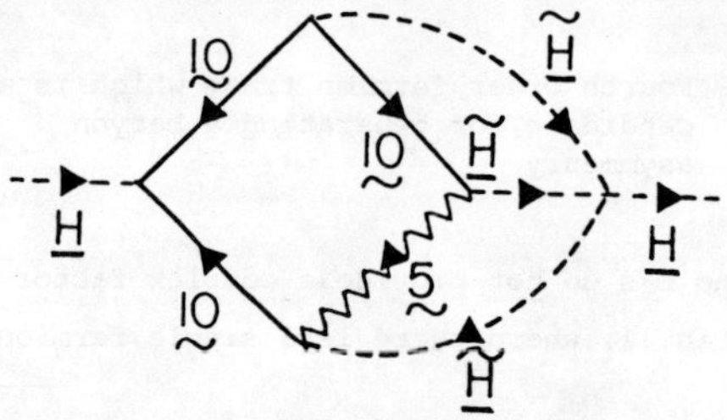

Fig. 12. Candidate sixth order diagram featuring a Higgs self-interaction.

the four-Higgs vertex, and one can then see easily that the same real catastrophe befalls more complicated diagrams with self-interactions of $\underline{H}$ and $\tilde{\underline{H}}$. [This is most easily seen by working in the original (H_1,H_2) basis.] On the other hand, the 6th order diagram of Fig.13, where one colour triplet Higgs

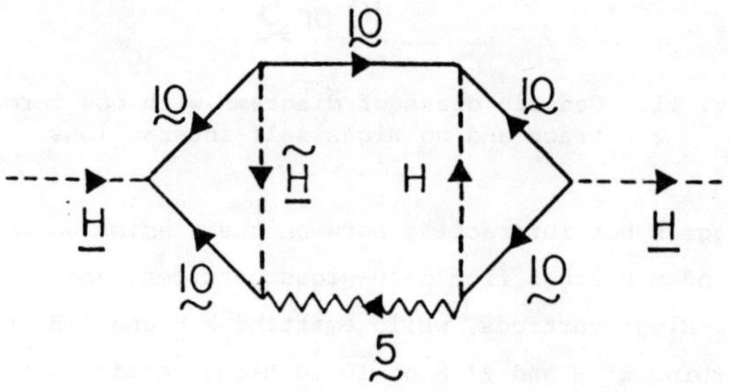

Fig. 13. Candidate sixth order diagram featuring the exchange of a Higgs triplet and a Higgs doublet.

and one SU(2) doublet Higgs is exchanged, does acquire a complex factor from the product of complex mixing parameters from Eq.(99). However, there is a mirror image diagram with the triplet and doublet exchanges interchanged, which has an equal and opposite imaginary part so that the sum is real and no (n_B/n_γ) is generated.

The above arguments demonstrate that there is no diagram before 8th order which contributes to (n_B/n_γ) in the minimal WGG SU(5) hypoaxion model[54]. Baryon number generation is therefore probably comparable to that estimated[59] in the minimal SU(5) model with a single $\underset{\sim}{5}$ of Higgs, which is far too small unless there is an implausible fourth generation of heavy "light" fermions with mass $O(m_W)$. However, one can easily construct richer SU(5) hypoaxion models which give (n_B/n_γ) large enough, e.g., with two 5 of Higgses coupled to each of the $\underset{\sim}{\bar{5}}$-$\underset{\sim}{10}$ and $\underset{\sim}{10}$-$\underset{\sim}{10}$ combinations of fermions, so this is not a mortal blow to the hypoaxion concept. Another example would be our version[27),55] of the SU(5) model described above.

Broaching the subject of cosmology, it is reasonable to ask whether the hypoaxion could produce any observable cosmological effects. Despite the weakness of its interactions, its presence would affect the rate of expansion during nucleosynthesis[60),61]. There is an upper limit from ^{4}He abundance on the ratio ξ of the Universe's actual expansion rate, corresponding roughly to the addition of one extra two-component light neutrino[61]:

$$\xi^2 - 1 < 0.15 \; . \tag{102}$$

By comparison, if the hypoaxion decoupled thermally from the rest of the Universe at a temperature T_A : $m_X > T_A > m_W$, one would expect[60]

$$\xi^2 - 1 \approx 4.4\times10^{-3} \tag{103a}$$

whereas if $m_X < T_A$ one would expect[61]

$$\xi^2 - 1 \lesssim 2.5\times10^{-3} \; . \tag{103b}$$

Both these estimates are way below the allowable limit (102), and it seems unlikely that such a small indirect effect of the hypoaxion could ever be inferred or excluded by cosmological observation.

5. A legitimate and experimentally relevant question to ask is the magnitude of θ (if different from zero) in hypoaxion GUT models. Certainly the estimates (73) and (75) do not apply to GUTs with a hypoaxion, in which bizarre cancellations are fated to occur. In such models there is a global Peccei-Quinn[49]

U(1) symmetry which is only broken by non-perturbative QCD effects - or possibly by effects associated with gravitation? We expect the QCD θ parameter to be zero[58),62)] up to contributions vanishing when $\Lambda_{QCD}\to 0$. It seems likely[58),62)] that the dominant contribution to θ may arise from a mismatch in phases between the effective chiral symmetry breaking $(\bar{\psi}\psi)$ and $(\bar{\psi}\psi)^2$ operators appearing when heavy non-strong degrees of freedom are integrated out. In this case one expects[58),62)]

$$\theta \approx O(\alpha^2)\Lambda^3_{QCD}\ m_q/m_W^4 \times \text{(product of phases and mixing angles)} \qquad (104)$$

which is certainly $\lesssim O(10^{-12})$. The coefficient and phase of one such $(\bar{\psi}\psi)^2$ operator, namely $(\bar{s}d)^2$, are known phenomenologically from the K_S-K_L system. They suggest that a reasonable estimate of Eq.(104) for θ may be more like

$$\theta \approx \left|\frac{\mathrm{Im}\ M_{12}}{m_K}\right| \approx O(10^{-17})\ . \qquad (105)$$

Inserting this estimate into Eq.(60) gives a contribution to d_n which is if anything smaller than the previous[40),41)] perturbative Kobayashi-Maskawa estimate, which is anyway too small to be seen in the foreseeable future.

We see one possible way of getting a contribution to θ which is larger than (104, 105). It is generally believed that Nature abhors global symmetries, and the Peccei-Quinn U(1) is surely no exception. We already know that it is broken by non-perturbative gauge theory effects, and may also suspect that it could be broken by (non-perturbative ?) phenomena occurring at the Planck scale. For example, one could imagine[63)] such effects generating fermion-fermion-multiHiggs or $(\mathrm{Higgs})^{n>4}$ interactions which did not respect the $U(1)_{PQ}$ symmetry. These could give rise to a hypoaxion mass larger than otherwise expected, and a contribution to of order $(\mathrm{phases})\times(m_X/m_{P\ell})^{n-4}$ which could in principle be much larger than (104,105). For this reason, it may well be necessary that any such interactions either respect the $U(1)_{PQ}$ symmetry or participate in a mechanism akin to Weinberg's "vacuum relaxation"[62)], in order to suppress θ below the experimental limit, while perhaps being larger than Eq.(105).

Barring this exotic possibility, detection of d_n in the next few years would be indirect evidence against an axion model, or indeed any other theory with a Peccei-Quinn U(1) symmetry. Conversely, non-observation of d_n at a level significantly below the order-of-magnitude "bound" (73) would be circumstantial evidence that the GUT does indeed possess some hitherto inapparent symmetry, very possibly of the $U(1)_{PQ}$ type. Independently of measurements of d_n, if one became convinced that a large entropy factor E must have been created after baryon number generation, e.g., during the Weinberg-Salam phase transition[44], one might become convinced already that the bound (73) was violated, in which case one could be forced into a hypoaxion model.

6. If, in accord with Dine, Fischler and Srednicki, we identify the scale M in Eq. (86) and (87) with the grand unification scale M_X (~10^{15} GeV) then we get

$$m_A \sim 10^{-8} \text{ eV}$$

$$f_A \sim 10^{-16} . \qquad (106)$$

As a result, it is important to find out the strength and the nature of its coupling to light particles at low energies. It has been noted already that the axion coupling to light fermions at the tree level is pseudoscalar in nature and has a strength ~f_π/M_X, where f_π is the pion decay constant and M_X is the grand unification mass (~10^{15} GeV). In the non-relativistic limit therefore, it gives rise to spin dependent couplings between matter with couplings, much smaller than the presently known bounds[64]. However, it is conceivable that in the presence of weak CP and P violating couplings present in the Lagrangian, one could generate a scalar coupling of axions to fermions. Such couplings would give rise to a modification of the gravitational potential GMm/r ($ae^{-(r/\lambda_a)}$), where λ_a is the Compton wavelength of the axion (~20 meters in this case) and could be detectable if a is not too small. R. Barbieri, R.N. Mohapatra, D. Wyler and myself have shown that[58], in the absence of instanton effects, the Goldstone boson nature of the axion forbids the appearance of scalar fermion couplings to all orders in perturbation theory in weak interactions. This implies that a scalar fermion coupling (if any) will arise only due to the presence of instanton effects and is therefore of order $(f_\pi/M_X)\times\theta \lesssim 10^{-25}$, if $\theta \lesssim 10^{-9}$.

Next, we point out that[58], again in the absence of instanton effects, the Goldstone boson nature of the axion prevents it from appearing in the Higgs potential. However, in the rest of the Lagrangian it appears only with derivative coupling and its couplings are suppressed by the superheavy mass M_X, thus making it invisible even in Higgs boson interactions.

The above mentioned properties of the grand unified axion justify its rich nomenclature: harmless[53], invisible[54], hypoaxion[57], phantom[65] etc. Also it strengthens even more the importance of the detection (or not) of d_n, as it may be the only evidence against (or for) a grand unified axion.

Epilogue

After our long march through some of the recent developments in grand unified theories, let us try to see things in perspective. I believe that the ideas presented (or reviewed) here on the generation of a hierarchical fermion mass spectrum hold water and I hope that they will be a major part in the understanding of such a perplexive subject as the fermion mass spectrum. Actually, the use of these ideas in models that they also try to explain the generation problem have already appeared[66] and looks encouraging. The solution of the strong CP-violation problem inside GUTs looks satisfactory. Afterall, the old Peccei-Quinn mechanism[49] was an attempt to solve the strong CP-problem by imposing constraints on the structure of weak interactions. It sounds then very natural to try to solve the problem inside a theory that unifies electroweak and strong interactions. It would even more satisfactory if the $U(1)_{PQ}$ symmetry appears <u>automatic</u>[65], i.e., gauge invariance, renormalizability and particle content entail that the theory possesses a $U(1)_{PQ}$ symmetry. That would be marvelous. On the other hand, if nature abhors global symmetries, then we saw that in the framework of grand unified theories, if, arbitrarily enough, we set $\theta=0$, the renormalized value of θ satisfies the present experimental limits. The detection (or not) of d_n within the next few years will clear up the air and somehow it will show us the way we have to follow. Of course, the solution of the strong CP-problem within the framework of GUTs is not immune to the "standard" disease of GUTs, the gauge hierarchy problem.

If recent efforts[19] to solve the gauge hierarchy problem in the framework of supersymmetric GUTs[67] have a happy end, then the whole picture makes much more sense. Then, it seems to me that the following proposed scenario should not be too far from being the real story.

<u>The Su-gar[*)] Scenario</u>

We start with N=8 supergravity with a dynamically realized SU(8) gauge symmetry. Notice that SU(8) is a big enough group so that has enough structure to accommodate:

1. The standard SU(5) group
2. Three families
3. An automatic $U(1)_{PQ}$
4. Enough superheavy particles (fermions, gauge and Higgs bosons) to trigger radiative fermion masses
5. Possibly a vanishing cosmological constant[68]

At the first symmetry breaking, presumably dynamical, SU(8) may break down to say $SU(5)\times U(1)_{PQ}(?)\times$ supersymmetry (SUSY), then

*) Any connection with the words Super and Grand is not accidental.

$$SU(5)\times U(1)_{PQ}\times SUSY \xrightarrow[10^{15}\ GeV]{} (SU(3)\times SU(2)\times U(1))\times SUSY \xrightarrow[10^{2}\ GeV]{} (SU(3)\times U(1)_{EM}).$$

Because supersymmetry remains unbroken up to say 1 TeV, fermion-boson symmetry and chiral invariance keeps the would be Higgs scalars massless. After supersymmetry breaking, presumably dynamical[19)] and thus making natural the appearance of a huge ratio like $\frac{M_W}{M_X}\sim 10^{-12}$, the Higgs particle picks up a negative mass and hence triggers an $SU(2)\times U(1)\rightarrow U(1)_{EM}$ breaking. The gauge hierarchy problem is solved naturally: no unchecked quadratic scalar self-energy divergence (fermion-boson symmetry), huge $\frac{M_W}{M_X}\sim 10^{-12}$ ratio because we tight together electroweak breaking with supersymmetry breaking which presumably and most probably seems to be dynamical[19)]. Needless to stress the fact that the severe constraints imposed on our Lagrangian because of supersymmetry, are going to help our program of a radiatively produced fermion mass spectrum by automatically keeping a lot of fermions massless at the tree level.

Ditto for the $U(1)_{PQ}$ symmetry. It is my strong belief and conviction that we are not far from the materialization of such a program. Extended supergravity shows us the way.

Aknowledgements

It is a great pleasure to thank Prof. Z. Maki, Prof. M. Konuma and the other members of the 4th Kyoto Summer Institute Organizing Committee, for their warm and generous hospitality, unbound kindness and wonderful atmosphere that they and the participants of the Summer Institute created and made the delivery of these lectures a real joy. Also, it is deeply appreciated the given chance to meet and discuss with our Japanese colleagues.

References

1) D.V. Nanopoulos, Nuovo Cimento Letters 8 (1973) 873.
   S. Weinberg, Phys. Rev. Letters 31 (1973) 494.

2) S.L. Glashow, A. Salam and S. Weinberg, Nobel Lectures, Stockholm, December 1979.

3) S. Weinberg, Phys. Rev. Letters 43 (1979) 1566;
   F. Wilczek and A. Zee, Phys. Rev. Letters 43 (1979) 1571.

4) J.C. Pati and A. Salam, Phys. Rev. Letters 31 (1973) 661; Phys. Rev. D8 (1973) 1240; Phys. Rev. D10 (1974) 275.

5) H. Georgi and S.L. Glashow, Phys. Rev. Letters 32 (1974) 438.

6) J. Learned, F. Reines and A. Soni, Phys. Rev. Letters 43 (1979) 907.

7) D.V. Nanopoulos, "Grand Unified Models", in the Proceedings of the XV Rencontre de Moriond, Vol.II, ed. by J. Tran Than Van, p.427 (1980).
   P. Langacker, Phys. Reports 72 (1981) 185.

8) S.L. Glashow and D.V. Nanopoulos, Nature 281 (1979) 464.
   J. Ellis and D.V. Nanopoulos, Nature 292 (1981) 436.

9) H. Georgi, H.R. Quinn and S. Weinberg, Phys. Rev. Letters 33 (1974) 451.

10) A.J. Buras, J. Ellis, M.K. Gaillard and D.V. Nanopoulos, Nucl. Phys. B135 (1978) 66.

11) D.V. Nanopoulos, "The Fermion Spectrum according to GUTs", in "Unification of the Fundamental Particle Interactions", ed. by S. Ferrara, J. Ellis, and P. van Nieuwenhuizen (Plenum 1980), p.435.

12) D.V. Nanopoulos and D.A. Ross, Nucl. Phys. B157 (1979) 273.

13) J. Ellis, M.K. Gaillard, D.V. Nanopoulos and S. Rudaz, Nucl. Phys. B176 (1980) 61.

14) Irvine-Michigan-Brookhaven Collaboration, M. Goldhaber et al., "Proposal for a Nucleon Decay Detector" (1979);
    Harvard-Purdue-Wisconsin Collaboration, J. Blandino et al., "A Decay Mode Independent Search for Baryon Decay Using a Volume Cerenkov Detector" (1979).
    University of Minnesota, H. Courant et al., "A Dense Detector for Proton Decay" (1979).
    Frascati-Milano-Torino Collaboration, G. Battistoni et al., "Proposal for an Experiment on Nucleon Stability with a Fine Grain Calorimeter" (1979).
    V.S. Narasimhan, "Test of Conservation of Baryon Number, Proposal for an Experiment at KGF" (1980).

15) H. Georgi and D.V. Nanopoulos, Nucl. Phys. B155 (1979) 52, and unpublished.
    M. Gell-Mann, P. Ramond and R. Slansky, in "Supergravity", Proceedings of the Supergravity Workshop at Stony Brook, ed. by P. van Nieuwenhuizen and D.Z. Friedman (North Holland, Amsterdam, 1979), p.315.

R. Barbieri, D.V. Nanopoulos, G. Morchio and F. Strocchi, Phys. Letters 90B (1980) 91.
16) F. Reines, H.W. Sobel and E. Pasierb, Phys. Rev. Letters 45 (1980) 1307.
V.A. Lubimov et al., Phys. Letters 94B (1980) 266.
17) H. Georgi, Nucl. Phys. B156 (1979) 126.
18) R. Barbieri and D.V. Nanopoulos, Phys. Letters 91B (1980) 369.
19) See for example, E. Witten, Nucl. Phys. B188 (1981) 513.
20) J. Ellis, M.K. Gaillard and B. Zumino, Phys. Letters 94B (1980) 343.
J. Ellis, M.K. Gaillard, L. Maiani and B. Zumino, LAPP preprint TH-15/CERN preprint TH.2481 (1980)
B. Zumino, CERN preprint TH.2954 (1980).
21) J.M. Cornwall, P.N. Levin and G. Tiktopoulos, Phys. Rev. Letters 30 (1973) 1268; Phys. Rev. D10 (1974) 1145.
22) C. Bouchiat, J. Iliopoulos and P. Meyer, Phys. Letters 38B (1972) 519.
23) For a review see:
D.V. Nanopoulos, "Cosmological Implications of Grand Unified Theories", in Progress in Particle and Nuclear Physics, Vol.6, ed. by Sir D. Wilkinson, (Pergamon Press, 1980), p.23.
24) D.V. Nanopoulos, D. Sutherland and A. Yildiz, Nuovo Cimento Letters 28 (1980) 205.
25) R. Barbieri and D.V. Nanopoulos, Phys. Letters 95B (1980) 43.
26) D.V. Nanopoulos, "A Solution to the Problem of the Fermion Mass", in the Proceedings of the XXth International Conference on High Energy Physics, Madison (July 1980), ed. by L. Durand and L.G. Pondrom, p.53.
27) R. Barbieri, D.V. Nanopoulos and D. Wyler, Phys. Letters B103 (1981) 433.
28) R. Barbieri, D.V. Nanopoulos and A. Masiero, Phys. Letters B104 (1981) 194.
29) See, for example, W. Hubschimd and S. Mallik, Univ. of Bern preprint, (December 1980) and references therein.
30) D.V. Nanopoulos, "Protons are not forever", in "High Energy Physics in the Einstein Centennial Year", ed. by A. Perlmutter, F. Krausz and L. F. Scott (Plenum Press, 1978), p.91.
H. Georgi and C. Jarlskog, Phys. Letters 86B (1979) 297.
H. Georgi and D.V. Nanopoulos, Nucl. Phys. B159 (1979) 16.
31) R. Barbieri, J. Ellis and M.K. Gaillard, Phys. Letters 88B (1979) 315.
32) H. Georgi and D.V. Nanopoulos, Nucl. Phys. B155 (1979) 52.
33) R. Barbieri, A. Masiero and D.V. Nanopoulos, Phys. Letters B98 (1980) 191.
34) D.V. Nanopoulos and S. Weinberg, Phys. Rev. D20 (1979) 2484.
35) J. Ellis, M.K. Gaillard and D.V. Nanopoulos, Phys. Letters 88B (1979) 390.
36) J. Ellis, M.K. Gaillard, D.V. Nanopoulos and S. Rudaz, Phys. Letters 99B (1981) 101; Nature 293 (1981) 41.

37) V. Baluni, Phys. Rev. D19 (1979) 2227.
R.J. Crewther et al., Phys. Letters 88B (1979) 193; 91B (1980) 487 (E).

38) D.N. Schramm and G. Steigman, Bartel Research Foundation preprint BA-80-19 (1980).

39) I.S. Altarev et al., Leningrad Nuclear Physics Institute preprint 636 (1981).

40) J. Ellis and M.K. Gaillard, Nucl. Phys. B150 (1979) 141.

41) D.V. Nanopoulos, A. Yildiz and P.H. Cox, Phys. Letters 87B (1979) 53; Ann. Phys. (N.Y.) 127, (1980) 126.

42) S. Weinberg, Phys. Rev. Letters 37 (1976) 657.

43) A.I. Sanda, Phys. Rev. D23 (1981) 2647.
N.G. Deshpande, Phys. Rev. D23 (1981) 2654.

44) E. Witten, Nucl. Phys. B177 (1981) 477.

45) J.R. Bond, E.W. Kolb and J. Silk, U. C. Berkeley preprint (1981).
J.D. Barrow and M.S. Turner, Nature 291 (1981) 469.

46) J. Ellis, M.K. Gaillard and D.V. Nanopoulos, Phys. Letters 90B (1980) 253.

47) R. Fabbri et al., Phys. Rev. Letters 44 (1980) 1563; 45 (1980) 401 (E).
S.P. Boughn et al., Astrophys. J. Letters 243 (1981) L113.

48) J. Silk and M.L. Wilson, Astrophys. J. Letters 244 (1981) L37.

49) R.D. Peccei and H.R. Quinn, Phys. Rev. Letters 38 (1977) 1440; Phys. Rev. D16 (1977) 1791.

50) S. Weinberg, Phys. Rev. Letters 40 (1978) 223.
F. Wilczek, Phys. Rev. Letters 40 (1978) 279.

51) T. Donnelly et al., Phys. Rev. D18 (1978) 1607.
A. Zehnder, ETH preprint June (1981).

52) K. Sato and H. Sato, Progr. Theor. Phys. 54 (1975) 1564.
D.A. Dicus, E.W. Kolb, V.L. Tepliz and R.V. Wagoner, Phys. Rev. D18 (1978) 1829; Phys. Rev. D22 (1980) 839.

53) M. Dine, W. Fischler and M. Srednicki, Phys. Letters B104 (1981) 199.

54) M.W. Wise, H. Georgi and S. Glashow, Phys. Rev. Letters 47 (1981) 402.

55) R. Barbieri, D.V. Nanopoulos and D. Wyler, CERN preprint TH-3119 (1981).

56) K. Symanzik, in Coral Gables Conference on Fundamental Interactions at High Energies II, ed. by A. Perlmutter, G.J. Iverson and R.M. Williams, (Gordon and Breach, 1970).

57) J. Ellis, M.K. Gaillard, D.V. Nanopoulos and S. Rudaz, CERN preprint TH-3121 (1981).

58) R. Barbieri, R.N. Mohapatra, D.V. Nanopoulos and D. Wyler, CERN preprint TH-3105 (1981) and paper in preparation.

59) J. Ellis, M.K. Gaillard and D.V. Nanopoulos, Phys. Letters 80B (1979) 360 and "Unification of the Fundamental Particle Interactions", ed. by S.

Ferrara, J. Ellis and P. van Nieuwenhuizen (Plenum Press, 1980), p.461.
S. Barr, G. Segre and A. Weldon, Phys. Rev. D20 (1979) 2494.

60) G. Steigman, K.A. Olive and D.N. Schramm, Phys. Rev. Letters 43 (1979) 239.
K.A. Olive, D.N. Schramm and G. Steigman, Nucl. Phys. B180 (1981) (FS 2) 497.

61) For a recent review and references, see
D.N. Schramm, Enrico Fermi Institute preprint EFI-81-03 (1981), presented at the Tenth Texas Symposium on Relativistic Astrophysics, Baltimore (1980).

62) S. Weinberg, Harvard preprint HUTP-78/A005 (1978).
H. Georgi, private communication (1981).

63) J. Ellis and M.K. Gaillard, Phys. Letters B88 (1979) 315 and Ref.31).

64) G. Feinberg and J. Sucher, Phys. Rev. D20 (1979) 1717.

65) D.V. Nanopoulos, CERN preprint TH-3116 (1981).

66) L.E. Ibanez, Oxford preprint TH-40/81.
P. Ramond, in these Proceedings.

67) H. Georgi, in these Proceedings.

68) R. Barbieri, S. Ferrara and D.V. Nanopoulos, CERN preprint TH-3159 (1981).

ON GRAND UNIFICATION

Pierre Ramond
Physics Department, University of Florida
Gainesville, Florida 32611, U.S.A.

in

Proceedings of the Fourth Kyoto Summer Institute on Grand Unified Theories and Related Topics, Kyoto, Japan, June 29 - July 3, 1981, ed. by M. Konuma and T. Maskawa (World Science Publishing Co., Singapore, 1981).

LECTURES ON GRAND UNIFICATION*+

P. Ramond

Physics Department, University of Florida,
Gainesville, FL 32611

Abstract

These lectures are devoted to the study of the mechanisms that Grand Unification provides for the study of the fermion mass matrix. We emphasize the difference between cubic and quartic Higgs couplings in this respect and point out that the $\Delta I_w = 0$ breaking at Grand Unified scales generates radiative $\Delta I_w = 1/2$ masses after electroweak breaking. We emphasize that in the standard unification picture there can be fermions with $\Delta I_w = 0$ masses in the so-called desert without significantly affecting the usual SU_5 predictions ($\sin^2\theta_w, \tau_p, \ldots$). We present a new model where the existence of low mass $\Delta I_w = 0$ fermions is linked to the PQ symmetry and the lightly coupled axion.

*Lectures delivered at the 4th Kyoto Summer Institute on Grand Unified Theories and Related Topics, June 29-July 3, 1981.

+Supported in part by DOE Contract DSR80136je7.

The predictive power of perturbative field theory has so far been found wanting for fermion masses and mixing angles. Indeed the perceived unification[1,2] of the gauge couplings into one,[3] which is the mainstay of Grand Unified Theories[4], has not been accompanied by a similar perception for the Yukawa coupling constants. In fact we are at present confronted with tantalizing patterns of fermion masses with no credible explanation. Still the only success in this direction, namely the mass ratio $\frac{m_b}{m_\tau}$[5], has come only with the help of Grand Unification. Thus it seems appropriate to survey in one place what insight the idea of unification may bring to the solution of the fermion masses and mixing angles.

We start with a semi-phenomenological description, using the electroweak theory as a basis for the classification of fermion masses. Specifically we use weak isospin I_w to classify the possible types of fermions one might consider. Thus at the level of the standard model we can study three main categories of fermions:

a) Fermions with $\Delta I_w = 1/2$ masses: these are the most plentiful ones at low energy. They are the common charged fermions (quarks and leptons) whose left-handed component interacts as a weak doublet ($I_w = 1/2$) and with a weak singlet ($I_w = 0$) right-handed component. This sector consists of several types:

leptons	e^-(.511),	μ^-(105.)	τ^-(1,784),	
quarks	u(5.)	c (1,500)	t (>18,000),	(1)
quarks	d(10.)	s (150)	b(4,700),	

where we have listed in parenthesis their masses in Mev's. The masses of the quarks are not directly interpretable since quarks are confined; these are the ultraviolet masses where quarks are almost free (asymptotic freedom); still for the light quarks the actual value of the mass is highly dependent on scale although quark mass ratios are practically scale invariant. Here the numerical values are at the scale of their constituent masses. As of this writing there only exists a limit on the mass of the t-quark since its effect has not yet been seen in the laboratory. For the remaining of these lectures we will assume its existence.

When the electroweak group SU_2xU_1 is broken, all these particles (and perhaps more) are supposed to become massive and one of the challenges of theory is to explain the numerical values of their masses and mixing angles. At first sight one is struck by the range of these values -- within each charge sector we find one particle much lighter than the other so it would seem natural to try to develop some sort of perturbative approach to obtain some of these masses.

b) <u>Fermions with $\Delta I_w = 1$ masses:</u> these are the left-handed neutrinos ν_e, ν_μ, ν_τ which are left-handed doublets, and have no right handed partners (C is maximally violated) but can have Majorana masses of the form (in the Weyl notation)

$$\nu_L^T \sigma_2 \nu_L . \tag{2}$$

So far there are only limits on these masses:

$$\nu_e(<.00006) \qquad \nu_\mu\ (<.57) \qquad \nu_\tau\ (<200). \tag{3}$$

These are extremely small on the scale of the ΔI_w breaking, for instance

$$\frac{m_{\nu e}}{m_w} < 6 \times 10^{-10}. \qquad (4)$$

However neutrinos can be prevented from becoming massive by lepton number conservation (Majorana masses have $\Delta L = 2$). If lepton number is violated, these neutrinos are expected to acquire Majorana masses, their smallness being explained by the weakness of the violation of lepton number.(6) This sector gives one example of mass unnaturally small on the scale of the symmetry breaking, because of a global conservation law.

c) Fermions with $\Delta I_w = 0$ masses: these are fermions with unusual or no weak interactions. At the level of the standard model there is no incentive for their existence but all but the simplest Grand Unified Theory have them. They fall into two classes -) I_w-singlets which can be either Majorana-like neutral lepton singlets or charged quarks and -) quarks and leptons with vector-like weak interactions which have conjugate I_w properties for their left- and right-handed components. Both types are supposed to become massive when the breaking of symmetries beyond the standard SU_2xU_1 occurs. In Grand Unified models, such breaking is supposed to happen at very large scales, characterized by a boson mass $m_x \sim 10^{15}$ GeV and one would expect the masses of $\Delta I_w = 0$ fermions to be of the same order of magnitude.(7) However, experience in the $\Delta I_w = 1/2$, 1 sectors shows us that this might not be true especially in the latter one where we have

$$\frac{m_{\nu e}}{m_w} < 6 \times 10^{-10}, \qquad (4)$$

as a result of a global conservation law. Later we will see that if there exists a global symmetry, broken at m_X, to prevent $\Delta I_w = 0$ fermion from getting large masses, then such fermions might well be within the range of the next generation of accelerators.[8] We will present one model where this global symmetry is precisely that introduced by Peccei and Quinn[9] to explain the absence of CP violation induced by the QCD anomaly. Thus we think it not unlikely that the famous desert of Grand Unified Theories might be populated by $\Delta I_w = 0$ fermions. We shall see later that their presence does not affect significantly the standard predictions of Grand Unified models.

At this stage it is well to discuss what tools the idea of Grand Unification puts at one's disposal. First it allows us to develop the idea of family. For instance in SU_5[2] each family consists of a $\bar{\underset{\sim}{5}} + \underset{\sim}{10}$ of left-handed fermions with the standard model decomposition

$$SU_5 \qquad SU_3^c \times SU_{2L} \times U_{1Y}$$

$$\bar{\underset{\sim}{5}} = (\underset{\sim}{1}^c, \underset{\sim}{2}) + (\bar{\underset{\sim}{3}}^c, \underset{\sim}{1}) \tag{5}$$

$$\underset{\sim}{10} = (\underset{\sim}{1}^c, \underset{\sim}{1}) + (\underset{\sim}{3}^c, \underset{\sim}{2}) + (\bar{\underset{\sim}{3}}^c, \underset{\sim}{1}). \tag{6}$$

In English this gives the family content to be a left-handed lepton doublet, a right-handed lepton singlet, a quark left-handed doublet and two quark right-handed singlets. Comparison with the $\Delta I_w = 1/2$, 1 sector tells us that we have three families: the e-family containing e, ν_e, u, d;
the μ-family with μ, ν_μ, c, s; and the τ-family with τ, ν_τ, t, b. The numerical value of the masses shows that the e-family central mass is much

less than the μ-family one which in turn is lighter than the τ-family. Further there are within each family mass splittings yet to be understood e.g. $t > b$, $c > s$ while $u < d$. Thus it appears that we have three families of "light" fermions; it is likely that any understanding of this triplication ("bureaucracy of nature") will prove of fundamental importance.

Together with the concept of family one may envisage that of the family group.[10,11] In fact it is tempting to organize the $\Delta I_w = 1/2$ masses in a family space basis as coming from a tree level contribution like[12]

$$\begin{bmatrix} 0 & 0 & 0 \\ 0 & 0 & 0 \\ 0 & 0 & 1 \end{bmatrix} \tag{7}$$

with the zero matrix elements being progressively filled by radiative corrections, thus explaining the smallness of the e- and μ- family masses with respect to the τ-family. In order to achieve this worthy goal we are faced with two types of problems — one technical and one conceptual.

The technical problem has to do with the actual implementation of this tree level mass matrix so that its zeros are natural at tree level and filled only at $\mathcal{O}(\hbar)$ or higher. This will lead us to a discussion of naturalness and calculability in perturbative field theory.[13]

The conceptual problem has to do with the use of a family group. The matrix (7) suggests an SU_3 family group structure with the mass matrix belonging to the sextet representation. This is because this matrix has a non-zero trace. (When analyzed in terms of SO_3 it corresponds to the reducible representation $\underset{\sim}{5} + \underset{\sim}{1}$ and when viewed as an SU_3 octet it is also reducible since it is not traceless.) Hence pure group theory would suggest that the oth order mass matrix is SU_4^f-invariant (here the superscript f stands uor family), with the effects of SU_2^f-breaking making its way to the fermion mass matrix only in higher orders of perturbation theory. This appealing picture is difficult to implement. However if we want to use SU_3 as a gauge family group we have to learn how to compensate for the anomalies it creates which we will show how to do in a simple way when we present models. On the other hand the experimental fact that interfamily processes such a $\mu \to e\gamma$, d-s mixings, etc... are very small shows that the family group must be broken at a large scale, perhaps at unification. (Could it be that the family group appears only as a global symmetry, with its Nambu-Goldstone bosons coupling ever so weakly to matter?)

Let us now review the notion of calculability in perturbative quantum field theory.[13] It is very appropriate to talk about this subject here in Kyoto because the mass matrix has to do with the structure of the Yukawa part of the Lagrangian. Schematically it is given by

$$\mathcal{L}_Y^{cl.} = yffh, \tag{8}$$

where f is a left-handed Weyl spinor and h is some scalar field (Higgs), and y is the dimensionless Yukawa coupling constant. For the purpose of this discussion we will assume that bare mass terms are excluded on symmetry grounds, as for instance in the standard model. When h gets a vacuum expectation value, vev, it generates a mass for the fermion f. Here y is an input parameter to the field theory, just like α in QED; it cannot be calculated within perturbative QFT since it suffers infinite renormalization. Thus if h gets a vev at tree level the resulting fermion mass has to be regarded as an undeterminable input parameter.

The aim of developing a successful theory is to minimize the number of these input parameters. Within this framework this can be achieved by making as many of the masses "calculable", which means that they can be expressed as finite, calculable, functions of the input parameters. The prime example of calculability in perturbative QFT is the anomalous magnetic moment of the electron which is calculated in terms of the undetermined input parameter α.

In QED one starts with the renormalizable interaction

$$\mathcal{L}^{\text{cl.}}_{\text{int.}} = e\bar{\Psi}\gamma^{\mu}\Psi A_{\mu}, \tag{9}$$

where e ($\alpha = \frac{e^2}{4\pi}$) is an input parameter. The radiative corrections of the field theory are found to generate an effective Lagrangian density containing among others a term of the form

$$\mathcal{L}^{\text{rad.}} \sim \bar{\Psi}\sigma_{\mu\nu}\Psi F^{\mu\nu}. \tag{10}$$

Dimensional analysis shows this term to be of dim 5, which means that its renormalization is finite as long as the infinite renormalization of the terms in the classical Lagrangian are taken care of. [Remember that renormalizability means that the ultraviolet infinities occur only in the terms appearing in the classical Lagrangian of dim 4 or less.] Hence the coefficient in front of (10) is calculated as a finite function of the input parameters α and m, the mass of the electron. This notion can be readily extended to the Yukawa part of the Lagrangian.

There are essentially two ways to do this. One is to arrange things so that h does not get a vev at tree level but only after radiative corrections are taken into account; we call this type (a). The other way is to arrange the classical Lagrangian in such a way that the direct coupling of some fermions with Higgs fields having tree level vev's is forbidden by symmetry but can occur in higher orders, that is, in the effective Lagrangian generated by perturbation theory; we call this type (b) mechanism. In this case we have to be careful that chiral symmetry of the non-coupling fermions not forbid masses to all orders; in practical terms it means that we can implement type (b) mechanism only when the starting fermion representation is reducible. If on the other hand the fermion representation is irreducible, we can rely only on type (a) mechanism.

Specifically, consider a Higgs classical renormalizable potential depending on two fields h and ϕ so that ϕ but not h gets a tree level vev. We write the potential schematically as

$$V_{cl}(\phi,h) = V_0(\phi) + \quad V_1(\phi) + \frac{h^2}{2}V_2(\phi) + \frac{h^3}{3}V_3(\phi) + \lambda\frac{h^4}{4}, \qquad (11)$$

where $V_i(\phi)$ is at most (4-i)th order in ϕ, since the theory is renormalizable. Minimization of (11) shows that a necessary condition for our required pattern of vev's

$\langle h\rangle_0 = 0,\ \langle\phi\rangle_0 \neq 0$

is that

$$V_1(\langle\phi\rangle_0) = 0. \tag{12}$$

It could be that this relation is true only for $\langle\phi\rangle_0$, thereby restricting enormously the form of $V_{cl.}$. In general such a constraint cannot be enforced by symmetry unless we have an organizing principle for the Higgs sector (perhaps supersymmetry?), and it would therefore be technically unnatural. It is less demanding to impose that (12) be true for any ϕ in which case terms of the form ϕ, ϕ^2 and ϕ^3 must be forbidden to appear in $V_{cl.}$ on grounds of symmetry. This means that no h-tadpole will be generated at tree level. Still we want h to eventually get a vev as a result of radiative corrections. The effective potential generated by perturbation theory is no longer restricted to be of at most 4th order in the field and it will, if allowed by the symmetries of the classical Lagrangian, contain terms of the form $h\phi^4$, $h\phi^5$, etc... These induced tadpole terms will give h a vev as a result of radiative corrections. Let us give several examples:

a-) Suppose our theory is invariant under SU_2 and that ϕ is a spin 1/2 field and h is a spin 2 field (not the usual spin, these are scalar fields!). Then clearly the lower term linear in h is $h\phi^4$ since it takes four spin 1/2 to make a spin 2 field. In this case, then, h will get an induced vev, calculable in terms of the input parameters of the theory.

b-) Suppose our theory is invariant under SU_N and that ϕ has N-ality m and h has N-ality n. Then our minimum requirement is that

$$n \pm pm \neq 0 \quad \text{mod } N \qquad p = 1,2,3, \tag{13}$$

but that there exists at least one positive integer $p > 4$ such that

$$n \pm pm = 0 \quad \text{mod } N \tag{14}$$

A solution is $N=8$ $m=1$ $n=4$, etc...[14] Actually these equations are overly strong: for example if ϕ and h are different representations of SU_N, we can have $n=m$. To be truly useful this analysis must be generalized to include the actual SU_N representations of ϕ and h; here we only give it as an illustration of the type (a) mechanism.

The type (b) mechanism can be implemented only with reducible fermion representations. The easiest way is to build the classical Lagrangian in such a way that a Higgs field which could couple directly to the fermions a la Yukawa is not present in $\mathcal{L}^{cl.}$ but can be generated by multi-Higgs.

Schematically this means that terms of the form

$$ffh^2, \ldots \tag{15}$$

will appear in the effective Lagrangian, which, when evaluated in the vacuum $\langle h \rangle \neq 0$ will give a contribution to the fermion masses, given as a finite function of the input parameters. Let us give an example: Suppose that we have a theory invariant under SU_2. The input fermions are an SU_2 singlet left-handed Weyl field F_L and an SU_2-doublet Weyl field f_L. The possible mass terms (two fermions scalar combinations) are

$$F_L F_L \sim 1 \text{ of } SU_2$$
$$F_L f_L \sim 2 \text{ of } SU_2 \tag{16}$$
$$f_L f_L \sim 3 \text{ of } SU_2.$$

The combination $f_L f_L$ contains only the triplet of SU_2 since f_L transforms as the (2,0) of the Lorentz group. The Higgs fields are taken to be a real singlet S and a complex doublet D. Then the classical Yukawa term reads (suppressing Lorentz indices and matrices)

$$\mathcal{L}_Y^{cl.} = F_L F_L S + F_L f_L D + c.c. \; . \tag{17}$$

It is invariant under a discrete symmetry

$$S \to -S, \quad F_L \to iF_L, \quad f_L \to -if_L \tag{18}$$

and a continuous chiral symmetry

$$F_L \to F_L, \quad f_L \to e^{i\alpha} f_L, \quad D \to e^{-i\alpha} D. \tag{19}$$

These allow for the triplet coupling to be generated in higher order by a term of the form

$$(f_L \tau_2 \vec{\tau} f_L)\; S\; D^T \tau_2 \vec{\tau} D, \tag{20}$$

which satisfies the symmetries (18) and (19). If S and D have a non zero vacuum value, the classical Lagrangian (17) gives a massless fermion and two massive ones; when radiative corrections are turned on, the massless fermion acquires a mass calculable in terms of the input parameters of the theory. This is an example of the type (b) mechanism; we will use it later on with SU_2 as a family group.[11]

One can also encounter intermediate situations where, even though the fermion mass itself appears at tree level, corrections to it are calculable. An example[15] of this type is present in the SU_5 Grand Unified Theory. For one family its fermion content has already been given by (5) and (6). The Higgs sector consists of a quintet $h_a \sim \underset{\sim}{5}$ and an adjoint $H^a{}_b \sim 24$. The renormalizable Yukawa coupling is

$$\mathcal{L}_Y^{cl.} = y\ \overline{5}^a\ 10_{bc}\ \delta^b{}_a h^c. \tag{21}$$

When h^a gets a $\Delta I_w = 1/2$ vev, it gives equal mass to the charge $-1/3$ quark and charge -1 lepton, so that at the Grand Unified scale we have

$$0 = m_b - m_\tau. \tag{22}$$

The Higgs adjoint representation $H^a{}_b$ gets a $\Delta I_w = 0$ vev at the scale of Grand Unification (as well as an induced tiny $\Delta I_w = 1$ vev). Now we see that the combination $H^b{}_a h^c - H^c{}_a h^b$ has the right index structure to couple to the two fermion combination. Thus if no symmetries restrict it, the radiative corrections will generate a term of the form

$$\overline{5}^a 10_{bc}\ \ H^b{}_a h^c. \tag{23}$$

(In fact this term is forbidden by the discrete symmetry $H \to -H$ but we can easily replace H by H^2.) The index structure of this calculable mass shows that in SU_5 language

$$(Hh)_a^{bc} \sim \overline{\underset{\sim}{45}} + \overline{\underset{\sim}{5}} \tag{24}$$

Thus the $\overline{\underset{\sim}{5}}$ part will just renormalize the undetermined coupling (21), while the strength of the $\overline{\underset{\sim}{45}}$ part, which did not appear in the renormalizable Lagrangian, will be calculable in terms of the input parameters. The end result will be a finite correction to the 0th order mass formula (22), yielding a new mass formula

$$0 = (m_b - m_\tau) + \varepsilon\,(m_b - 3m_\tau), \tag{25}$$

where ε is a finite function of the input parameters.

This example, taken from the simplest Grand Unified model, illustrates a new mechanism at our disposal to generate radiative $\Delta I_w = 1/2$ masses. Let us characterize the two scales of Grand Unified models as follows: H is a Higgs which gets a $\Delta I_w = 0$ vev and breaks the unified group at scale $\sim m_x \sim 10^{15}$ Gev. Call h another Higgs field which gets a $\Delta I_w = 1/2$ vev at a scale $\sim m_w \sim 10^2$ GeV. The discrepancy between these scales leads to the gauge hierarchy problem about which we have nothing to say. We see therefore that in a Grand Unified framework we have two ways of generating $\Delta I_w = 1/2$ masses: by an undetermined amount via the renormalizable Yukawa coupling

$$ffh, \tag{26}$$

and by unrenormalizable couplings of the form[16,17]

$$\frac{z}{m_x^n} ffH^n h \qquad n > 1, \tag{27}$$

since $H^n h$ has the quantum numbers $\Delta I_w = 1/2$; z is dimensionless, and calculable. In fact one can regard the breaking as appearing sequentially: first perform the $\Delta I_w = 0$ breaking and then obtain an effective field theory. In that theory, the terms of the form (27) will appear as regular Yukawa couplings when H is replaced by its vev

$$\frac{z}{m_x^n} ffH^n h \quad \text{--->} \quad y_{eff} ffh, \tag{28}$$

where y_{eff} is a function of the input parameters of the initial theory and of logarithms involving m_x/μ, μ being the renormalization point. These will therefore appear as <u>determined</u> from the initial (unbroken) theory although, at the level of the effective theory, they are regarded as arbitrary input parameters, which are more numerous than those of the initial unbroken theory. In the absence of quantum numbers to distinguish h from $H^n h$, the induced terms (27) or (28) will not produce anything new. However, if we give h and H family quantum numbers, h and $H^n h$ will fill in different matrix elements of the family mass matrix (7). This mechanism perhaps provides the key to obtaining a realistic mass matrix. At any rate it is a new mechanism that we have at our disposal in theories with two scales of breaking, and we should use it. If we wish to use it to fill in all the zeroes of the matrix (7), we cannot use the minimal 3 family SU_5 because we must couple the first two families in the Yukawa term to avoid chiral symmetries -- this can be done in the charge 2/3 sector since we have

$$\underset{\sim}{10} \times \underset{\sim}{10} = (\underset{\sim}{\overline{5}} + \underset{\sim}{50})_S + \underset{\sim}{\overline{45}}_A, \tag{29}$$

and we can use the $\underset{\sim}{50}$ which has no $SU_3^c \times U_1^Y$ neutral components -- but in the charge -1/3, -1 sector the same cannot be done without introducing direct Higgs couplings with $\Delta I_w = 1/2$ vev unless one can implement the type (a) mechanism. Thus the only hope at the level of SU_5 is to have more fermions with $\Delta I_w \neq 1/2$ masses, e.g. $\Delta I_w = 0$. Hence we see that the desire to implement the observed mass matrix in perturbative QFT leads us to adding new fermions to the theory or, in other words, to go beyond SU_5.

To conclude this general discussion let us say a word about evaluating the strength of these calculable contribution to the mass matrix. For this purpose it is easiest to draw Feynman diagrams in the unbroken theory. All-we need to remember is that a fermion mass term involves a transition from left-handed to right-handed components, that a spin 0 exchange does a L-R transition and a spin 1 exchange does not, i.e. preserves handedness. (In the unbroken theory all particles are massless, which enables us to use the concept of helicity interchangeably with that of handedness.) The renormalizable vertices are thus[16]

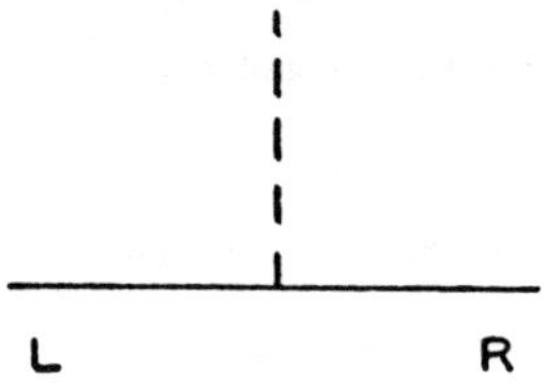

and

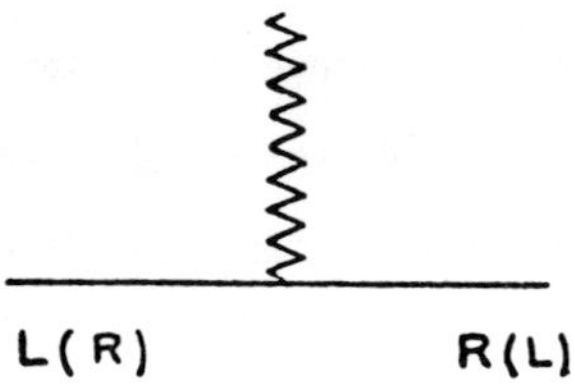

(30)

Thus it is easy to see that

ffh^2: 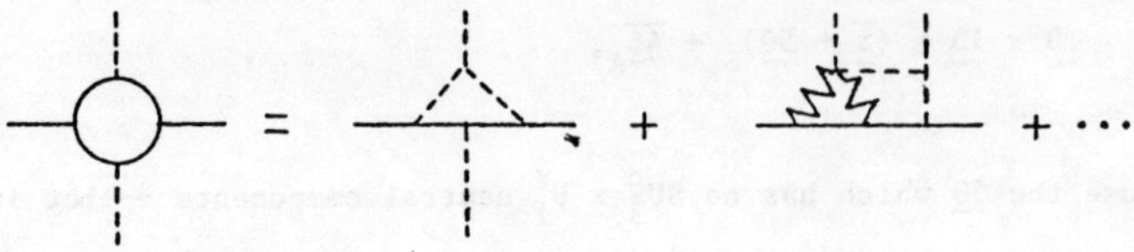 (31)

which can be generated only if a Higgs cubic self coupling exists. Also

ffh^3: 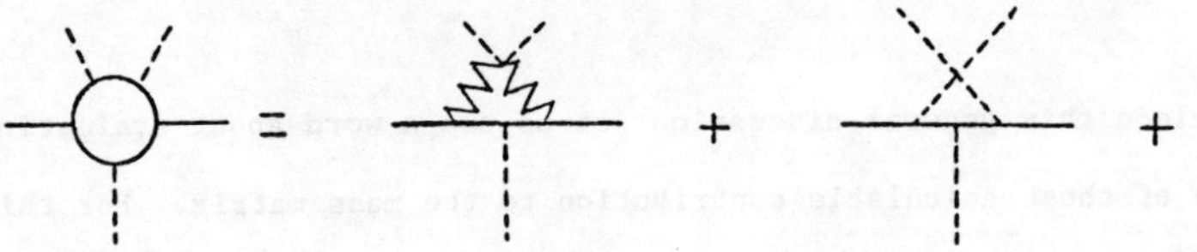 (32)

etc... We see that the higher the n in the induced ffh^n term the more powers of coupling constants appear and the smaller the contribution. Hence for practical purposes it is sufficient to consider only (31) and (32). In a theory with a gauge hierarchy problem one might worry about large logarithms which might spoil the concept of a low energy effective theory and therefore the estimates obtained from the diagrams of (31) and (32). However there is a consistent way[18] of devising an effective theory even when a hierarchy is present. This concludes our brief, but general, discussion on calculable masses. The main lessons are: a-) if we want to obtain claculable masses in a theory with one fermion irreducible representation (of the gauge group) we have to rely on the type (a) mechanism, i.e. prevent, in a technically natural way, some Higgs from getting tree level vev's; b-) if the fermion representation is reducible, then one can use the easier to implement type (b) mechanism. c-) theories with two scales, $\Delta I_w = 1/2$ and $\Delta I_w = 0$ say, have a natural radiative mechanism, the "small h, big H" mechanism to contribute to $\Delta I_w = 1/2$ masses. The ideal scenario would be to have the family matrix (7) at tree level and gradually fill its zeroes by means of this mechanism. No such model has yet been presented.[19]

We now turn to the presentation of several Grand Unified models with a non-trivial radiative structure.

We start by considering a one family E_6 model[20] which provides a good example of calculability. In E_6, the left-handed fermions belong to the complex 27-dimensional representation, and the mass matrix stems from the product

$$\underset{\sim}{27} \times \underset{\sim}{27} = (\underset{\sim}{\overline{27}} + \underset{\sim}{\overline{351}}')_S + \underset{\sim}{\overline{351}}_A. \tag{33}$$

In the one family case only the symmetric part if Lorentz invariant. Thus the Higgs that couple to the fermion masses either belong to the $\underset{\sim}{27}$ or the $\underset{\sim}{351}'$. The adjoint representation is 78-dimensional and is abstracted from the product

$$\underset{\sim}{27} \times \underset{\sim}{\overline{27}} = \underset{\sim}{1} + \underset{\sim}{78} + \underset{\sim}{650}. \tag{34}$$

The physical content of the E_6 family can be deduced from the decompositions

$$E_6 \supset SO_{10} \times U_1 \supset SU_5 \times U_1 \times U_1 \supset SU_3 \times SU_2 \times U_1 \times U_1 \times U_1, \quad (35)$$

In SO_{10} language

$$\underset{\sim}{27} = \underset{\sim}{16} + \underset{\sim}{10} + \underset{\sim}{1}, \quad (36)$$

$$\underset{\sim}{78} = \underset{\sim}{45} + \underset{\sim}{1} + \underset{\sim}{16} + \overline{\underset{\sim}{16}}, \quad (37)$$

and in SU_5 language

$$\underset{\sim}{27} = (\overline{\underset{\sim}{5}} + \underset{\sim}{10} + \underset{\sim}{1}) + (\underset{\sim}{5} + \overline{\underset{\sim}{5}}) + \underset{\sim}{1} \quad (38)$$

$$\underset{\sim}{78} = (\underset{\sim}{24} + \underset{\sim}{10} + \overline{\underset{\sim}{10}} + \underset{\sim}{1}) + (\overline{\underset{\sim}{5}} + \underset{\sim}{10} + \underset{\sim}{1}) + (\underset{\sim}{5} + \overline{\underset{\sim}{10}} + \underset{\sim}{1}) + \underset{\sim}{1}. \quad (39)$$

From (36) and (38) we see that a fermion family consists of the usual SU_5 family $\overline{5} + 10$ plus two singlet neutral leptons and a $5 + \overline{5}$ pair. Thus its physical content is: one charge 2/3 quark, two charge −1/3 quarks, two charge −1 leptons and five neutral leptons of a given handedness. It is one example where we have fermions with $\Delta I_w = 0$ masses, given by $5.\overline{5}$ and 1.1, $\Delta I_w = 1/2$ masses given by $\overline{5}.10$ and 10.10, and $\Delta I_w = 1$ masses appearing in both $\overline{5}.5$ and $5.\overline{5}$ (as a mass splitting). It is easy to obtain the U_1 assignment, generated by T_6, the coset E_6/SO_{10}. It is: 1 for the 16, −2 for the 10 and +4 for the 1 appearing in (36). Hence we see that of the possible SO_{10} invariant mass terms

10.10 has value −4

1.1 has value +8,

so that the neutral lepton mass term cannot be in the $\overline{27}$, but is rather in the $\overline{351}'$. Hence if we take our Higgs fields to be only along the 27, the neutral lepton SO_{10} singlet will necessarily be massless at tree level.

We start from Yukawa terms of the form

$$\mathcal{L}_Y^{cl.} \sim 27_f 27_f 27_H, \tag{40}$$

where the subscript stands for fermion (f) or Higgs (H). Since the $\underset{\sim}{27}$ is a complex representation, (40) has a global phase symmetry generated by X; we assign X-value of +1 to 27_f and -2 to 27_H.

We will need several Higgs fields to break E_6 down to $SU_3^c \times U_1^Y$. If we use exclusively 27_H's we cannot break SU_5 without breaking SU_{2L}; thus we need at least one 78_H to break SU_5 to the standard model. The pattern goes as follows:

$$E_6 \rightarrow SO_{10} \qquad \text{by means of } 27_H$$

$$SO_{10} \rightarrow SU_5 \qquad \text{by means of } 27'_H$$

$$SU_5 \rightarrow SU_3^c \times SU_{2L} \times U_{1Y} \qquad \text{by means of } 78_H$$

$$SU_{2L} \times U_{1Y} \rightarrow U_1^Y \qquad \text{by means of } 27''_H. \tag{41}$$

(It is intriguing to note "en passant" that the number of 27_H's is the same as the number of fermion families.) In the above we have not shown the U_1 factors except in the last stages. We have noted that the Yukawa term has a global phase symmetry, which can be broken or preserved by the Higgs potential with various phenomenological consequences.

Suppose first that the phase symmetry generated by X is kept by $V_{cl.}$.

If the 78_H is real, then one easily finds that the extra neutrinos cannot get a mass until the weak isospin is broken, with the result that the left-handed neutrino gets a mass of the order of the t-quark mass, if we take our one-family model to represent the τ-family! The initial X-symmetry survives all stages of symmetry breaking by mixing with a local U_1 in the same way that B-L is formed in the minimal SU_5. There results a global symmetry at the level of the standard model. We say that the global symmetry has been "'t Hoofted". We leave it to the reader to discover the meaning of this left-over global symmetry. This case is clearly unrealistic.

If the 78_H is complex (i.e. doubled) then the situation becomes more interesting. One can have in the potential a term of the form

$$78_H 27_H 27_H 27_H, \tag{42}$$

which assigns an X-value of +6 to 78_H. Now when 27_H and $27'_H$ break E_6 down to SU_5, the X symmetry survives by "'t Hoofting". At the SU_5 level it looks exactly like the U_1 of Peccei and Quinn, and when 78_H gets a vev, it breaks it spontaneously, leading to the "invisible axion".[21] Thus in this model the U_1 of P.Q. has its origin in the fact that the starting fermion representation and Higgs representations are all complex. The existence of a U_1 P.Q. might be linked to the non-vector like nature of the initial Lagrangian!

Furthermore, X symmetry allows induced terms of the form

$$27_f 27_f \overline{78}_H \overline{27}_H \overline{27}_H, \tag{43}$$

which give an effective, calculable, 351 contribution to the mass matrix when 27_H and 78_H get their vev. This is how the "right-handed" neutrinos get their masses. Let us detail the process:

a-) When 27_H gets a vev along the SO_{10} singlet direction (see (36)), it gives a tree level mass to the fermions in the 10 of SO_{10}. The 16_f and 1_f are massless, and the theory is now $SO_{10} \times U_1'$ invariant, where U_1' is generated by

$$X' = T_6 + 2X; \tag{44}$$

recall that T_6 generates the coset E_6/SO_{10}.

b-) We assume that $27_H'$ gets a vev in the 16 of SO_{10}, along the SU_5 singlet, breaking $SO_{10} \times U_1'$, where U_1'' is generated by

$$X'' = 5X' - 3T_5; \tag{45}$$

T_5 generates the coset SO_{10}/SU_5; its values in the 16 are +3 for the $\overline{5}$, -1 for the 10 and -5 for the singlet. Its effect on the fermions is to mix the two $\overline{5}_f$ present in each 27_f so as to leave one massless combination in each of the charge -1 and -1/3 sectors. We are therefore left at this stage with a massless standard SU_5 family and two massless Majorana (Weyl) neutrinos.

c-) When 78_H gets a vev along the 24 of SU_5, it breaks SU_5 down to the standard model and breaks the 't Hoofted X", leading to the "invisible axion". At the same time it induces mass terms for the neutrinos via the effective interaction (43), because the SO_{10} and SU_5 singlets in the 78_H automatically get tree level vev once X" is broken. So we are left with the standard model plus a weakly coupled axion.

d-) Finally assume that $27''_H$ gets a vev so as to break I_w by $\Delta I_w = 1/2$. In principle this can be done in two ways, along the 5_H or along the $\overline{5}_H$. Unfortunately it does not seem to be possible to choose among those: if one starts with the vev along the 5_H only, there one finds in the classical potential tadpoles induced in the $\overline{5}_H$ direction with the result that both (b,τ) and t quark masses occur at tree level. Since there are several tadpole terms one can require the relation (12) to hold at tree level (in the absence of symmetries this is an unnatural act!), in which case we would obtain by means of the Hh-mechanism a relation of the form

$$\frac{m_b}{m_w} \sim \mathcal{O}(\alpha), \tag{46}$$

generated at the one-loop level by interchange of heavy fermions, a mechanism previously proposed. This concludes the study of this case where one obtains radiative calculable masses for the "heavy" neutrinos, together with a lightly coupled axion.

Now suppose that the X-symmetry is explicitly broken by the classical Higgs potential.(16,22,23) This can occur in several ways:

a-) By means of a quartic term of the form $78_H 27_H 27_H 27_H$ with a real 78_H. Then the analysis proceeds in the same way as before except that there is no PQ symmetry since the global X-symmetry has been broken to a discrete symmetry.

b-) By means of a cubic term of the form

$$27_H 27_H 27_H, \tag{47}$$

which breaks X to another discrete symmetry. This case is different from the ones just described because now the extra neutrinos can get radiative masses by means of induced terms of the form

$$27_f 27_f \overline{27}_H \overline{27}_H, \tag{48}$$

which are generated by the diagrams of (31). This mechanism has also been proposed before at the level of SO_{10}.[24] Here it occurs automatically. It is amusing to see that this simple one family E_6 model displays a non-trivial radiative structure, in either case. The only thing it does not do is to display a mechanism which produces radiative masses for particles with vector-like weak interactions, ($\Delta I_W = 0$ fermions), but a three-family generalization of this model will show such a mechanism for the "heavy" fermions associated with the lightest family.

The generalization of these models to include three families will proceed via the introduction of a family group, either SU_2^f or SU_3^f, as suggested by the structure of the tree level matrix (7). In these notes we do not discuss the possibility of enforcing the family structure by means of Abelian (global or local) symmetries.[25]

Let us start by analyzing a model based on $E_6 \times SU_2^f$, where SU_2^f is the family group.[11,22] We assign two families to form one SU_2^f doublet, the other a singlet. Since SU_2^f has no d^{ABC} coefficient, we can gauge it without fear of anomalies. The classical Yukawa couplings are taken to be

$$\mathcal{L}_Y^{cl.} = y_1(27,1)_f(27,1)_f(27,1)_H + y_2(27,2)_f(27,1)_f(27,2)_H . \tag{49}$$

It has two global phase symmetries. As before we may or may not choose to break them in the Higgs potential.

We first consider the option of having no cubic terms in the Higgs potential, and keep one global symmetry (the same as before) by introducing terms like

$$(78,1)_H(27,1)_H(27,1)_H(27,1)_H , \tag{50}$$

and

$$(78,1)_H(27,2)_H(27,2)_H(27,1)_H , \tag{51}$$

where $(78,1)_H$ is a complex field carrying the X-value +6. This last term is made possible by the fact that in E_6[26]

$$\underset{\sim}{78} \times \underset{\sim}{27} = \underset{\sim}{27} + \underset{\sim}{351} + \underset{\sim}{1728} \tag{52}$$

so that (51) is E_6 invariant through the use of the $\underset{\sim}{351}$. In order to break the initial symmetry down to the standard model, we have to introduce yet another type of Higgs which we take to be a complex family doublet $(1,2)_H$ with the couplings

$$(1,2)_H(27,1)_H(27,1)_H(27,1)_H, \tag{53}$$

and

$$(1,2)_H(27,2)_H(27,2)_H(27,2)_H, \tag{54}$$

so that $(1,2)_H$ also has X = +6. The symmetry breaking proceeds as

$$
\begin{array}{lll}
E_6 \times SU_2 \times U_{1X} \rightarrow SO_{10} \times SU_2 \times U_1' & \text{via} & (27,1)_H \\
SO_{10} \times SU_2 \times U_1 \rightarrow SU_5 \times U_1 \times U_1'' & \text{via} & (27,2)_H \\
SU_5 \times U_1 \times U_1'' \rightarrow SU_3^c \times SU_{2L} \times U_{1Y} \times U_1 & \text{via} & (78,1)_H \\
SU_3^c \times SU_{2L} \times U_{1Y} \times U_1 \rightarrow SU_3^c \times SU_{2L} \times U_{1Y} & \text{via} & (1,2)_H \\
SU_{2L} \times U_{1Y} \rightarrow SU_3^c \times U_1^Y & \text{via} & (27,1)_H'
\end{array} \tag{55}
$$

It is interesting to note that in this model even after the breaking of SU_5 there persists a local U_1 which is a linear combination of flavor and family quantum numbers. This is why we need the $(1,2)_H$ to break it for it causes flavor changing effects and cannot be broken (or so it seems) at the m_W scale by, say, the extra $(27,1)'_H$. The induced mass terms come from the dimension 6 invariant couplings

$$(27,2)_f(27,1)_f(\overline{27},2)_H(\overline{27},1)_H(\overline{78},1)_H$$

$$(27,2)_f(27,1)_f(\overline{27},1)_H(\overline{27},1)_H(1,\overline{2})_H$$

$$(27,3)_f(27,3)_f(\overline{27},2)_H(\overline{27},2)_H(\overline{78},1)_H$$

$$(27,3)_f(27,3)_f(\overline{27},2)_H(\overline{27},1)_H(1,\overline{2})_H, \tag{56}$$

as well as (43). So we have all the masses of the lighest family generated radiatively, and we have a PQ symmetry broken by $(78,1)_H$. The interested reader might want to work out the fermion spectrum at each stage of breaking. Let us note that if the scale at which $(1,2)_H$ gets its vev is depressed over unification, it can lead to low $\Delta I_w = 0$ masses. This model provides an example where all different types of masses are radiative, $\Delta I_w = 0$ masses for the lightest family for instance as well as the e- u-d masses. It is interesting to see if the sign of the u-d mass difference can be understood in this context. (We leave it as an exercise!)

Alternatively, one can break the global symmetries of the Yukawa coupling (49) by taking $(78,1)_H$ to be real. Then the quartic terms (50) and (51) leave behind only discrete symmetries. The analysis proceeds in the same way except that there is no PQ symmetry. (Note that $(1,2)_H$ has to be taken to be complex because $(1,2)_H$ is pseudoreal, and would not have a kinetic term: Higgs fields belonging to pseudoreal representations are necessarily auxiliary fields). In both of the above cases, when there are no cubic terms present in the Higgs potential, induced masses can come only from terms of even dimensions in the induced Lagrangian.

On the other hand, we can break the global symmetries by introducing cubic Higgs couplings, leaving behind discrete symmetries. In that case the analysis proceeds along the lines of the one-family case with the induced masses coming from dim. 5 invariants such as, among others,

$$(27,2)_f(27,2)_f(\overline{27},2)_H(\overline{27},2)_{H'}, \tag{57}$$

which can give mass to different types of fermions. For instance as long as I_w has not been broken this induced term can give masses only to $\Delta I_w = 0$ fermions. Then, when I_w is broken by means of $(27,1)_H'$, the $(27,2)_H$ gets an induced tree level vev as well. Hence (57) will generate a $\Delta I_w = 1/2$ mass by taking the cross product between the $\Delta I_w = 0$ and $\Delta I_w = 1/2$ vev's: this is the hH mechanism described earlier. This discussion holds as well for the induced terms (56).

It is not clear what the best way of generating radiative masses actually is. There are some apparent disadvantages to using cubic Higgs couplings since they introduce explicit scale breaking at the classical level and do not allow for the very beautiful symmetry breaking scheme of Coleman and E. Weinberg(27). They also tend to break global symmetries into discrete symmetries, which is a good thing unless one wants to keep the U_1 of Peccei and Quinn to solve the strong CP problem. Finally from the point of view of evaluating the strength of the calculable masses it would seem that the cubic couplings are less favored since they allow a direct pure Higgs one loop contribution (31) while in their absence the lowest order contribution (32) depends primarily on gauge couplings.

On the other hand cubic self couplings of the Higgs tend to appear in effective "low" energy theories, and Grand Unified models probably are themselves some sort of effective theories. In short one does not know what is preferable, and so we present both possibilities. As usual, search for systematics in the Higgs sector is very frustrating, unless one is willing to introduce supersymmetry. Thus on three grounds: allowing for the Coleman, E. Weinberg mechanism, estimating calculable masses in terms of gauge couplings, and perhaps in order to get a global U_1 à la P.Q., it would seem better to avoid the cubic Higgs couplings.

We now proceed to introduce SU_3 as a family group. Unlike SU_2, SU_3 has anomalies and since we are dealing with left-handed fermions we will introduce anomalies. If we want to gauge the family group we have to produce an anomaly-free set of fermions. Let us show one easy way to calculate the anomaly number of SU_n representation (SU_n $n>2$ and U_1 are the only groups with anomalies). Let A_r be the anomaly number associated with a left-handed fermion transforming as the r-dimensional representation of SU_n. We know that

$$A_r = -A_{\bar{r}}, \qquad (58)$$

so that if r is real, or pseudoreal, the associated fermion produces no anomaly. Diagrammatically, we have

$$A_{\underline{r}} \quad \longleftrightarrow$$  , (59)

where the solid line represents the fermion ($\sim\underline{r}$) and the outside legs are the SU_n currents. Now consider a Weyl fermion transforming as the reducible representation $\underset{\sim}{r} \times \underset{\sim}{r}'$; represent it by two lines one corresponding to $\underset{\sim}{r}$, the other to $\underset{\sim}{r}'$. Its associated anomaly $A_{\underline{r}x\underline{r}'}$ is given in terms of diagrams by

 (60)

This is so because the SU_n currents must couple all to the same line -- if one couples to one line and the other two to the other we get no contribution since the quantum numbers of the single current cannot couple to a closed fermion line [Technically $Tr(T^a)=0$]. Hence the two terms, which are easily understood. The first one is the anomaly of the outside line, call it $\underset{\sim}{r}$, multiplied by the closed loop, which is the dimension of the other representatin $\underset{\sim}{r}'$, and vice versa. Hence we have

$$A_{rxr'} = d_{r'}A_r + d_r A_{r'}, \tag{61}$$

where d_r is the dimension of the representation.[28] This formula can be easily generalized to an arbitrary number of representations, yielding

$$F_{\underset{\sim}{r}_1 x..x \underset{\sim}{r}_K} = F_{\underset{\sim}{r}_1} + F_{\underset{\sim}{r}_2} + \ldots + F_{\underset{\sim}{r}_K}, \tag{62}$$

where

$$F_{\underset{\sim}{r}} = \frac{A_r}{d_r} \tag{63}$$

plays the role of a logarithm for anomalies. We leave it to the reader to prove, using diagram techniques that

$$A_{(\underset{\sim}{r} x \underset{\sim}{r})_s} = (d_r + 4)\, A_{\underset{\sim}{r}}. \tag{64}$$

Using (58) and (61) it is easy to compute the anomaly number of low lying SU_3 representations. For instance, since

$$\underset{\sim}{3} \times \underset{\sim}{3} = \underset{\sim}{6} + \underset{\sim}{\overline{3}}, \tag{65}$$

we find, using (61) and setting $A_{\underset{\sim}{3}} = +1 = -A_{\overline{\underset{\sim}{3}}}$

$$A_{\underset{\sim}{3} x \underset{\sim}{3}} = 3.1 + 1.3 = 6$$

$$= A_{\underset{\sim}{6}} + A_{\overline{\underset{\sim}{3}}},$$

so that

$$A_{\underset{\sim}{6}} = 7 \tag{66}$$

Similarly, using

$$\underset{\sim}{6} \times \underset{\sim}{3} = 10 + \underset{\sim}{8}, \tag{67}$$

and (61),

$$A_{\underset{\sim}{6}\times\underset{\sim}{3}} = 3.6 + 6.1 = 27, \tag{68}$$

since $A_{\underset{\sim}{8}} = 0$, it follows that

$$A_{\underset{\sim}{10}} = 27, \tag{69}$$

and so on.

If we want to put the three families in a triplet of SU_3^f, we owe 15 units of anomaly in the SU_5 case (or 5 if the $\overline{5}$ and 10 are put in $\overline{3}^f$ and 3^f respectively), 16 units with SO_{10} and 27 units with E_6. In view of (69) the SU_3^f generalization of E_6 has a particularly simple anomaly-free fermion content

$$(\underset{\sim}{27},\underset{\sim}{3}^f) + (\underset{\sim}{1},\overline{\underset{\sim}{10}}^f) \tag{70}$$

We take the Higgs fields to transform as $(27,\overline{6})_H$ and, say $(1,27)_H$, with the Yukawa term

$$\mathcal{L}_Y^{cl.} = (\underset{\sim}{27},\underset{\sim}{3})_f(\underset{\sim}{27},\underset{\sim}{3})_f(\underset{\sim}{27},\overline{\underset{\sim}{6}})_H + (\underset{\sim}{27},\underset{\sim}{3})_f(\underset{\sim}{1},\overline{\underset{\sim}{10}})_f(\overline{\underset{\sim}{27}},\underset{\sim}{6})_H'$$

$$+ (\underset{\sim}{1},\overline{\underset{\sim}{10}})_f(\underset{\sim}{1},\overline{\underset{\sim}{10}})_f(\underset{\sim}{1},\underset{\sim}{27})_H. \tag{71}$$

We can take $(\underset{\sim}{1},\underset{\sim}{27})_H$ to be real in which case (71) displays only one global symmetry. This global symmetry can be maintained in the Higgs potential; if it is, it will lead to a PQ symmetry. The symmetry breaking pattern can proceed in the following way:

a) $(27,\overline{6})'_H$ breaks $E_6 xSU_3 xU_{1X} \to SO_{10} xSU_2 xU_1 xU'_1$;

b) $(27,\overline{6})_H$ then breaks the above to $SU_5 xSU_2 xU_1 xU''_1$;

c) $(1,\overline{27})_H$ breaks it further to $SU_5 xU_1 xU''_1$, where U_1 is a local symmetry which is a linear combination of flavor and family quantum numbers and U''_1 is the PQ symmetry.

d) In order to break to the standard model we need a $(78,1^f)_H$! and a $(1,8)_H$ to break the left-over U_1. Interestingly one reconstructs in this way the adjoint representation at the level of the Higgs [Could it come from a theory in higher dimensions?].

e) Finally one needs yet another $(27,\overline{6})''_H$ to break SU_{2L}. In this theory since all fermion masses couple to some Higgs at tree level, calculable masses can be generated only if type (a) mechanism is enacted, and a superficial analysis of the potential indicates that it cannot be done in a technically natural way. Still we thought this model simple enough to present it in these lectures.

Our last topic is more phenomenological in nature. Suppose that SU_5 in its minimal form is essentially correct and that over the next two years the proton is found to decay according to plan. What does this imply for physics beyond SU_5? All Grand Unified models beyond the minimal SU_5 predict the existence of new types of fermions with masses in the $\Delta I_w = 0$ and $\Delta I_w = 1$ sectors. These have been briefly discussed at the beginning of these lectures. It is thought that when the $\Delta I_w = 0$ breaking occurs at scale $m_x \sim 10^{15}$ Gev these will get commensurate masses, out of reach by earth accelerator experiments. In the following we want to explore the possibility(29) that in some interesting cases these need not be as heavy as one naively thought because there may exist global conservation laws which make those particles massless until they are broken.(30) We have already discussed the example of neutrinos whose masses are less than 10^{-10} m_w. Can one duplicate this situation with these $\Delta I_w = 0$ fermions?

To fix our ideas let us try to formulate a model with $\Delta I_w = 0$ radiative masses. To make it as simple as possible we start with one SU_5 family $\overline{5}_f + 10_f$ and an extra $\overline{5}'_f + 5_f$ set of Weyl fermions. The two fermion combinations

$$\overline{5}_f 5_f \text{ and } \overline{5}'_f 5_f \tag{72}$$

have $\Delta I_w = 0$ and $\Delta I_w = 1$ masses. The extra set of fermions in the absence of mixings describes a lepton weak isodoublet with charges (-1,0) and vector-like weak interactions, that is, a purely vectorial neutral current, and a weak singlet charge -1/3 quark. At the level of SU_5, there is nothing to prevent bare mass terms for the combinations (72). Hence one needs an additional symmetry. We now present a new model[29,31] where this symmetry is the phase symmetry introduced by Peccei and Quinn to solve the strong CP problem! On the other hand if the Yukawa term includes only the usual SU_5 two fermion terms $\overline{5}_f 10_f$, $\overline{5}'_f 10_f$ and $10_f 10_f$, the 5_f never appears so that the resultant chiral symmetry $5_f \rightarrow e^{i\alpha} 5_f$ forbids $\Delta I_w = 0$ masses to all orders of perturbation theory: 5_f must appear in the Yukawa term.

The simplest way to have it appear is to introduce a new Higgs field 10_H, which transforms as a $\underset{\sim}{10}$ of SU_5. The resultant term

$$5_f 10_f 10_H \tag{73}$$

links the extra chiral symmetry to a phase transformation on 10_H. We now want to connect it to the usual fermions.

One way is to introduce a Higgs cubic coupling

$$10_H \overline{5}'_H \overline{5}''_H \tag{74}$$

where $\overline{5}'_H$ and $\overline{5}''_H$ are Higgs fields that couple to $10_f \cdot 10_f$. The Yukawa part of $\mathcal{L}$ is now given by

$$\mathcal{L}_Y^{cl.} \sim (x\overline{5}_f + x'\overline{5}'_f) 10_f \overline{5}_H + 10_f 10_f (y5'_H + y'5''_H) + 5_f 10_f 10_H \tag{75}$$

and the Higgs potential contains the cubic term (74). We have at this level two global symmetries X and Y with the following assignments

	$\overline{5}_f(5'_f)$	5_f	10_f	5_H	$5'_H(5''_H)$	10_H	
X	-3	3	1	-2	-2	-4	
Y	1	0	0	+1	0	0.	(76)

We see that X is the global symmetry already present in the standard SU_5 model; it leads to B-L conservation by 't Hoofting. The extra Y symmetry gives a non zero value for the combinations $5_f\overline{5}_f$ and $5_f\overline{5}'_f$; it therefore prevents bare mass terms from developing. However we want these fermions to get masses in higher order. The Higgs field capable to couple to $5_f\overline{5}_f$ is the 24_H already present in the standard SU_5, but it cannot have any Y value since it is real. We thus extend the model by making it complex so as to give it a Y-value. As $5_f\overline{5}_f$ has Y value +1, the 24_H must have Y value of -1/2 so as to forbid the renormalizable coupling $5_f\overline{5}_f\ 24_H 24_H$, allow the induced coupling $5_f\overline{5}_f\ 24_H 24_H$, and therefore produce a calculable radiative $\Delta I_w = 0$ mass. This can be achieved by quartic terms in the Higgs potential of the form

$$24_H 24_H 5_H \overline{5}'_H,\ 24_H 24_H 5_H \overline{5}''_H. \tag{77}$$

When 24_H gets a vev at a scale $\sim m_x$, it spontaneously breaks the Y symmetry, producing a Nambu-Goldstone boson which couples weakly to matter. The Y-symmetry is recognized as the PQ symmetry and the N.G. boson is nothing but the axion! This model resembles closely a model recently proposed(32) to incorporate the invisible axion into the standard SU_5 model. It is amusing that one arrives at the axion by requiring low $\Delta I_w = 0$ masses!!

In this model, because we have chosen to use a cubic Higgs coupling (74) the calculable mass comes from diagrams of the form (31) at the two loop level with only Higgs exchange.

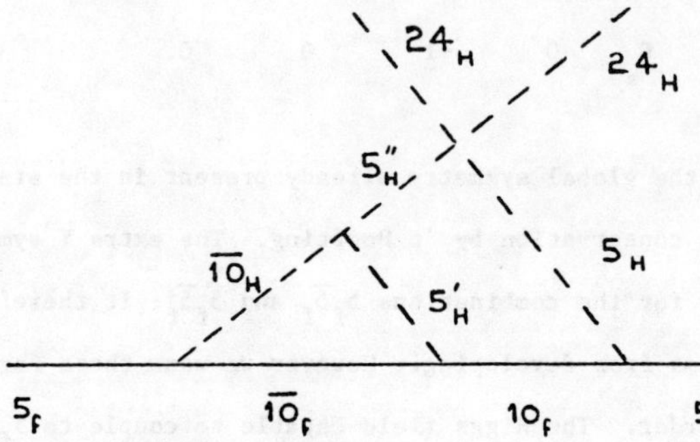

(78)

Thus the diagram is depressed over the usual two loop suppression by the smallness of the Yukawa couplings. In fact note that the contribution is proportional to the mass of the charge 2/3 quark, i.e. for the lightest family m_u! Hence we can achieve at least a 10^{-10}-10^{-13} suppression over the unification scale!

This raises the exciting possibility that there might be such fermions with anomalously low mass. If so, what is their effect[29] on low energy phenomenology as well as on the successful SU_5 predictions?

The most striking feat of SU_5 is the close agreement of its value of $\sin^2\theta_w$ with experiment. However, it is easy to see that at the one loop level the addition of extra fermion representations does not change the value of $\sin^2\theta_w$ as long as the extra fermions come in SU_5 representations. For instance in the model we just presented the extra fermions consist of an extra $5_f + \overline{5}_f$ and they give to the one-loop β-function for the strong, weak and hypercharge coupling constants the same contribution, as ordinary $\overline{5} + 10$ fermions do.[3] Hence unless the new fermions within each SU_5 multiplet get masses at widely different values, the standard calculation of $\sin^2\theta_w$ will not be affected.

Similar considerations show that the unification mass (at the one-loop level) will not be changed by the presence of these fermions. However, since extra fermions tend to change the slope of the coupling constant curve, their effect will be to increase the value of the coupling constant at unification, and therefore make the proton a little bit more short lived. For example, adding one extra pair of $5_f + \overline{5}_f$ multiples the standard SU_5 prediction for τ_p by ~ .7.[29] Hence the precise determination of τ_p will be an important guide as to the number of such particles, as well, of course, a crucial test of the idea of unification itself!

The next question of importance is the stability of these particles: how do they decay? When he electroweak group is broken, the weak eigenstates will mix with the result that these heavy particles will decay via the usual weak interactions, but their decay width will be much suppressed by mixing angles.

Take as one example our model. It has two mass eigenstates in the charge -1 sector, call them e^- and E^-, two eigenstates in the charge -1/3 sector, d and D, one charge 2/3 quark u, one massless neutrino ν_L and a Dirac neutrino N. It is a simple matter to diagonalize the mass matrix obtained by electroweak breaking and to rewrite the currents in terms of the mass eigenstates. In the quark sector, one finds the charge current density to be

$$u_L^+(\cos\alpha d_L + \sin\alpha D_L), \tag{79}$$

where α is a mixing angle given approximately by

$$\tan^2\alpha \sim \frac{m_d}{m_D}. \tag{80}$$

The quark neutral current density is

$$\frac{1}{3}\sin^2\theta_w(d_L^+ d_L + D_L^+ D_L) + \frac{1}{2}(\cos\alpha d_L^+ + \sin\alpha D_L^+)(\cos\alpha d_L + \sin\alpha D_L), \tag{81}$$

showing flavor-changing neutral current suppressed by the mixing angle α. This simple example shows that when all three families are included we expect to find induced flavor changing neutral current effects. Since we have only mixing angle suppression, we naively expect that the absence of such effects will put a constraint on the magnitude of the masses of these exotic fermions, e.g.

$$m_D > 10^4 m_d, \tag{82}$$

in the absence of a GIM mechanism, which is not too severe.

In the lepton sector, the situation is a bit more complicated because of the neutral lepton mass matrix. Nevertheless the same systematics appears: the heavy particles decay into light ones via ordinary charge-changing and neutral weak interacting their decay being slowed down by mixing angle effects. Again, in the realistic case of three families, these will induce lepton flavor changing effects, thus putting a limit on the mixing angles.

Let us remark that the mass difference between the numbers of the heavy doublet (E^-,N) is exceedingly small. The reason is that its quantum number is $\Delta I_W = 1$, which in the usual Grand Unification framework is twice suppressed by the hierarchy. In this picture one can expect to have particles with masses of hundreds of Gev, separated from one another by ev's!!

In one simple model, the extra particles have the same color and charge quantum numbers as ordinary particles. The same would be true if we included an extra $10_f + \overline{10}_f$. Then these heavy particles can mix with the ordinary light ones and decay reasonably fast, as far as cosmological contraints are concerned. However the situation would be different with, for instance, an extra $45_f + \overline{45}_f$ of fermions. It is easy to see that it contains quarks with exotic quantum numbers. These would be quasi stable on cosmological scale. Hence we must restrict our fermions to representations that contain no exotic quarks or leptons. Details of the phenomenology of these new fermions will be presented elsewhere.

To conclude, let us hope that the proton decays and that new types of fermions will be found by the new generation of accelerators.

Acknowledgements

I would like to thank Professors Konuma and Maki for their superb hospitality and kind invitation, as well as the audience for its receptiveness and inquisitiveness. I would also like to thank Prof. A. Sanda and Mark Bowick who have collaborated with me on various aspects of these lectures.

References and Footnotes

(1) J. C. Pati and A. Salam, Phys. Rev. D8 (1973) 1240; D10 (1974) 275.

(2) H. Georgi and S. L. Glashow, Phys. Rev. Lett. 32 (1974) 438.

(3) H. Georgi, H. R. Quinn, and S. Weinberg, Phys. Rev. Lett. 33 (1974) 451.

(4) For a recent review, see P. Langacker, SLAC-PUB 2544 (1980), to be published in Phys. Reports C; this review contains an exhaustive list of references.

(5) A. J. Buras, J. Ellis, M. K. Gaillard and D. V. Nanopoulos, Nucl. Phys. B135 (1978) 66.

(6) For a specific mechanism see, M. Gell-Mann, P. Ramond, R. Slansky, as mentioned in P. Ramond "The Family Group in Grand Unified Theories" invited talk at Sanibel Symposia, Feb. 1979, CALT-68-709, unpublished and in "Supergravity" ed. by P. Van Nieuwenhuizen and D. Z. Freedman (North-Holland, 1979, Amsterdam). See also T. Yanagida in Proc. of Workshop on the Unified Theory and the Baryon Number in the Universe, ed. by O. Sawada and A. Sugamoto KEK (1979). The smallness of the neutrino masses can also be regarded as stemming from the fact that their quantum number $\Delta I_w = 1$ is different from the main breaking mechanism $\Delta I_w = 1/2$.

(7) H. Georgi, Nucl. Phys. B156 (1979) 126. This formulates the so-called "survival hypothesis" by which fermions get masses commensurate with the scale at which the symmetry that allows them is broken.

(8) M. Bowick, P. Ramond, and A. Sanda in preparation. See also M. Bowick and P. Ramond, UFTP-81-7, to be published in Physics Letters; P. Ramond, invited talk at the Johns Hopkins Workshop on Current Problems

in Particle Theory 5, May 25-27, 1981. UFTP-81-8.

(9) R. Peccei and H. Quinn, Phys. Rev. Lett. 38 (1977) 1440, and Phys. Rev. D16 (1977) 1791.

(10) F. Wilczek and A. Zee, Phys. Rev. Lett. 42 (1979) 421
K. Akama and H. Terazawa, INS-report-257, Tokyo 1976.

(11) P. Ramond, Sanibel Symposia Talk, CALT-68-709, 1979, unpublished.

(12) M. Gell-Mann, P. Ramond, and R. Slansky, in "Supergravity" op. cit.

(13) Calculability is reviewed in J. C. Taylor, "Gauge Theories of Weak Interaction", Cambridge University Press, Cambridge 1976. See especially chapter 15 and references contained therein.

(14) I thank M. Ramond for a general analysis of these equations.

(15) P. Ramond, unpublished 1979.

(16) P. Ramond, in Proceedings on "Weak Interactions as Probes of Unification" 1980. G. B. Collins, L. N. Chang, J. R. Ficenec, eds., American Institute of Physics, N.Y. 1981.

(17) R. Barbieri and D. V. Nanopoulos, Phys. Lett. 91B (1980) 369. These authors consider this mechanism in the context of an E_6 model, and point at that at the one loop level diagram (32) can occur by exchange of a heavy fermion.

(18) Y. Kazama, D. Unger, and Y. P. Yao, Fermilab preprint 1980.

(19) For an interesting attempt, see the lectures of Dr. D. V. Nanopoulos at this summer school.

(20) F. Gürsey, P. Ramond, and P. Sikivie, Phys. Lett. B60 (1975) 177.

(21) J. E. Kim, Phys. Rev. Letters 43 (1979) 103.
M. Dine, W. Fischler, A. Srednicki, IAS preprint (1981).

(22) M. Bowick and P. Ramond, UFTP-81-7, to be published in Physics Letters.

(23) F. Gürsey, Private communication; and F. Gürsey, M. Serdaroglu, Yale University preprint 1981.

(24) E. Witten, Phys. Lett. 91B (1980) 81.

(25) See for instance, R. Barbieri, D. V. Nanopoulos and A. Masiero, CERN preprint TH-3048 (1981).

(26) For more Kronecker products, see R. Slansky, Los Alamos Preprint 1981, to be published in Physics Reports.

(27) S. Coleman and E. Weinberg, Phys. Rev. D7 (1973) 1888.

(28) This formula is not new; it appears in many papers. For many more, see J. Patera, Montreal Reprint 1980.

(29) M. Bowick, P. Ramond, and A. Sanda, in preparation.

(30) The lessons of ref. (21) show that one could have many global conservation laws spontaneously broken at unification scales. The resulting Nambu-Goldstone bosons would then couple very lightly to matter, thus having escaped detection.

(31) See also P. Ramond, invited talk at INS Symposium on Quark and Lepton Physics, Tokyo 1981.

(32) M. B. Wise, H. Georgi, and S. S. Glashow, HUTP-81/A019 (1981).

THE CASE FOR AND AGAINST NEW DIRECTIONS IN GRAND UNIFICATION

Howard Georgi

Lyman Laboratory of Physics, Harvard University,

Cambridge, MA 02138, U.S.A.

in

Proceedings of the Fourth Kyoto Summer Institute on Grand Unified Theories and Related Topics, Kyoto, Japan, June 29 - July 3, 1981, ed. by M. Konuma and T. Maskawa (World Science Publishing Co., Singapore, 1981).

Contents

1. SUPERSYMMETRY -- A PRACTICAL PRIMER

I review the rules for constructing supersymmetric gauge theories.

2. SUPERSYMMETRIC GUTs -- PROMISE AND PROBLEMS

I discuss a supersymmetric SU(5) theory.

3. EFFECTIVE THEORIES BELOW A SYMMETRY BREAKING SCALE

I give and discuss a precise definition of what I mean by the title.

Howard Georgi†

Lyman Laboratory of Physics, Harvard University

Cambridge, MA 02138 U.S.A.

†Supported in part by the National Science Foundation under Grant Number PHY77-22864.

1. Yesterday, I had the great pleasure of sightseeing in your beautiful city. I saw shrines and gardens and castles and in all of them I saw what I assumed was the influence of Zen, a compulsion to modify nature to make the pieces fit together better. In a Zen garden, I had the feeling that the individual trees and rocks and flowers were not very important. They were sacrificed to the whole, and if the rocks had to be moved and the trees had to be twisted, that was okay.

I do not think that I will take up Zen. I enjoy the separate pieces of the world too much. The point I want to make, however, is that while the Zen approach is all right in philosophy or aesthetics, it does not work in physics. No matter how beautiful you may think your theory is in the whole, if the pieces make no sense, the theory is no good. This is the trouble with subquark models. The grand idea is fine, but the pieces do not work. In such a case, a serious physicist must either abandon the entire program or spend his time developing new pieces that actually work.

This bit of philosophy may help you to understand some of the things I talk about this week. If I have a choice between talking about a slightly contrived model in which I understand how to calculate and a seemingly more beautiful model in which calculation is beyond our abilities. I will pick the contrived model every time. In the second case, you should not be building models at all, but figuring out how to calculate.

Today, I want to talk about supersymmetry. Not the fancy SO(8) supergravity which has been glorified by the Zen masters at CERN, but simple $N = 1$ supersymmetry. I will discuss it in the context of a realistic SU(5) model incorporating supersymmetry, constructed by Savas Dimopoulos and me.[1] The model itself is not very interesting. I certainly do not believe it. But it will be a useful vehicle for a review of supersymmetry.

I am not going to start from the beginning of supersymmetry. All I want to do is to teach you how to use it. I refer you to the review by Fayet and Ferrara[7] for the mathematical details.

In a supersymmetric theory each fermionic degree of freedom is associated with a degenerate bosonic degree of freedom in a supermultiplet. Renormalizable theories involve essentially only two kinds of multiplets, scalar and gauge. In each type, the fermionic particles are the two states of a chiral fermion, like a neutrino and antineutrino. In the scalar supermultiplet, the bosons are the two real scalars in a complex scalar field. In the gauge supermultiplet, the bosons are the two spin states of a gauge field. We will take all our fermions to be left-handed for convenience.

Suppose we have N scalar supermultiplets. The following notation is very convenient: Let Ψ^a (A=1 to N) be the left-handed fermion fields associated with the scalar supermultiplets and ϕ^a be the corresponding complex spinless meson fields. Let A^μ_α be the gauge fields (where μ is a vector index and α labels the generators of the gauge group) and χ_α be their supersymmetric partners which we can take to be left-handed fermion fields. The covariant derivative on the Ψ and ϕ fields is

$$D^{\mu a}_{\ \ b} = \partial^\mu \delta^a_b + i\, g_\alpha A^\mu_\alpha T^a_{\alpha b} \tag{1.1}$$

where T_α is the gauge generator and g_α is the gauge coupling constant. The gauge invariant function is $v(\phi)$. We will also need the derivatives:

$$\begin{aligned} v_a(\phi) &= \frac{\partial}{\partial\phi^a}\, v(\phi), \\ v_{ab}(\phi) &= \frac{\partial^2}{\partial\phi^a\partial\phi^b}\, v(\phi), \\ v_{abc}(\phi) &= \frac{\partial^3}{\partial\phi^a\partial\phi^b\partial\phi^c}\, v(\phi). \end{aligned} \tag{1.2}$$

Gauge invariance of v implies

$$v_a(\phi)\, T^a_{\ b}\phi^b = 0 \tag{1.3}$$

(summation over repeated indices implied). The penultimate pieces of notation we need is for the complex conjugates of ϕ^a;

$$\bar{\phi}_a = \phi^{a*}$$

$$\bar{v}(\phi) = v(\phi)^*$$

$$\bar{v}^a(\phi) = v_a(\phi)^* \tag{1.4}$$

etc.

Finally, we define the functions

$$K_\alpha = g_\alpha \bar{\phi}_a T^a_{\alpha b} \phi^b \tag{1.5}$$

(no sum on α) with derivatives defined as in (1.2 and 4).

To put all this together into the Lagrangian for the supersymmetric theory, it is convenient to work in a Majorana basis for the fermion fields. The kinetic energy terms and the gauge couplings are the usual ones with the covariant derivatives given by (1.1). The Yukawa couplings and fermion mass terms are

$$-\frac{1}{2}\psi^{aT}\gamma^0 v_{ab}(\phi)\psi^b + \text{h.c.}$$

$$-\chi_\alpha\gamma^0 K_{\alpha a}(\bar{\phi})\psi^a + \text{h.c.} \tag{1.6}$$

The scalar meson potential is

$$v(\phi, \bar{\phi}) = v_a(\phi)\bar{v}^a(\bar{\phi}) + \frac{1}{4}K_\alpha K_\alpha. \tag{1.7}$$

The vacuum expectation value (VEV) of the ϕ's is determined by minimizing (1.7). In most cases, the minimum will occur when

$$v_a(\phi) = \bar{v}^a(\bar{\phi}) = K_\alpha = 0. \tag{1.8}$$

If (1.8) can be satisfied, then although the gauge symmetries may be spontaneously broken, the supersymmetry remains unbroken. It is sometimes possible to arrange $v(\phi)$ so that (1.8) is not allowed, so that (1.7) has its minimum for nonzero values of $v_a(\phi)$ and K_α. In this case the supersymmetry is broken spontaneously.

Now I can write down the mass matrices. These depend only on the VEV's of the functions $v(\phi)$, $v_a(\phi)$... and $K_\alpha(\bar{\phi}, \phi)$, $K^a(\phi)$... . We will denote the

VEV's by the same symbols without explicit ϕ dependence. In this notation, the condition that the VEV extremizes the potential is

$$v_{ab}\bar{v}^{b} + \frac{1}{2} K_{\alpha a} K_{\alpha} = 0. \qquad (1.10)$$

The mass squared matrix of the vector bosons is given by

$$\frac{1}{2}\left\{K_{\alpha a} K^{a}_{\beta} + K^{a}_{\alpha} K_{\beta a}\right\}. \qquad (1.11)$$

The mass matrix of the left-handed fermions is the symmetric matrix

$$\begin{bmatrix} v_{ab} & K_{\alpha b} \\ K_{\alpha a} & 0 \end{bmatrix}. \qquad (1.12)$$

The mass squared matrix of the spinless bosons is

$$\begin{bmatrix} \bar{v}^{ab} v_{bc} + \frac{1}{2} K^{a}_{\alpha} K_{\alpha c} + \frac{1}{2} K^{a}_{\alpha c} K_{\alpha} & \bar{v}^{abc} v_{b} + \frac{1}{2} K^{a}_{\alpha} K^{c}_{\alpha} \\ v_{abc} \bar{v}^{b} + \frac{1}{2} K_{\alpha a} K_{\alpha c} & v_{ab} \bar{v}^{bc} + \frac{1}{2} K_{\alpha a} K^{c}_{\alpha} + \frac{1}{2} K^{c}_{\alpha a} K_{\alpha} \end{bmatrix} \qquad (1.13)$$

The signal of spontaneous supersymmetry breakdown is a nonzero value of v_a or K_α. If one or both of these is nonzero, there is a Goldstone fermion, in the direction of the vector

$$\begin{bmatrix} v_{a} \\ \frac{1}{2} K_{\alpha} \end{bmatrix} \qquad (1.14)$$

It is easy to check that this annihilates (1.12), using (1.10) and (1.3) which is equivalent to

$$v_{a} K^{a}_{\alpha} = 0. \qquad (1.15)$$

Note that if $v_a = K_\alpha = 0$, then

$$v^{ab} K_{\alpha b} = 0 \qquad (1.16)$$

and

$$K_{\alpha a} K^{a}_{\beta} = K_{\beta a} K^{a}_{\alpha}. \qquad (1.17)$$

With these relations you can easily show that in the supersymmetric limit, the theory falls apart in the correct way into degenerate supermultiplets.

A nonzero $K_{\alpha a}$ is a signal that the gauge symmetry has been spontaneously broken. We will be breaking SU(5) while preserving supersymmetry in the next lecture, so I will examine this case in more detail.

If $v_a = K_\alpha = 0$, then (1.16) breaks the fermion and scalar mass matrices apart into two parts. We can write the matrix $K_{\alpha a}$ as a sum,

$$K_{\alpha a} = \sum_j m_j \chi_\alpha^j Z_a^j \tag{1.18}$$

where the m_j are positive. Then it is clear from (1.11) and (1.17) that the m_j are the masses of the vector bosons associated with the broken gauge symmetries.

The vectors

$$\begin{bmatrix} \bar{Z}^{jc} \\ \\ -Z^j \end{bmatrix} \tag{1.19}$$

annihilate (1.13). These correspond to the Goldstone bosons which are eaten by the Higgs mechanism. But

$$\begin{bmatrix} \bar{Z}^{jc} \\ \\ Z_c^j \end{bmatrix} \tag{1.20}$$

is an eigenvector of (1.13) with eigenvalue m_j^2. Thus, there is a real scalar field degenerate with each massive vector meson. Furthermore, the $K_{\alpha a}$ part of the fermion mass matrix is a Dirac mass matrix which combines the partners of the massive gauge bosons with the partners of the Goldstone bosons into four component fermions. These, together with the massive gauge and scalar bosons comprise what are called massive vector supermultiplets.

Thus, the supersymmetric broken gauge theory consists of scalar supermultiplets, massless gauge supermultiplets for the unbroken gauge symmetries and massive vector multiplets for the broken gauge symmetries. In the next lecture, we will see how to use these ideas in an SU(5) model.

In the next lecture, we will see how to use these ideas in an SU(5) model.

REFERENCES

1) S. Dimopoulos and H. Georgi, Harvard Preprint HUTP-81/A022. The same thing has been done by N. Sakai, Tohoku University Preprint TU/81/225.

2) P. Fayet and S. Ferrara, Phys. Rep. 32C, 249 (1977).

2. The original motivation, I think, for constructing supersymmetric grand unified theories was the hope that the supersymmetry could help with the hierarchy puzzle. This seemed reasonable. Because the scalar fields are associated with fermions in supermultiplets, they can carry chiral symmetries which keep them massless. This sounds like just what we need for the Higgs doublet. Of course, to use supersymmetry, at least in any obvious way, we must keep the supersymmetry unbroken down to a scale of the order of 1 TeV.

To see whether such a thing is possible, we must examine the structure of an SU(5) model in detail. Clearly, the model must contain 10's and $\bar{5}$'s of scalar supermultiplets describing the usual left-handed fermion fields. The supersymmetric partners of the leptons include an SU(2) x U(1) doublet of scalar fields with the right gauge structure to be the Higgs. But it cannot be! The most serious problem with such an identification is that it gives rise to far too much baryon number violation. I want to show you how this comes about in gory detail.

Let the supermultiplets containing the quark and lepton fields be

$$M_j^{xy} = - M_j^{yx}, \quad M'_{jx} \tag{2.1}$$

where x = 1 to 5 is an SU(5) index and j is a generation index (j = e, μ, τ). M_j^{xy} is a 10 and M'_{jx} is a $\bar{5}$. The SU(3) x SU(2) x U(1) symmetry is embedded such that the components of the 5 ($\bar{M}'^{x}_{e}$ say) correspond to quarks and leptons as follows:

$$\bar{M}'^{a}_{e} = d^a \quad \text{for} \quad a = 1 \text{ to } 3 \tag{2.2}$$

where a = 1 to 3 are the three colors

$$\bar{M}'^{4}_{e} = e^+, \; \bar{M}'^{5}_{e} = \bar{\nu}_e . \tag{2.3}$$

An example of the identification of Higgs with lepton partners is the following SU(5) invariant term in v,

$$M_e^{xy} \, M'_{ex} \, M'_{\mu y} . \tag{2.4}$$

If the ν_μ component of M'_μ gets a VEV, this term gives rise to a mass for the d and e.

The trouble with (2.4) is that the coupling of the triplet components of M'_μ, the SU(5) partners of the putative Higgs, mediate proton decay. This is probably obvious, because the scalar part of M'_μ acts like the usual 5 of Higgs in SU(5), whose triplet component is well known to give baryon number violation. But let us see it explicitly. $M'_{\mu a}$ couples to (ignoring Cabibbo-KM mixing)

$$\sum_{b=1}^{3} M^{ab}_{\;} M'_{eb} + M^{a4}_e M'_{e4} + M^{a5}_e M'_{e5}$$

$$= \varepsilon^{abc} \bar{u}_b \bar{d}_c + u^a e^- + d^a \nu_e .$$

This coupling to antiquark-antiquark and also quark-lepton obviously produces proton decay.

There are a host of other reasons why the lepton partners cannot be the Higgs. They cannot give mass to the u quark, because the term

$$M^{xy}_e M^{wz}_e \bar{M}^r_\mu \varepsilon_{xywzr} \tag{2.5}$$

is not allowed by the supersymmetry (only unbarred fields are allowed in v -- the barred fields carry the wrong handedness). They produce degenerate lepton masses, because of the antisymmetry of M^{xy}_j. They cause funny neutral currents, because the leptons mix with the gauge fermions.

All of these difficulties are avoided if there are two light Higgs doublets which are contained in supermultiplets (distinct from M^{xy}_j and M'_{jx}) in which the colored components are all superheavy, with mass of the order of the unification scale. I will construct the simplest such model (based on my work with S. Dimopoulos) and show that it fixes all these problems.

In the simplest model, the Higgs doublets are in a 5 and $\bar{5}$, H^x and H'_x. In addition, we need a Higgs to develop a large VEV and break SU(5) down to SU(3) x SU(2) x U(1). We take this to be a 24, Σ^x_y. ($\Sigma^x_x = 0$). The function v with the right properties is

$$\lambda_1\left|\frac{1}{3}\,\Sigma^x_y\Sigma^y_z\Sigma^z_x + \frac{m}{2}\,\Sigma^x_y\Sigma^y_x\right|$$

$$+ \lambda_2\, H'_x\left|\Sigma^x_y + 3m\delta^x_y\right|H^y$$

$$+ f_{jk}\,\varepsilon_{uvwxy}\,H^u\,M^{vw}_j M^{xy}_k$$

$$+ g_{jk} H'_x N^{xy}_j M'_{ky}. \tag{2.6}$$

Applying the formalism of the first lecture, we find that the potential is minimized for

$$\langle H^x\rangle = \langle H'_x\rangle = \langle M^{xy}_j\rangle = \langle M_{jx}\rangle = 0, \tag{2.7}$$

and one of three VEV's for Σ;

$$\langle\Sigma^x_y\rangle = 0, \tag{2.8a}$$

$$\langle\Sigma^x_y\rangle = \frac{m}{3}\,\delta^x_y - \frac{5m}{3}\,\delta^x_5\delta^5_y, \tag{2.8b}$$

$$\langle\Sigma^x_y\rangle = 2m\,\delta^x_y - 5m\left(\delta^x_4\delta^4_y + \delta^x_5\delta^5_y\right). \tag{2.8c}$$

Supersymmetry is not broken. (2.8a) does not break SU(5). (2.8b) breaks it down to SU(4) x U(1). (2.8c) breaks it down to SU(3) x SU(2) x U(1). At this point, these three physically inequivalent vacua are completely degenerate. This bizarre situation is not uncommon in supersymmetric theories (it may be relevant in the very early universe).

At any rate, (2.8c) is what we want. If we assume it, it is easy to see that (2.6) gives a mass of order m (the unification scale) to the color triplet components of the Higgs fields $H^{1,2,3}$ and $H'_{1,2,3}$. But it leaves the SU(2) doublet components, $H^{4,5}$ and $H'_{4,5}$, massless. All the components of Σ are either eaten by the Higgs mechanism or have large mass. Thus, we are left with only the quark, lepton and Higgs doublet supermultiplets and the SU(3) x SU(2) x U(1) gauge supermultiplets massless.

I will return shortly to what I don't like about this model. But first let me note how it solves the problems we discussed above. Clearly, the terms in (2.6) proportional to f_{jk} and g_{jk} contain the usual couplings of the Higgs doublets to quarks and leptons. If the Higgs get a VEV, the quark and lepton masses will not be a problem. Furthermore, as the discussion in lecture 1 shows, if the Higgs get a VEV, the gauge fermions will mix with the supersymmetric partners of the Higgs, the Higgs fermions. But as long as the partners of the leptons do not get a VEV, the leptons will not mix with gauge fermions. The neutral currents will be normal. Finally, we have solved the problem of proton decay mediated by the SU(5) partners of the Higgs by the same brutal expedient used in standard SU(5) models--pushing up their mass. The color triplet Higgs still mediate proton decay, but because they are superheavy, the rate is small. Note that we have also eliminated several other potential sources of fast proton decay. With the triplet Higgs out of the picture, the remaining couplings of the scalar supermultiplets in the superpotential conserve baryon number and the separate lepton numbers, where the superpartner bosons are defined to have the same baryon and lepton number as their corresponding fermions.

There are still, however, several sources of baryon number violation from the couplings of the leptoquark gauge bosons and their partners. One is, of course, the standard gauge boson exchange diagram, (an example is shown in Figure (2.1)). This effect is obviously of order $1/M^2$ where M is the unification scale because of the vector boson propagator. But there are also baryon number violating effects from the diagrams shown in Figures (2.2) and (2.3), where the dashed lines and capital letters refer to the supersymmetric partners of the quarks and leptons. Figure (2.2) represents exchange of the fermion partner of the leptoquark boson. Figure (2.3) represents the $K_\alpha K_\alpha$ term in the scalar potential (α in the leptoquark direction) which in some

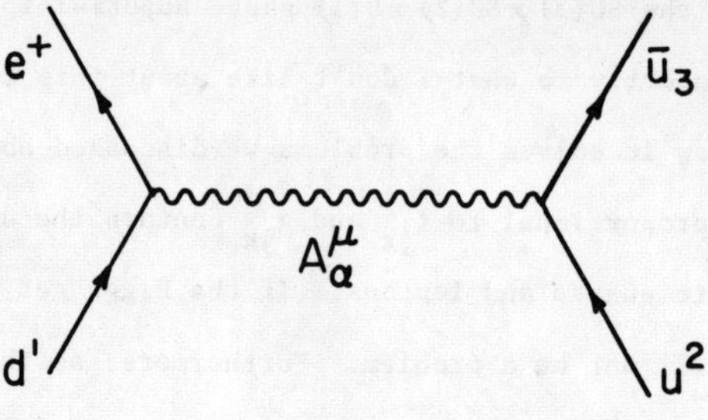

Fig. (2.1)

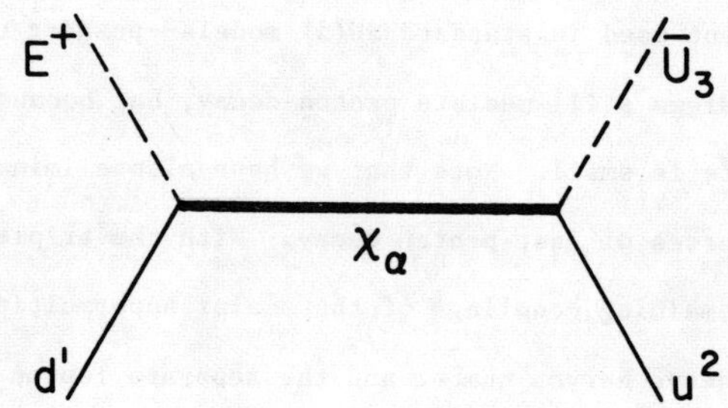

Fig. (2.2)

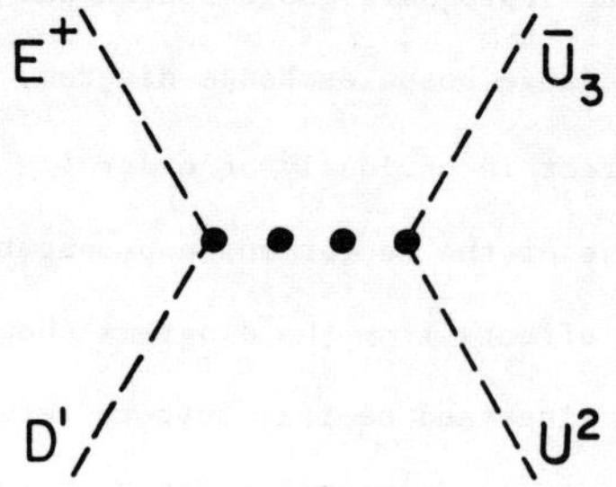

Fig. (2.3)

sense results from exchange (with "propagator" 1) of an auxiliary field in the leptoquark gauge boson supermultiplet. The diagram in Figure (2.2) is naively of order 1/M and that in Figure (2.3) is naively of order 1. You might well ask, what is going on?

In fact, all of these diagrams are actually order $1/M^2$. The leading 1/M term in Figure (2.2) does not contribute to baryon number violation because it involves a change in chirality. But the massive fermion is built (as shown in lecture 1) out of the chiral gauge fermion and a chiral fermion in the Σ supermultiplet. The mass term changes one to the other. But the Σ field does not couple to the quark and lepton supermultiplets. Thus, the leading term cancels, leaving only a $1/M^2$ effect.

The diagram in Figure (2.3) is even more subtle. The same $K_\alpha K_\alpha$ term which produces it also produces a coupling of the external particles to $m\phi_\alpha$ where m is the leptoquark boson mass and ϕ_α is its scalar partner (in the massive vector supermultiplet). Thus, there is a precisely analogous diagram in which this scalar is exchanged. In this diagram, the auxiliary field "propagator" (which is just 1) is replaced by $m^2/(q^2-m^2)$ where q is the momentum transfer carried by the virtual line. Thus, the order 1 contribution is cancelled, leaving only a $1/M^2$ effect. A more formal way to see the cancellation is to notice that the ϕ_α field can be shifted so that the entire $K_\alpha K_\alpha$ term is just the ϕ_α mass term. This induces effects elsewhere, but they are all of order $1/M^2$.

We have saved the proton, but at what cost? The trouble with this theory (at this stage, at least) is that just as in the unsymmetric version, we must make accurate adjustment to keep the Higgs doublets light. We must adjust the two m's which appear in (2.6) to be equal to a high degree of accuracy. Actually, in a rather technical and not very satisfying sense, this adjustment is more natural than the analogous adjustment in the simpler nonsupersymmetric theory because the ratio of the two mass parameters is not renormalized. We set it equal to one at the beginning and forget about it.

But still, we had no reason to make this adjustment except our desire to make the doublet light.

To put the problem in different terms, we had hoped at the beginning to keep the Higgs doublets light by means of an unbroken chiral symmetry. We could do that very simply in (2.6) by setting $\lambda_2 = 0$. This is completely natural but it doesn't give physics, because not only are the SU(2) doublets, $H^{4,5}$ and $H'_{4,5}$, massless when λ_2 goes to zero, but the color triplets, $H^{1,2,3}$ and $H'_{1,2,3}$, are massless as well. As shown above (in boring detail) this is unacceptable because it leads to fast proton decay. Thus, we must break the chiral symmetry explicitly to give large mass to the color triplets.

Are there any ways out of this embarrassing situation? Not in SU(5). In other groups, I know of two possible escape routes, both suggested to me originally by Savas Dimopoulos. One is to embed the SU(3) x SU(2) x U(1) in a unifying group in which proton decay can be avoided entirely. It is then possible to let the analog of the color triplet be light, so that the global chiral symmetry which keeps the Higgs doublets light can be unbroken.

I have constructed such a model (on the plane on the way to Japan) based on the group SU(3) x SU(3) x SU(3) with a discrete symmetry that cyclically permutes the factors. This is a subgroup of E(6), as discussed by Ramond, but the model <u>cannot</u> be extended to E(6). That would introduce proton decay and spoil the whole purpose. At any rate, the interesting representations are 27's, as in E(6), and I have built a theory in which the pieces (except for neutral leptons) of five 27's are all light. Three of them contain the quarks and leptons. The other two contain the Higgs. All are light because of an unbroken chiral symmetry.

If I decide to write up this model, I will call it "A Supersymmetric GUT with No Fine Tuning Puzzle, No Proton Decay and No Redeeming Value". The last because in this theory, there is too much stuff below the unification scale so all the gauge couplings get strong before unification, and I do not really know what the model means. Incidentally, I think that this model can be improved to the point where it makes sense by breaking the 27's into pieces and keeping some of the pieces light with chiral symmetries. But that brings me to the next subject.

The other way out is to give the color triplet a large mass and break the original chiral symmetry spontaneously, but in such a way that a linear combination of the chiral symmetry generator and a spontaneously broken gauge generator combine to generate an unbroken global chiral symmetry. This sort of phenomenon is familiar from B-L in SU(5). There, above the SU(2) x U(1) breaking scale, there is a global chiral symmetry of the SU(5) Lagrangian. It is not B-L. But when SU(2) x U(1) is broken, the global symmetry combines with the broken U(1) to give unbroken B-L. Ramond referred to this as 't Hoofting a symmetry.

Here, the idea would be to use this hybrid global symmetry to keep the Higgs doublets light. Unfortunately, this cannot happen in SU(5) (or SO(10) or E(6)). The relevant gauge symmetries are not broken at the unification scale. But it can be done in an SU(7) (or O(14)) model first considered by Kim.[1]

In this model, the 7 of SU(7) contains the 5 of SU(5) and two <u>charged</u> SU(5) singlets, with charges ±1. Thus, this is a really different embedding of SU(3) x SU(2) x U(1) into a unified group. So different in fact that "$\sin^2\theta$" at the unification scale is 3/20 instead of 3/8 as it is in SU(5). You might think that would get you into a lot of trouble. But, it turns out that if you break SU(7) down to SU(4) x SU(3) (with no U(1)) at the unification scale and then further break it to SU(3) x SU(2) x U(1) at a few TeV, you get an acceptable $\sin^2\theta$. Amazing!

Anyway, a U(1) subgroup of SU(7) (with the right properties--it distinguishes SU(2) from SU(3)) gets broken in the first stage of symmetry breaking. This can combine with a global symmetry of the original theory to give an unbroken chiral symmetry which keeps the Higgs light.

I infer from conversations with Savas that he and Frank Wilczek are working out just such a model. Personally, I don't think it can work. At the very least, it will be subject to the same objection that I applied to my SU(3) x SU(3) x SU(3) model. There is so much stuff in it that the couplings get strong before unification.

As I mentioned earlier, I believe that this idea of combining symmetries can be incorporated in my SU(3) x SU(3) x SU(3) model to produce a model that might work. I have not worked out all the details because I am not fantastically impressed with its beauty. But for what it is worth, here is the structure I envision. The 27 consists (in an inappropriate but familiar SU(5) langauge) of a $10 + \bar{5}$ of quarks and leptons, a $5 + \bar{5}$ and two singlets. I will start with the same five 27's, three matter and two Higgs. But in the matter multiplets, I will make the $5 + \bar{5}$ and singlets very heavy (of the order of the unification scale). In the Higgs 27's, I will make the $10 + \bar{5}$ very heavy. Since this is a chiral multiplet with respect to the unbroken SU(3) x SU(2) x U(1), I can only do this by adding two $\overline{27}$'s and coupling the $10 + \bar{5}$ to the $\overline{10} + 5$ in the $\overline{27}$ (this is the part I am really not sure how to do).

The result of these imaginings would be a model with the usual three generations of quarks and leptons and their superpartners, plus four 5 and $\bar{5}$ supermultiplets containing the Higgs and lots of friends. In such a model, the couplings get large around the unification scale, so it is just at the edge of respectability.

Believe it or not, in the SU(5) model, or any of the generalizations, to produce an acceptable low energy physics we must do further violence to the theory. We must break the supersymmetry explicitly (but softly) by adding various mass terms of the order of a TeV.[2] These push the unobserved gauge and Higgs fermions and matter (quark and lepton) bosons up in mass and also generate VEVs for the Higgs doublets. There is no alternative to explicit symmetry breaking in an SU(5) model. Spontaneous symmetry breaking always leaves an unrealistic mass spectrum with light color triplet scalar bosons. It may be possible to avoid this embarrassment by enlarging the group to include either supercolor[3] or an extra U(1).

I hope that this lecture has betrayed a certain lack of enthusiasm on my part for theories with low energy supersymmetry. It is clear to me that supersymmetry can help with the fine tuning problem in GUTs. But, it is also clear that the cure is worse than the disease. Fortunately, we may soon have incontrivertible evidence that supersymmetry is not relevant at low energies. This good news can come in the form of proton decay. As Dimopoulos first pointed out to me, supersymmetry kills proton decay. The superpartners of the $SU(3) \times SU(2) \times U(1)$ gauge bosons form an incomplete SU(5) multiplet with the same group structure as the gauge bosons. They cancel some of the effect of the gauge bosons in the β functions so that the three coupling constants approach each other more slowly. Thus, the unification scale is increased. In fact, the unification scale in a supersymmetric model is 10^{17} GeV[2] or 10^{18} GeV[4] depending on whether the effect of the Higgs is included. In either case, proton decay from leptoquark gauge boson exchange is too slow to be observable. Thus, when proton decay is shown to exist with the rate predicted by the nonsupersymmetric SU(5) model, low energy supersymmetry will be decisively ruled out.

REFERENCES

1) J. Kim, Phys. Rev. Letters 45, 1916 (1980).
2) S. Dimopoulos and H. Georgi, Harvard Preprint HUTP-81/A022; N. Sakui, Tohoku Univ. Preprint TU/81/225; K. Harada and N. Sakai, Tohoku Univ. Preprint TU/81/226.
3) S. Dimopoulos and S. Raby, SLAC Preprint SLAC-PUB-2719; E. Witten, Princeton Univ. Preprint (1981); M. Dine, W. Fischler and M. Srednicki, Princeton IAS preprint (1981).
4) S. Dimopoulos, S. Raby and F. Wilczek, Santa Barbara Preprint NSF-ITP-81-31.

3. I will introduce today's talk by describing two amusing models of neutrino masses which are superficially similar, but which are in fact very different. Both these models can be unified, but unification is not the point in either case, so I will describe them in the context of $SU(2) \times U(1)$.

The first model is due to Chikashige, Mohapatra and Peccei (CMP).[1] They wanted a model in which B-L symmetry was broken spontaneously. They introduced right-handed neutrinos ν_R which couple to the leptons doublets Ψ_L as follows:

$$\bar{\nu}_R \phi^\dagger \psi_L + h.c. \tag{3.1}$$

where ϕ is the usual Higgs doublet field. Because (3.1) would give a Dirac mass to ν_L, CMP wanted to give ν_R a large Majorana mass. But the obvious mass term, $\nu_R \gamma^0 \nu_R$, violates B-L, so they introduced a complex scalar field σ with the Yukawa interaction

$$\sigma \nu_R \gamma^0 \nu_R + h.c. \tag{3.2}$$

Thus σ carries B-L. If σ develops a VEV

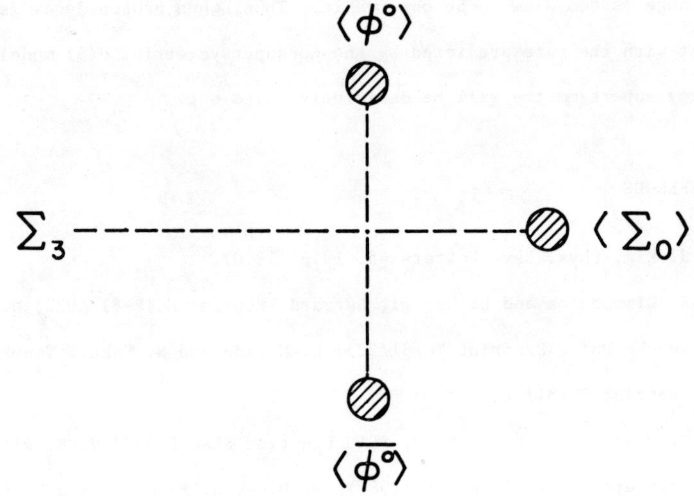

FIGURE 3.1

$$\langle\sigma\rangle = f/\sqrt{2}, \tag{3.3}$$

then the ν_R masses are of order f and the ν_L masses are of order m^2/f where m is a lepton mass.

Of course, (3.3) breaks B-L spontaneously, so there is a Goldstone boson. CMP called it the Majoron. If f is large enough to suppress ν_L masses to an acceptable level, the Majoron is invisible.

Gelmini and Roncandelli (GR)[2] arrived at spontaneous breakdown of B-L in a different way. They did not introduce ν_R. Instead they coupled the lepton doublets

$$\Psi_L = \begin{pmatrix} \nu \\ \ell^- \end{pmatrix}_L \tag{3.4}$$

to SU(2) triplet Higgs fields

$$\chi = \begin{bmatrix} \chi^0 & \chi^-/\sqrt{2} \\ \chi^-/\sqrt{2} & \chi^{--} \end{bmatrix} \tag{3.5}$$

as follows:

$$\Psi_L^T \gamma^0 \chi^\dagger \Psi_L + \text{h.c.} \tag{3.6}$$

Now the χ field carries B-L, which can be preserved in its Higgs couplings if the SU(2) x U(1) invariant terms

$$\phi^\dagger \chi \phi^* \text{ and h.c.} \tag{3.7}$$

are left out. If χ^0 develops a VEV

$$\langle\chi^0\rangle = v/\sqrt{2}, \tag{3.8}$$

B-L is spontaneously broken and the ν_L get mass of order v. In this case, v must be very small to give small enough ν_L masses. Again, there is a massless Goldstone boson. Again, it is called the Majoron. Again, it is very difficult to see.

Both of these models have a spontaneously broken B-L symmetry, but the similarity ends there. The CMP model has very little to offer low energy particle physics. Only the almost invisible Majoron shows that B-L was spontaneously broken. But the GR model has many implications for low energy physics. The χ^{--} and χ^{-} fields must exist with masses of a few hundred GeV or less, and interesting couplings to leptons. The Majoron is associated to a light neutral Higgs scalar with mass of order v. The Z^0 decays into Majoron plus this light Higgs. The effect of this decay on the Z^0 width should be observable (it is just like having two extra flavors of neutrino). It is very reasonable to ask, why are these two models so different?

In today's talk, I will introduce a set of organizing principles which, I hope, will help you to understand not only the difference between GR and CMP, but many other things as well.

The principles are based on the obvious fact that there are several different momentum scales in the theories that interest us at which symmetries are broken. I will try to make the idea of <u>an effective field theory below a symmetry breaking</u> scale as precise as I can.[3]

Suppose that some symmetries, gauge and/or global are broken (spontaneously or even explicitly) at/or above some mass scale M. I will define what I mean by the effective theory below M by telling you what is not in it. To do this, I imagine that the parameters in our theory are modified as follows:

(1) The dimensional parameters which characterize the symmetry breaking are increased so that $M \to \infty$;

(2) The dimensional parameters which characterize all smaller scales of symmetry breaking are held fixed;

(3) All dimensionless parameters (except ratios of (1)/(2)) are held fixed.

Then a particle is not in the effective theory below M if:

(A) Its mass goes to infinity in the limit (1-3); or

(B) It becomes a free particle in the limit (1-3).

Everything else is in the effective theory.

I hope that the meaning of the limit (1-3) and the conditions (A-B) are clear. In (1-3) I am trying to push up the scale M while leaving the structure of the low energy theory fixed. The motivation is just that I hope to make distinctions between low energy and high energy sharper by considering the limit $M \to \infty$.

It will also be useful to define the following concept: If the theory has a symmetry or other property which becomes exact only in the limit (1-3), I will say that the symmetry or property is approximately satisfied in the effective theory below M.

Now I claim that these principles will help us to understand what is going on. Let us discuss some examples.

First, I should discuss the CMP and the GR model. Clearly the scales of the B-L symmetry breaking are very different. This is the essential difference between the two models. In the CMP model the breaking scale is large, and I can apply my definitions. The scale M is f, the VEV of the σ field. The scalar part of σ is a heavy Higgs with mass of order f, so it does not belong to the low energy theory by condition (A). The right-handed neutrinos are likewise not in the effective theory because their masses are of order f. The pseudoscalar part of σ is the Majoron. Because it is a Goldstone boson, all its couplings to light matter are suppressed by f^{-1}. Thus, the Majoron is a free particle in the limit (1-3) and even though it is massless, it does not belong in the low energy theory by condition (B). So the effective theory below f contains the same particles as in the standard SU(2) x U(1) models. Furthermore, the effective theory has an approximate B-L symmetry because the ν_L masses are of order m^2/f and vanish as f goes to infinity. Thus, in the CMP model, the effective theory below f is just the standard SU(2) x U(1) model. If f is large enough, therefore, we cannot tell the difference between CMP and standard. The same holds for theories in which B-L is broken explicitly at a large scale f.

In a sense, the original B-L symmetry of CMP is a fake. It has little to do with the structure of the low energy theory. The interesting B-L is the approximate symmetry of the effective theory below f.

The GR model, on the other hand, has no large symmetry breaking scale. B-L is a real symmetry of the low energy theory, and yet the ν_L masses are nonzero. I wish I had more time to discuss this fascinating model, but I must push on and refer you to the original GR paper and to an exegisis of the model by Glashow, Nussinov and me.[4]

The fact that the Majoron in CMP does not appear in what I call the effective low energy theory may bother some people. To me, it seems very reasonable because the Majoron completely decouples from everything in the limit $f \to \infty$. It lives, therefore, in its own world, not in ours.

Now, let us apply these ideas to GUTs, starting with the simplest SU(5) theory. Here I want to show you that the effective theory language simplifies the minimization of the Higgs potential. The point is that in the $M \to \infty$ limit which defines the effective theory below M, all lower scales of symmetry breaking are irrelevant and can be ignored. In SU(5), this corresponds to ignoring the SU(2) x U(1) breaking at the unification scale. If SU(2) x U(1) breaking is ignored, the 24 of Higgs breaks SU(5) down to SU(3) x SU(2) x U(1) at the unification scale, M. The components of the 24, Σ, are either Goldstone bosons which are eaten by the Higgs mechanism or Higgs mesons with mass of order M. The color triplet components of the 5 of Higgs likewise has mass of order M. Thus, the effective theory below M is just the standard model, with quark and lepton fields and a doublet of Higgs. Notice that we have not mentioned the VEV of the SU(2) triplet component of the Σ field, Σ_3.[5] We have assumed that only the SU(2) singlet component Σ_0 gets a VEV.

This is perfectly consistent. You can find the VEV of the Higgs doublet, $\langle\phi^0\rangle = u/\sqrt{2}$, in the effective theory below M, without reference to the Σ field. But then, if you are interested, you can go back and consider the effect of $\langle\phi^0\rangle$ on the full theory. It induces a nonzero VEV for Σ_3 of order u^2/M through tadpole diagrams such as that shown in Figure (3.1). (There is a $1/M^2$ from the Σ_3 propagator). This perturbative approach gives

VEV's which are correct to leading order in 1/M, and it is usually much easier than minimizing the full potential.

In the simplest SU(5) model, B-L is an exact symmetry. But if we modify the theory in such a way that B-L is broken, spontaneously or explicitly, the effective theory is the same, except that B-L is only an approximate symmetry. It is broken by Majorana ν_L masses of order u^2/M. I emphasize again that this is independent of the nature of the B-L breaking. This is my translation of the discussion in Tony Zee's talk of ν_L masses in terms of higher dimension operators.

I next want to apply this language to the complex of ideas discussed here by Roberto Peccei and Demetres Nanopoulos; strong CP violation; Peccei-Quinn symmetry; and the axion.[6] I will use an approach which is different from Roberto's. It may be less rigorous (though I don't think so) but I find it much easier to understand. I will treat θ as the coefficient of a parameter in the Lagrangian,

$$\frac{\theta g^2}{32\pi^2} \varepsilon_{\mu\nu\alpha\beta} G_a^{\mu\nu} G_a^{\alpha\beta}, \tag{3.9}$$

and restrict myself to chiral transformations associated with gauge invaraint currents. I will forget about the vacuum completely except to note that it is such that the $G\tilde{G}$ term in (3.9) contributes to the action even though it is a total divergence. In this language, the axial U(1) transformation associated with the current

$$j_5^\mu = \bar{\Psi} \gamma^u \gamma^5 \Psi \tag{3.10}$$

is not a symmetry, because it is not conserved.

$$\partial_\mu j_5^\mu \propto G\tilde{G} \tag{3.11}$$

(I give up on the factors of 2 and π). The change in the Lagrangian under a transformation is related to the divergence of the associated current, thus the U(1) chiral transformation changes θ.

The Peccei-Quinn "symmetry" is a U(1) transformation on the Higgs and quark fields which is a symmetry of the classical action. But the U(1) rotation on the quark fields has an axial U(1) component, so it is not a true symmetry of the theory. We can define a modified Peccei-Quinn transformation which consists of the same transformation on the Higgs fields and a different transformation on the quark field. The transformation on the quark fields is constructed so that it is the usual Peccei-Quinn transformation on the heavy quark fields, but on the light quark fields u and d, it is chosen to cancel the axial U(1) anomaly. This is then not a symmetry. However, it becomes an exact symmetry if the Yukawa couplings (and therefore the masses) of the light quarks go to zero. And the important thing is, it doesn't change θ! Thus, it is an approximate symmetry of the theory, broken only by the light quark mass terms and Yukawa couplings.

The axion is the pseudo-Goldstone boson associated with the spontaneous breakdown of this modified Peccei-Quinn symmetry. Suppose, for simplicity, the breakdown is due to the VEV of a single complex scalar field σ, which transforms into $e^{i\theta}\sigma$ under the modified Peccei-Quinn. If σ gets a VEV of order f

$$\langle\sigma\rangle = f/\sqrt{2}, \tag{3.12}$$

then the phase of the σ field is just the axion field divided by f, A/f. The axion field is thus just the phase in the Peccei-Quinn transformation. Any nonderivative term which is invariant under the transformation is obviously independent of A. In the limit of exact symmetry A appears in the Lagrangian only through derivative terms, $\partial^{\mu}A/f$. All of this is just the usual Goldstone theorem.

Because the axion field always appears multiplied by 1/f in its coupling to light stuff (where no additional powers of f can arise in other ways) it is not in the effective theory below f by condition B. It is also clearly difficult to see if f is finite but large. One might worry, however, that even if its couplings were small enough to escape detection in particle physics experiments, its effect might still show up as a long range force competing with gravity.

If the axion were a true Goldstone boson, such a long range force would be impossible because the axion would have only derivative couplings. A Goldstone boson can never mediate a $1/r^2$ force. This is not obvious if you plunge into perturbation without thinking. You might expect, for example, that in a P and CP violating theory, radiative corrections could induce a flavor conserving scalar coupling for a Goldstone boson. But the Goldstone theorem is very smart and any such contribution you find is doomed to cancel with something else. The reason is just the derivative couplings. All the nonderivative coupling must cancel because of the symmetry. But a derivative coupling can never be equivalent to a flavor conserving scalar coupling, because flavor conserving vector currents are divergenceless.

But the axion is not a true Goldstone boson. There is an induced flavor conserving scalar coupling of the axion to quarks. This coupling is intimately related to the strong CP violation in the theory. To see what it is, we must examine in detail how the Peccei-Quinn (PQ) approximate symmetry solves the strong CP problem.

We will begin by going to the effective theory below not only the scale of PQ symmetry breaking but also below the $SU(2) \times U(1)$ symmetry breaking. In this region, there are only quarks, leptons, photons and gluons. Although the axion does not really belong in this theory, we must keep track of it too. The reason is that nothing at the higher scales determined the phase of the VEV of σ, which is just the VEV of the axion field A. Before we start, we can make a chiral U(1) transformation to make $\theta = 0$ in the effective OCD theory. We can then make a chiral SU(N) transformation leaving $\theta = 0$ but making the heavy quark masses real, and the u and d masses relatively real. Then the only phases in the effective QCD theory are an overall phase in u and d quark mass matrix and phases in more complicated terms generated by higher order effects (i.e., of the weak interactions). These are the same terms which break the modified PQ symmetry. Furthermore, because these terms are all small (the light quark masses are small compared to the QCD scale parameter Λ), the QCD theory is approximately parity invariant. We will therefore use parity invariance in evaluating QCD matrix elements.

Now consider the u and d quark mass term. This has the form

$$m_o \, e^{i(\xi+A/f)} \, \bar{\Psi}_L \, \Psi_R + \text{h.c.}, \tag{3.13}$$

where I have included the axion coupling because the $\bar{\Psi}_L \, \Psi_R$ term breaks the PQ symmetry so the phase appears explicitly. I have also split the axion field into a VEV ξ and a field A with no VEV by definition.

To determine ξ, we must minimize the Hamiltonian for the effective theory as a function of ξ. All the other terms in the classical Lagrangian are real and independent of ξ, so to first approximation we need only minimize the VEV of (3.13)

$$|m_0| \langle\bar{\Psi}\Psi\rangle \cos(\xi + \arg m_0). \tag{3.14}$$

I haven't been careful about signs, but it is clear that (3.14) is minimized when

$$m = m_0 \, e^{i\xi} \text{ is real.} \tag{3.15}$$

This is the light quark mass matrix in this approximation. Thus to this order, the quark masses are real at the same time that the θ parameter is zero. There is no strong CP violation at this level.[6]

Now if I put (3.15) back into (3.13), you see that in the same approximation, the axion has purely pseudoscalar couplings to the quark mass term. And this is the only nonderivative coupling, at this level, so we have yet to find any long range force.

Note, also, that expanding (3.13) to second order in A gives an axion mass term of order

$$m\langle\bar{\Psi}\Psi\rangle/f^2 \simeq m_\pi^2 \, f_\pi^2/f^2, \tag{3.16}$$

which is just what Roberto obtained.

Finally, let us consider the effect of higher dimension operator which are induced in the effective theory by the weak interactions. Of course, the interesting operators are those that break the modified PQ symmetry. Suppose that 0 is such an operator. It will appear in terms of the form

$$K\, e^{i\lambda(\xi+A/f)}\, 0 + \text{h.c.} \tag{3.17}$$

where K is a calculable coefficient and λ is a constant which depends on how 0 transforms under PQ. In general, there will be many such operators, but we will restrict our attention to the one which makes the largest ξ dependent contribution to the vacuum energy. Then we must minimize the sum of (3.14) and

$$|K| \langle 0\rangle \cos(\lambda\xi + \arg K). \tag{3.18}$$

Thus

$$\arg m = \xi + \arg m_0 \simeq \frac{|K|\langle 0\rangle}{|m|\langle\bar{\Psi}\Psi\rangle} (\lambda \arg m_0 - \arg K). \tag{3.19}$$

This phase is a real strong CP violating effect. How big is it? The first operator which seems to contribute significantly in (3.19) is $(m\,\bar{\Psi}\,\Psi)^2$. Lower dimension operators involving gluons exist (like a color electric dipole moment interaction) but their contribution is small. They tend to have the same phase as the mass term for reasons similar to those discussed by Demetres Nanopoulos in his talk on the neutron dipole moment. If the operator

$$0 = (m\,\bar{\Psi}\,\Psi)^2 \tag{3.20}$$

is generated by the weak interactions, we expect on dimensional grounds that

$$K \simeq G_F^2. \tag{3.21}$$

Further, we can roughly estimate <0>,

$$\langle 0\rangle \simeq \langle m\,\bar{\Psi}\Psi\rangle^2 \simeq (m_\pi^2\, f_\pi^2)^2. \tag{3.22}$$

Then

$$\arg m \simeq \varepsilon\, G_F^2\, m_\pi^2\, f_\pi^2 \simeq 10^{-14}\, \varepsilon \tag{3.23}$$

where ε measures the relevant phase in (3.19). Thus, the strong interactions violate CP, but the induced θ parameter is very small, less than 10^{-14}. Furthermore, we have finally induced a flavor conserving coupling for the axion. But it is very, very small, of order $\theta m/f$.

There is a great deal more to be said about the axion, but I am running out of time. I will just mention some new ideas which are causing excitement. Kim[7] and Dine, Fischler and Srednicki[8] taught us that if the PQ symmetry is carried by an $SU(2) \times U(1)$ singlet scalar field, then f can be made very large. Conventional particle physics experiments do not put severe bounds on f, but Dicus, Kolb, Teplitz and Wagoner[9] have argued that astrophysical considerations (of red supergiants) suggest $f > 10^9$ GeV. This immediately suggests that f should be of the order of the unification scale in a GUT. I refer you to a paper by Shelly Glashow, Mark Wise and me[10] for a more thorough discussion. But clearly, if $f \simeq 10^{15}$ GeV, the axion is invisible. The one question that remains about strong CP violation in such a theory is "Who imposed the PQ symmetry?" Larry Hall, Mark Wise and I[11] have shown that in a certain type of theory it need not be imposed. It is an automatic consequence of gauge invariance and renormalizability.

What does this solution to the strong CP puzzle look like in the effective theory below PQ symmetry breaking? The answer is rather amusing. No remnant of the PQ symmetry remains. It looks like a perfectly standard KM model in which someone has simply adjusted θ to be very small.

My final topic is the so-called survival hypothesis.[12] Using the effective theory language, I can give a precise statement of it. Consider a set of scalar meson fields transforming according to an irreducible representation of the symmetry group of the effective theory. Or a set of Dirac fermions, with both left and right-handed parts transforming according to the same irreducible representation. In both cases, a mass of order M is allowed by all the symmetries. Because nothing forbids it, it will always be induced in some order. Thus, such a field is not in the effective theory unless an unnatural adjustment is made to cancel the order M mass.

For each irreducible representation of this kind that we want in the effective theory, we need an unnatural condition. Alas, we seem to need one such condition to keep the true Higgs doublet light if the effective theory below the grand unification scale is to make any sense at all. But I would guess that the number of such unnatural conditions should be kept to a minimum. If so, all other such particles, and in particular all Dirac fermion fields should have mass of order M and thus not be in the effective theory.

However, that does not necessarily mean that such fermions are uninteresting. As emphasized in the talk by Pierre Ramond, it may happen that some of these fermions get mass only through radiative corrections. In that case, their masses, while proportional to M, can be suppressed by powers of small coupling constants. They may be very light. Nanopoulos argued that such particles may be cosmologically interesting even if they are too heavy to be seen in particle physics experiments.

But, can they be really light, with mass of 100 GeV or less? I call such a particle an interloper because if they exist, they are refugees from the high energy world. Interlopers must probably have the color and electromagnetic charge of quarks or leptons. If they did not, they would be quite stable and cause cosmological embarrassment. But if they do have the same quantum numbers, they can mix with the quarks and leptons and that destroys the GIM mechanism.

We know that the mixing is of order u/M because the interlopers go away as M goes to infinity. The effective theory has an approximate GIM mechanism. But for finite M, we have to be careful about where the coupling constants go. If the mass of the interloper is more suppressed than its mixing, flavor changing neutral current processes may be a problem. However, it frequently happens that the mixing and the mass are equally suppressed.[11] Then the mixing of the interlopers with the appropriate quarks or leptons is of order

$$u/M \simeq 10^{-13}. \tag{3.24}$$

I agree with the other speakers that the observation of such interlopers would be great fun. Though unstable on a cosmological time scale, they would be very stable by our usual standards. Some might even live long enough (seconds? years?) to be useful.

I hope I have convinced you that the idea of an effective field theory below a scale of symmetry breaking is an indispensible tool in analyzing unified (and nonunified) theories. In my opinion, however, it would be a mistake to get so caught up in the low energy effective theory that you stop wondering what is happening at the unification scale. Just as today, most of us are certain of the validity of SU(2) x U(1) even though we have not seen a W or Z, so also if proton decay is seen at the expected rate, we can presume that unification involves SU(5). Indeed, our faith in SU(2) x U(1) is not based entirely in hard experimental evidence. There is an infinite class of theories which give the same neutral current structure. But compared to SU(2) x U(1), the alternatives are complicated, unnatural and ugly. My faith in SU(5) is likewise based on my belief that the world is simple and beautiful.

Finally, I would like to thank the organizers and the participants in KSI '81 for making my stay in Japan memorable and pleasant. I may not understand Zen, but I know warm hospitality when I feel it. Thank you all.

REFERENCES

1) Y. Chikashige, R.N. Mohapatra and R.D. Peccei, Phys. Letters 98B, 265 (1981).

2) G.B. Gelmini and M. Roncandelli, Phys. Letters 99B, 411 (1981).

3) The idea of effective low energy theories in the modern context has developed gradually. It has roots in nonlinear chiral theories, see S. Weinberg, Phys. Rev. Lett. 18, 188 (1967) and Physica 96A, 327 (1979). For the connection to renormalization, see K. Wilson, D 2, 1438 (1970); T. Appelquist and J. Carazzone, Phys. Rev. D 11, 2856

(1975); H. Georgi, H.R. Quinn and S. Weinberg, Phys. Rev. Letters 33, 451 (1974). See, also, E. Witten, Nucl. Phys. B104, 445 (1976); S. Weinberg, Phys. Lett. 82B, 387 (1979); H. Georgi, Nucl. Phys. B156, 126 (1979); H. Georgi and D.V. Nanopoulos, Physics Lett. 82B, 95 (1979); S. Weinberg, Phys. Rev. Lett. 43, 1566 (1979); F. Wilczek and A. Zee, ibid. 43, 1571 (1979).

4) H. Georgi, S.L. Glashow and S. Nussinov, Harvard Preprint HUTP-81/A026 (1981).

5) A.J. Buras, J. Ellis, M.K. Gaillard and D.V. Nanopoulos, Nucl. Phys. B135, 66 (1978).

6) R. Peccei and H.R. Quinn, Phys. Rev. Letters 38, 1440 (1977); S. Weinberg, Phys. Rev. Letters 40, 223 (1978); F. Wilczek, Phys. Rev. Letters 46, 279 (1978).

7) J. Kim, Phys. Rev. Letters 43, 103 (1979).

8) M. Dine, W. Fischler and M. Srednicki, "A Simple Solution to the Strong CP Problem with a Harmless Axion", Princeton Preprint, 1981.

9) D.A. Dicus, E.W. Kolb, V.L. Teplitz and R.V. Wagoner, Phys. Rev. D 18, 1829 (1978).

10) M.B. Wise, H. Georgi and S.L. Glashow, Harvard Preprint HUTP-81/A019 (1981).

11) H. Georgi, L.J. Hall and M.B. Wise, Harvard Preprint HUTP-81/A031 (1981).

12) H. Georgi, Nucl. Phys. B156, 126 (1979).

GRAND UNIFICATION AND GRAVITY —— SELECTED TOPICS

Anthony Zee

Physics Department, University of Washington,

Seattle, Washington 98195, U.S.A.

in

Proceedings of the Fourth Kyoto Summer Institute on Grand Unified Theories and Related Topics, Kyoto, Japan, June 29 - July 3, 1981, ed. by M. Konuma and T. Maskawa (World Science Publishing Co., Singapore, 1981).

Contents

GRAND UNIFICATION AND GRAVITY—SELECTED TOPICS

A. Zee

University of Washington, Seattle, WA 98195, USA

Abstract

The material given here was presented in lectures delivered at the 4th Kyoto Summer Institute on Grand Unification and Related Topics. It consists of six sections. The sections are completely independent of each other and may be read in any order the reader chooses. We list here the section titles.

1. The Family Problem.
2. Fermion Mass Hierarchy.
3. Maximal Local Symmetry.
4. Operator Analysis of New Physics.
5. Dynamically Generated Gravity.
6. Kaluza Theory and Grand Unification.

The last section contains a (hopefully) pedagogical introduction to Kaluza theory. For pedagogical completeness we have added several appendices briefly reviewing some elementary notions of differential geometry.

1. <u>The Family Problem</u> 家庭問題

When three generations live together, we have inevitably a family problem. We do not understand the repetitive grouping of quarks and leptons into families. "Who ordered the muon?" "Why does Nature repeat Herself?" We have as yet no satisfactory answer to those questions. Over the last two decades we have witnessed the unification of the fundamental forces, but the fundamental fermions are yet to be truly unified. Yang-Mills theory does not provide a strong enough restriction on the fermion sector. Even in grand unified theories the fermion representation is simply repeated as many times as desired.

At the moment one can imagine two possible approaches to the family problem. Perhaps quarks and leptons are composites and the observed family structure merely reflects some kind of excitations in the composite system. One might also adopt a group theoretic approach, associating with fermion families some group structure without inquiring on the deep origin of the underlying group structure. These two approaches are of course not mutually exclusive and one would hope that any proposed family symmetry would eventually be understood on a deeper level.

Here we restrict ourselves to discussing the group theoretic approach. As a first step, one might imagine a gauge group orthogonal to the usual gauge group, the so-called horizontal symmetry group or family group, under which the different families transform into each other[1)]. It is then possible, with certain assumption, to relate fermion masses and mixing angles. This sort of game has met with a measure of success and is still actively pursued by some people. However, this limited approach suffers from an unacceptable amount of arbitrariness. The ultimate ambition of the group theoretic approach is to unify the horizontal group with the "vertical" group into a simple group $\mathfrak{g}$ and to place all the fermions into a single irreducible representation $\mathfrak{R}$ of $\mathfrak{g}$.

In my opinion, a truly unifying theory of the strong and electroweak interactions should have all the fundamental fermions placed in a single irreducible representation of the gauge group. It seems to me that this is a fairly minimal esthetic requirement.

We envisage thus the following scenario to explain the observed repetitive family structure. Fermions are unified into an irreducible representation $\mathfrak{R}$ of a simple gauge group $\mathfrak{g}$. When the group $\mathfrak{g}$ is broken down into some subgroup G, the group theory is such that $\mathfrak{R}$ decomposes into exact replicas:

$$\mathfrak{R} \to R + R + R \qquad \text{when } \mathfrak{g} \to G \quad .$$

Here R denotes a representation of G.

This clearly poses a rather peculiar restriction on $\mathfrak{g}$ and $\mathfrak{R}$. Remarkably enough, as we will see, we are led to an essentially unique choice of $\mathfrak{g}$ and $\mathfrak{R}$. To begin with, the tensor representations of SO(n) and SU(n) are clearly no good. Theories based on the SU(n) groups have the rather unattractive feature that freedom from anomalies has to result from the adjusted cancellation between several irreducible representations, as is the case in the well-known Georgi-Glashow SU(5). However, the cancellation in this case may be understood "more deeply" by passing to SO(10). (Another example is E(6) $\to$ SU(6), 27 $\to$ 15 + $\overline{6}$ + $\overline{6}$.) In general, however, the arranged cancellation of anomalies appears to have no group theoretic significance. While there are in fact complex anomaly-free representations of SU(n) (found by Okubo and by Bars and Balantekin) they are realized by such high ranking tensors of mixed symmetry that it appears unlikely that they are relevant to the real world. This probably precludes our putting all the fermions into one single irreducible representation of SU(n) unless the representation is real. But, as is well-known the observed fermion representation in SU(3,2,1) is complex. In contrast, the orthogonal groups are automatically anomaly-free (except for SO(6) ≃ SU(4)) and admit spinorial representations.

Amazingly enough, the spinor representations of SO(2(n+m)) decomposes into the direct sum of 2^m spinor representations of SO(2n). This may be just what we need to incorporate the observed repetitive fermion structure particularly since we have a ready-made SO(10) theory[2)] with fermions in each family assigned to the 16-dimensional spinor representation. This rather attractive possibility was first noted independently[3)] by Gell-Mann, Ramond, and Slansky, and by Wilczek and Zee.

Unfortunately, as we will see, the group theory does not quite work out and this approach is afflicted with some serious difficulties. Nevertheless, I consider the group theoretic structure so suggestive that it is probably worthwhile to continue investigating in this direction in the hope that these difficulties might be resolved. The following discussion is based on papers by Wilczek and Zee[3)] to which we refer for further details.

Let us give a brief review of the relevant group theory. Because of their familiarities with the Lorentz group, particle physicists should have no trouble with the existence of spinor representations. Suppose that a set of 2n matrices

γ_i i=1,...,2n, dimension 2^n, exists with the property that $\{\gamma_i,\gamma_j\} = 2\delta_{ij}$. This is known as a Clifford algebra C_{2n}. Define $\sigma_{ij} = \frac{1}{2}[\gamma_i,\gamma_j]$ and $u(R) = e^{i\theta_{ij}\sigma_{ij}}$ where $\theta_{ij} = -\theta_{ji}$ is a set of c-numbers. Then it is easy to see that

$$u^\dagger(R)\gamma_k u(R) = R_{k\ell}\gamma_\ell$$

where R denotes the rotation in 2n-dimensional space through angle θ_{ij} in the i-j plane. The mapping $R \rightarrow u(R)$ defines a unitary representation of SO(2n). The representation is reducible however. Let $\psi \rightarrow u(R)\psi$ so that ψ is a spinor. Now consider the matrix

$$\gamma_{FIVE} \equiv \gamma_1\gamma_2\cdots\gamma_{2n}$$

Then clearly $\{\gamma_{FIVE},\gamma_i\} = 0$ and $[\gamma_{FIVE},\sigma_{ij}] = 0$ and so

$$\psi_\pm \equiv \frac{1}{2}(1 \pm \gamma_{FIVE})\psi$$

transform independently. There are thus two irreducible spinor representations $s_\pm$ of SO(2n) with dimensions 2^{n-1}. For SO(4N+2) $s_\pm$ are complex and conjugate of each other. For SO(4N) $s_\pm$ are real. Thus, we should consider only the SO(10), SO(14), SO(18), etc.

The existence of Clifford algebras may be proven by explicit iterative construction. For n=1, take $\gamma_1^{(1)} = \tau_1$ and $\gamma_2^{(1)} = \tau_2$. Given $\gamma_i^{(n)}$ i=1,...,2n we construct

$$\gamma_i^{(n+1)} = \gamma_i^{(n)} \times \tau_3 = \begin{pmatrix} \gamma_i^{(n)} & 0 \\ 0 & -\gamma_i^{(n)} \end{pmatrix} \qquad i=1,\ldots,2n$$

$$\gamma_{2n+1}^{(n+1)} = 1 \times \tau_1$$

$$\gamma_{2n+2}^{(n+1)} = 1 \times \tau_2 \quad .$$

This explicit iteration makes clear the group theory property which motivates this whole discussion. When we break SO(2n+2) → SO(2n) we restrict ourselves to generators $\sigma_{ij}^{(n+1)}$ i,j=1,...,2n which have the form

$$\begin{pmatrix} \sigma_{ij}^{(n)} & 0 \\ 0 & \sigma_{ij}^{(n)} \end{pmatrix} .$$

Thus, a spinor ψ of SO(2n+2) has the form

$$\begin{pmatrix} \psi_1 \\ \psi_2 \end{pmatrix}$$

with ψ_1 and ψ_2 spinors of SO(2n).

Unhappily for us, a spinor does not decompose into exact replicas as desired; some of the replicas are reversed. This follows since

$$\gamma_{FIVE}^{(n+1)} = \gamma_{FIVE}^{(n)} \times \tau_3 = \begin{pmatrix} \gamma_{FIVE}^{(n)} & 0 \\ 0 & -\gamma_{FIVE}^{(n)} \end{pmatrix}$$

from our explicit construction. Thus, the two spinors ψ_1 and ψ_2 introduced above have opposite γ_{FIVE} chirality. In group theoretic language, a complex representation of a group may well be real when restricted to a subgroup. The complex spinor representations of SO(10 + 4n) are real when restricted to SO(10).

More generally, when SO(2n + 2m) → SO(2n) × SO(2m) we have

$$\underset{\sim}{2}_{+}^{n+m-1} \rightarrow \left(\underset{\sim}{2}_{+}^{n-1}, \underset{\sim}{2}^{m-1}\right) + \left(\underset{\sim}{2}_{-}^{n-1}, \underset{\sim}{2}^{m-1}\right) \quad ;$$

we denote the spinor representations by their dimensions. This decomposition is disastrous for physics. If the observed fermions with their V-A weak interactions are assigned to the 16_+ of SO(10), then the fermions corresponding to the 16_- will have V+A weak interactions. These so-called V+A fermions have not been observed experimentally.

We are thus faced with several possibilities. Firstly, the idea of deriving repetitive family structure from orthogonal groups may be altogether wrong, of course. However, we feel that the group theoretic structure is sufficiently suggestive to warrant some serious efforts to somehow coneal the V+A fermions from view. The most obvious way is to introduce Higgs to give large masses to the unwanted V+A fermions. But the detailed Higgs structure necessary appears somewhat ad-hoc and artificial, as is almost unavoidable. A more ingenious scheme was suggested by Gell-Mann et al.[3)] who utilize the idea of technicolor to conceal the V+A fermions. After breaking SO(10 + 4n) down to SO(10) x SO(4n) they assume that some subgroup (call it TC) of SO(4n) remains unbroken and is to be identified as technicolor. The point is that the two spinor representations $\underset{\sim}{S}_{\pm}$ of SO(4n) in general decompose quite differently under the subgroup TC and, in particular, will contain different numbers of TC singlets. The fundamental hypothesis of technicolor, that TC non-singlets are confined by techni-strong forces, is then invoked to conceal the V+A fermions.

We will explore the technicolor scheme in detail here. Gell-Mann et al. chose a symplectic group for TC. We will show that a whole sequence of orthogonal groups may be used. As we will see, this scheme suffers from serious difficulties. To begin with, this scheme is not without its share of arbitrariness stemming from our ignorance of strong interaction dynamics. Starting with a gauge symmetry SO(10 + 4n) one can in general break the symmetry down step-wise into $SU(3)_{color} \times U(1)_{em}$ in many different ways. Without a detailed theory of symmetry breaking we do not know which symmetry breaking chain is favored. Of course, we do not even know if we could really get away without using explicit Higgs.

Another problem is the proliferation of fermions, thus threatening to modify the renormalization group analysis of Georgi, Quinn, and Weinberg[4]. This actually constrains the possible symmetry breaking chain.

SO(14) is not large enough to contain more than two V-A families. Thus, the smallest theory we may consider is SO(18). (Incidentally, an SO(22) theory with fermions in the spinor representation is not asymptotically free.) Upon breaking SO(18) → SO(10) x SO(8) we have the spinor $\underset{\sim}{S}_+$ decomposing into $(16_+,8_+)$ + $(16_-,8_-)$.

We are now faced with a well-defined group theory problem. What is a subgroup TC of SO(8) such that under TC, 8_- does not contain singlets while 8_+ does?

A "natural" subgroup of SO(8) is SU(4). Upon breaking SO(8) to SU(4) we have

$$\underset{\sim}{8}_+ \rightarrow [0] + [2] + [4] = [0] + [2] + [\bar{0}]$$

$$\underset{\sim}{8}_- \rightarrow [1] + [3] = [1] + [\bar{1}] \quad .$$

The notation [k] ($[\bar{k}]$) denotes the representation of SU(n) realized by an antisymmetric tensor with k upper (lower) indices. If we identify SU(4) as TC we would then have two nonconfined V-A families. Suppose we go further and consider the breaking SO(8) → SU(4) → SU(2) x SU(2), then we have

$$\underset{\sim}{8}_+ \rightarrow [0] + [2] + [\bar{0}] \rightarrow [0,0] + [0,2] + [1,1] + [2,0] + [0,0]$$

$$\underset{\sim}{8}_- \rightarrow [1] + [\bar{1}] \rightarrow [0,1] + [1,0] + [0,\bar{1}] + [\bar{1},0] \quad .$$

(Obviously, under SU(n+m) → SU(n) x SU(m)

$$[k] \rightarrow \sum_{j+\ell=k} [j,\ell] \qquad \text{with } [j,\ell] \equiv [j] \times [\ell] \quad .)$$

For SU(2), the representation [2] is equivalent to [0]. Thus, if we identify SU(2) x SU(2) as TC then this theory contains four V-A fermion families. All V+A fermions are confined. Thus, the theory predicts one additional family beyond the τ family.

Incidentally, Gell-Mann et al. identified TC as Sp(4). The group Sp(4) contains SU(2) x SU(2) and is itself contained in SU(4). Under the breaking chain SU(4) → Sp(4) → SU(2) x SU(2) we have the decomposition

$$\underset{\sim}{4} \rightarrow \underset{\sim}{4} \rightarrow [1,0] + [0,1]$$

$$\underset{\sim}{6} \rightarrow \underset{\sim}{5} + \underset{\sim}{1} \rightarrow [1,1] + [0,0] + [0,0] \quad .$$

Thus, if one stops the symmetry breaking at Sp(4) as Gell-Mann et al. does, one would have three V-A families.

This discussion underscores the fact that, due to our ignorance of symmetry breaking, the physical predictions of the scheme depend completely on the unbroken subgroup chosen. In the present state of the art, this choice is largely arbitrary. We could, for instance, break down further into the diagonal SU(2) subgroup of SU(2) x SU(2). In this case we end up with five V-A families. It is amusing to note the local isomorphisms SU(4) ≃ SO(6), Sp(4) ≃ SO(5), SU(2) x SU(2) ≃ SO(4), and SU(2) ≃ SO(3). Thus, in the present discussion, the number of V- families is "predicted" to be two, three, four, or five respectively according to whether SO(8) is broken down to SO(6), SO(5), SO(4), or SO(3). (These orthogonal subgroups are not embedded in SO(8) in the obvious way, however.)

Our discussion of the theory is incomplete without an analysis of the behavior of the gauge coupling constants under the renormalization group. As noted before, this imposes certain restrictions on the allowed chain of symmetry breaking. The group SO(8) may be broken down in many different ways. For example, we may have the chain SO(18) → SU(9) → SU(5) x SU(2) → SU(3) x SU(2) x U(1) x SU(2). Without a detailed dynamical theory of symmetry breaking we can only examine each of the many possible chains in turn and find the one which works best.

We suggest the symmetry breaking occurs through SO(18)

$$\xrightarrow{m} SU(3) \times SU(2) \times U(1) \times SO(8)$$

$$\xrightarrow{M} SU(3) \times SU(2) \times U(1) \times SU(2) \times SU(2) \ldots \quad (9) \quad .$$

We know of no dynamical reason why the symmetry should break down in this (somewhat peculiar) fashion.

A straightforward analysis shows that the values for $\sin^2\theta$ and M are not changed. The input $8\alpha_8(M) \sim 1$ fixes $18\alpha_{18}(M)$ to be roughly of order 1 and M to be of the order 10^{10} GeV. This value for M may pose a serious difficulty <u>if</u> we take it to mean that the $16\_$ fermions all have masses of the order 10^{10} GeV. As is well-known, to give masses to all sixteen fermions in the $16\_$ one has to break the SU(2) x U(1) subgroup of SO(10). We certainly do not want to break SU(2) x U(1) at a scale of 10^{10} GeV!

With explicit Higgs fields we could of course give the $16\_$ fermions any mass value we choose; in particular we can choose a mass scale of a few hundred GeV. Unfortunately, this suggestion also meets with difficulties. Above a few hundred GeV all 256 fermions of the SO(18) spinor will contribute to the beta functions.

In summary, we are faced with a rather peculiar situation. The group theory of orthogonal groups strongly suggests a natural incorporation of the observed repetitive family structure. But our attempts to implement the group structure have met apparently insurmountable difficulties. We believe that the group theory structure is mathematically so elegant that the existence of fermion families should really have something to do with the orthogonal groups[5)]. Further investigations to circumvent the difficulties discussed above will be most worthwhile. Also, we believe that experimenters should keep their eyes out for V+A fermions.

References: Section 1

1. T. Yanagida, F. Wilczek and A. Zee, Phys. Rev. Lett 42, 421 (1979); K. Akama and H. Terazawa.
2. H. Georgi, Nucl. Phys. B156, 126 (1979).
3. M. Gell-Mann, P. Ramond, R. Slansky, unpublished; F. Wilczek and A. Zee, "Spinor and Families," 1979 Princeton preprint and 1981 Santa Barbara preprint. We have learned that H. Georgi and E. Witten have also considered spinors in the present context (private communication from H. Georgi).
4. H. Georgi, H. Quinn, and S. Weinberg, Phys. Rev. Lett. 33, 451 (1974).
5. A partial list of work on the orthogonal group includes H. Sato, Phys. Rev. Lett. 45, 1997 (1980); A. Davidson et al., Phys. Rev. Lett. 45, 1335 (1980); Z. Ma et al., preprint BIHEP-TH-3 (1980); R. Mohapatra and B. Sakita, Phys. Rev. D21, 1062 (1980); J. Maalampi and K. Enquist, Phys. Lett. 97B, 217 (1980); M. Ida, Y. Kayama and T. Kitazoe, preprint; R. Cahn and H. Harari, Nucl. Phys. B176, 135 (1980); J.E. Kein, Phys. Rev. Lett. 45, 1916 (1980); I. Umemura and K. Yamamoto, Kyoto preprint NEAP-28; S. Nandi, A. Stern, and E.C.G. Sudarshan, Fermilab-PUB-81/53, (1981); and K. Enquist et al., Helsinki preprint (1981).

2. Fermion Mass Hierarchy

Quark and lepton masses span an enormous range. There is a factor of at least 10^4 between the mass of the yet-to-be observed top quark and the electron mass. In the standard theory this large number is totally unexplained. It is incorporated into the theory by simply allowing vastly different Yukawa couplings and/or vacuum expectation values of Higgs fields.

A natural explanation for this enormous disparity in fundamental fermion masses suggests itself: members of the top family become massive in tree approximation while the masses of the bottom and up family are due to radiative effects. This hope is hardly new and has long been bolstered by various numerological formulas such as $m_e/m_\mu = (3\alpha/\pi)\log 2$. In the standard theory, all three families are treated on the same footing, and thus to realize this radiative mass scenario one must complicate the standard picture, unfortunately.

In this subsection we sketch the work of S. Barr. The reader is referred to his papers[1)] for further details. Related work has been done by a number of other people and will be discussed by other lecturers at this school.

Let us now explain Barr's approach. In the standard SU(5), each of the three families belongs to a $\bar{\underset{\sim}{5}} + \underset{\sim}{10}$ of SU(5). In order to distinguish the three families group theoretically we might consider embedding SU(5) into a larger gauge group G. The fundamental theory is to be based on G which breaks down to SU(5) whereupon various representations of G decompose into SU(5) representations. Now the $\bar{5} + 10$ containing e,d,u may well have "descended" from a totally different representation of G than the

$\overline{5} + 10$ containing τ, b and t. The Yukawa couplings of the Higgs fields to the fermions must respect G of course. So it is quite possible that the group theory is such that Higgs fields may couple to the τ family but not to the electron family. But in this case members of the electron family are "protected" only by G from having masses. After all, in SU(5), it looks the same as the other families. Thus, when G is broken, members of the electron family will in general become massive through radiative effects.

It is unfortunately awkward and rather complicated to actually incorporate this extremely simple idea. The situation is complexified by the presence of three families, rather than two. We illustrate this mechanism here with G = SO(10). In the standard version of SO(10) fermions are assigned to three (left-handed) $\underset{\sim}{16}_L$'s and Higgs fields to $\underset{\sim}{45}_H$, $\underset{\sim}{10}_H$, $\underset{\sim}{126}_H$. Of the Higgs fields only the $\underset{\sim}{10}_H$ can couple to fermions, viz.,

$$a_{ij}(16^i_L 16^j_L)10_H \qquad i,j = 1...3 \qquad . \qquad (1)$$

The Yukawa couplings a_{ij} have to be vastly different in magnitude in order to account for the disparity in fermion masses. In Barr's version of SO(10), the fermion representation content is chosen to be $\underset{\sim}{16}^i_L$ $i = 1,2,3$, $\underset{\sim}{10}_L$, $\underset{\sim}{45}_L$ while the Higgs representation content is $\underset{\sim}{45}_H$ and $\underset{\sim}{16}^\alpha_H$. (The number of $\underset{\sim}{16}_H$'s needed will not be discussed here.) This differs from the standard version in two crucial ways: (1) the $\underset{\sim}{10}_H$ Higgs is banished and (2) two real fermion representations $\underset{\sim}{10}_L$ and $\underset{\sim}{45}_L$ are added. The following mass and Yukawa terms are allowed by the representation content:

$$m_1(10_L 10_L) \tag{2a}$$

$$m_2(45_L 45_L) \tag{2b}$$

$$b_{i\alpha}(45_L 16_L^i)\overline{16}_H^\alpha \tag{2c}$$

$$a_{i\alpha}(10_L 16_L^i) 16_H^\alpha \ . \tag{2d}$$

According to Georgi's survival hypothesis[2] the real fermion representations $\underset{\sim}{10}_L$ and $\underset{\sim}{45}_L$ should have masses of the order of the grand unification scale. Here we see how the hypothesis actually works in practice. The reality of the $\underset{\sim}{10}_L$ and the $\underset{\sim}{45}_L$ allows bare mass terms to be written down for them as in Eqs. (2a) and (2b). We adopt the "natural" principle of taking all dimensional parameters in the theory such as m_1 and m_2 and $\langle 16_H^\alpha \rangle$ to be the same order of magnitude as the grand unification mass scale. Since 16 x 16 = 10 + 126 + 120, there is no $(16_L 16_L)$ coupling to Higgs fields at the tree level. Instead, the $\underset{\sim}{16}_H^\alpha$ Higgs fields couple the $\underset{\sim}{16}_L^i$ to the real fermion representations $\underset{\sim}{10}_L$ and $\underset{\sim}{45}_L$ as in Eqs. (2c) and (2d). At the SU(5) level the sector contains four $\overline{5}$'s, one $\underset{\sim}{5}$, four $\underset{\sim}{10}$'s, one $\overline{10}$, one $\underset{\sim}{24}$, and four $\underset{\sim}{1}$'s. According to Georgi's hypothesis, real representations such as $\underset{\sim}{1}$, $\underset{\sim}{24}$, $\underset{\sim}{5} + \overline{\underset{\sim}{5}}$, $\underset{\sim}{10} + \overline{\underset{\sim}{10}}$ should all get superheavy masses leaving three $\underset{\sim}{10}_L + \overline{\underset{\sim}{5}}_L$'s in the low energy theory. However, because of the couplings in Eq. (2c) and (2d) the three $\underset{\sim}{10}_L + \overline{\underset{\sim}{5}}_L$ are not just the $\underset{\sim}{10}_L + \overline{\underset{\sim}{5}}_L$ found in the $\underset{\sim}{16}_L$'s. Lest we get confused it is extremely convenient to adopt the notation $\underset{\sim}{r}(\underset{\sim}{R})$ with $\underset{\sim}{r}$ denoting an SU(5) representation and $\underset{\sim}{R}$ the SO(10) representation from which $\underset{\sim}{r}$ "descends." We call $\underset{\sim}{R}$ the ancestor of $\underset{\sim}{r}$. Thus, of the four $\underset{\sim}{10}_L$'s in SU(5) one of them has for ancestor the real

$\underset{\sim}{45}_L$ of SO(10) while the others descend from the $\underset{\sim}{16}_L$ spinor of SO(10).

Quark and lepton masses break SU(2) x U(1) and as usual come from $\underset{\sim}{5}_H$ at the SU(5) level. In this theory the ancestors of the $\underset{\sim}{5}_H$'s are the $\underset{\sim}{16}_H^{\alpha}$. We see from Eqs. (2b) and (2c) that the $\overline{10}_L(45)$ and one linear combination of $10_L(16^i)$ and $10_L(45)$ mate to become ten superheavy Dirac fermions. There remain three $\underset{\sim}{10}_L$'s. However, not all of them are allowed to couple to $5_H(16)$ and $5_H(\overline{16})$. That privilege is reserved for those who can boast of having the right ancestor. Thus, the couplings $10_L(16^i)10_L(45)5_H(\overline{16^{\alpha}})$ are allowed while, in contrast, the couplings $10_L(16^i)10_L(16^j)5_H(\overline{16^{\alpha}})$ are not. It turns out that one of the three charged +2/3 quarks does not get a mass at this level. We identify it as the up quark.

A similar discussion may be given for the $(10_L\overline{5}_L)\overline{5}_H$ couplings. At tree level, it happens that only t,b,τ, and c are massive. However, since SO(10) is in fact broken, even members of those families with the "wrong" SO(10) ancestors will all eventually become massive through radiative effects. The mechanism could be discussed using operator analysis (see Section 3). The point is that although the group theory does not allow any operator which couples 16_L to 16_L with dimension less than or equal to four one can readily construct such operators with dimensions greater than four. For instance, consider the dimension five term (written schematically)

$$\frac{1}{M}(16_L 16_L)\overline{16}_H\overline{16}_H \quad .$$

Here M is a superheavy mass needed for dimensional reasons. This term will be generated radiatively since it is not forbidden by group theory. The $\overline{16}_H$ when reduced to SU(5) contains 5_H and 1_H. Thus the operator in Eq. (3) contains at the SU(5) level an effective term

$$\frac{\langle 1_H \rangle}{M} 10_L 10_L \langle 5_H \rangle \quad . \tag{4}$$

We assumed $\langle 1_H \rangle$ and M are of the same order. The reader is referred to ref. (1) for further details.

The trouble with the sort of approach outlined here is that the theory necessarily involves a number of parameters which cannot be fixed in the present state of the art. While it is possible, as shown by Barr, to obtain a fermion mass spectrum more or less in accord with the real world, it is impossible, in the absence of explicit numbers, to establish any one scheme as correct. Also, in this approach, the fact that there are three families is put in, and it appears to us that an understanding of the number of families is surely an essential ingredient in a true explanation of the fermion mass spectrum.

References: Section 2

1. S. Barr, Phys. Rev. D21, 1424 (1980); 1981 preprint (to appear in Phys. Rev.).

2. H. Georgi, Nucl. Phys. B156, 120 (1979).

3. Maximal Local Symmetry

We all believe in gauge theories. But, in a sense to be explained presently, the known gauge groups are strikingly small. Why?

The free Lagrangian for N_f two-component Weyl fields has a global $SU(N_f) \times U(1)$ symmetry. With 45 "known" fermion fields (we are including the top quark) we have available a global symmetry of $SU(45) \times U(1)$, which is enormous compared to SU(5) or SO(10).

Why does Nature gauge only a tiny subgroup of the available symmetry? What are the principles which determine which subgroup Nature gauges?

These questions may well lie outside the province of physics at present: Nevertheless, it may be amusing to explore them. In fact, we can answer the first question quite readily. If the entire available global symmetry is gauged the theory would suffer from anomalies and would be non-renormalizable. Note that the question is unanswerable before the discovery of the axial anomaly. Physics does progress, and apparently unaswerable questions eventually will be answered. Ultimately, physicists will have to understand why Nature chooses to have a certain number of fermions and to gauge a certain group.

The question we are asking is in a sense the reverse of the so called family problem. The family problem asks why, for a given gauge group, are there so many fermions. For SU(5), 15 fermions suffice; why does Nature order 45? We are asking here why, for a given number of fermions, are there so few gauge bosons.

We would like to explore what principles Nature might possibly use in determining the appropriate gauge group for a given number of fermions. We propose that the absence of bare masses might be an important criterion.

We pose the following group theoretic question. What is the largest simple subgroup G of $SU(N_f) \times U(1)$ such that the resulting gauge theory based on G is (1) free from anomalies and (2) free from bare masses?

Let us briefly explain the second condition. Starting with the "empirical" observation that the standard SU(3) × SU(2) × U(1) theory has no bare mass terms, Glashow, Georgi, and others have suggested that in gauge theories all bare masses have values of the order of the mass scale at which the gauge symmetry is broken. Thus, for example, if the fermions in SU(5) consist of $5+\bar{5}+\bar{5}+10$, the conjecture just mentioned would assert that the 5 and one of the $\bar{5}$'s would "mate" to form five Dirac particles of masses of order 10^{15} GeV. Of course, we can imagine a theory with a large number of fermions, most of which are "mated" off, leaving only the observed fermions. Theories of this type have been envisaged by Georgi[1]. Here we assume that Nature is not so profligate. Of course this is purely an assumption, open to doubt. In fact, SO(10) may be an example in which some fermions, namely the right-handed neutrinos, are removed to high masses.

In any case, we adopt the principle that the theory should be free from bare masses. Mathematically, a representation D(g) of some group G is said to be real if

$$D^*(g) = S^{-1}\, D(g) S \qquad \text{for all } g \in G$$

and if S is symmetric. (If S is anti-symmetric, the representation is said to be pseudo-real.) The assertion is that if ψ_L transforms according to a real representation, then a G invariant mass term for ψ_L exists and reads

$$\psi_L C S^{\dagger} \psi_L \ .$$

Note that if S is antisymmetric then this term vanishes identically by Fermi statistics. We exclude by assumption the pseudo-real representations from consideration. This eliminates the symplectic groups, for instance.

We could thus formulate our problem mathematically as follows. Given an integer N_f find the largest simple group G such that G has a set of representations R whose dimensions add up to N_f, which is anomaly free, and which does not contain any real (in general, reducible) representations.

This question was posed and analyzed in ref. 2, which readers should consult for details and references.

Armed with a table of anomalies and dimensions (we mention later some tricks for calculating anomalies), one can readily answer the above question for any given N_f. A table is given in ref. 2. This exercise actually yields several amusing results.

(1) For $N_f = 15$, 30 and 45 the solution to our problem is the SU(5) of Georgi and Glashow. For $N_f = 60$, however, the solution turns out to be SU(8) with the fermions assigned to $4[\bar{1}]+[2]$. Could this be a clue on why there are three families?

Similarly, suppose there are sixteen fermions per family. For $N_f = 16$, 32, 48, the solution is SO(10). For $N_f = 64$, the solution is SO(14).

(2) Ultimately, physicists may have to understand how the number of fermion fields used to construct the world is chosen. Our program cannot answer the basic question; however, a small clue might be furnished by the fact that for some numbers no solution exists. For instance, were the top quark not to exist the known fermions would make 39 fields. There is no solution for $N_f = 39$.

For further details, assorted caveats and disclaimers, see ref. 2.

Our insistence on simple groups is necessary. Otherwise, for 45 fields the largest group will be SU(5) x SU(5) x SU(5) and for 48 fields, SO(10) x SO(10) x SO(10). This poses an amusing way to approach the family problem. The gauge group is in fact $(SU(5))^3$; each family has its "own gauge group." The "observed" SU(5) is then the diagonal SU(5). The remaining gauge bosons, supposed to be heavier, will be responsible for small deviations from μe universality.

We begin by asking why Nature is so miserly with local symmetries, gauging only a tiny subgroup of an enormous global group. The answer turns out to be that Nature is not miserly; on the contrary, Nature gauges the largest subgroup which frees Nature from the twin scourges of anomalies and bare masses.

It is tempting to quantify the preceding remark. Consider the ratio of the number of generators in G to the number of generators in $SU(N_f) \times U(1)$. It turns out that this ratio is maximized (to be $\approx$ 0.18) by the solution $N_f = 16$, G = SO(10). But then we do not understand why there are families.

References: Section 3

1. H. Georgi, Nucl. Phys. B156, 126 (1979).

2. A. Zee, Phys. Lett. 99B, 110 (1981).

Appendix III-1

General expressions for the anomaly have been given by a number of authors[1,2]. In practice, however, one usually encounters only relatively simple representations of low-ranking groups and it is often easier to figure out the anomalies by various recursion techniques[3], especially since, normally, one can never remember the general formula.

If we embed SU(N-1) into SU(N) so that the fundamental representation $[1]_N$ of SU(N) decomposes into $[1]_{N-1} + [0]_{N-1}$, then the anomaly A(R) of a representation R of SU(N) is clearly equal to the sum of the anomalies of the representations of SU(N-1) which R decomposes into. The anomaly A([1]) is normalized to 1. The antisymmetric tensor with indices $[m]_N$ decomposes into $[m]_{N-1}$ + $[m-1]_{N-1}$ and so we have the recursion relation

$$A([m]_N) = A([m]_{N-1}) + A([m-1]_{N-1}) \quad . \tag{1}$$

This enables us to construct Table I. To start the recursion we note that for N=3 [2] = $[\bar{1}]$ and so $A([1]) = -A([\bar{2}]) = +1$. In particular, this gives

$$A([2]_N) = N-4 \quad . \tag{2}$$

To remember this formula, one need only recall that it gives the well-known Georgi-Glashow cancellation for N=5.

Similarly, we have, for the symmetric tensor with m indices $\{m\}_N$ the recursion

$$A(\{m\}_N) = \sum_{k=0}^{m} A(\{k\}_{N-1}) \tag{3}$$

or equivalently $A(\{m\}_N) = A(\{m-1\}_N) + A(\{m\}_{N-1})$. This enables us to construct Table II. Notice that with our normalization $A(\{1\}) = A([1]) = +1$ so that the entries for N=2 are to be disregarded. They should be zero since all representations for SU(2) are either real or pseudoreal. What is important is the ratio between the anomalies of different representations for a given group.

Another useful fact is to note that for two representations R_1 and R_2 of a given group

$$A(R_1 \times R_2) = d(R_1)A(R_2) + A(R_1)d(R_2) \tag{4}$$

where d(R) denotes the dimension of R. To prove this, we note that if a generator is represented on R_i by the matrix T_i, then it is represented on $R_1 \times R_2$ by

$$T = T_1 \times 1 + 1 \times T_2 \tag{5}$$

Then,

$$T^3 = T_1^3 \times 1 + 3T_1^2 \times T_2 + 3T_1 \times T_2^2 + 1 \times T_2^3 \ . \tag{6}$$

Taking the trace of Eq. (6) we obtain Eq. (4).

As an application of Eq. (4) let us consider for SU(N)

$$\begin{aligned} A([1] \times [1]) &= 2NA([1]) = 2N \\ &= A([2]) + A(\{2\}) \\ &= N - 4 + A(\{2\}) \end{aligned} \tag{7}$$

Thus,

$$A(\{2\}) = N + 4 \tag{8}$$

Compare with Eq. (2).

We remark that dimensions of representations satisfy the same recursion relations. For instance,

$$d(\{m\}_N) = \sum_{k=0}^{m} d(\{m\}_{N-1}) \quad . \tag{3"}$$

For convenience, we also list the dimensions in the two tables.

To start the recursion in Table II, we exploit the following trick. We "analytically" continue to the group SU(0) for which the anomalies are easier to calculate. Note that for SU(0) all representations have dimension zero except for the singlet. We calculate the anomalies by evaluating $\mathrm{Tr}Y^3$ over $\{m\}_N$ choosing Y to be the diagonal matrix with elements given by (-N+1,1,1,...,1). Then on $\{m\}_N$

$$\mathrm{tr}\, Y^3 = \sum_{k=0}^{m} d(\{k\})_{N-1} [(m-k)(-N+1) + k]^3 \quad . \tag{9}$$

The kth term in the sum corresponds to those tensor components with k indices not equal to 1. We note that if we continue this expansion to $N \to 0$ we obtain, using Eq. (3"),

$$\lim_{N\to 0} m^3 d\{m\}_N \quad . \tag{10}$$

This tells us that the cube of a zero-by-zero traceless Hermitean matrix has the same value, namely m^3, for each component of the tensor {m} in SU(0)! Thus, $\mathrm{tr}Y^3$ is just m^3 times the dimension of {m} which is well-known to be given by

$$d(\{m\}_N) = \frac{(N - 1 + m)!}{(N - 1)!\ m!} \quad . \tag{11}$$

Continuing Eq. (11) to N=0 we find $d(\{m\})_{N\to 0} \to N/m$ so that $\mathrm{tr}Y^3 \to m^2 N$. Finally, normalizing our convention $A(\{1\}_N) = 1$ we find $A(\{m\}_0) = m^2$ which is just what is given in the N=0 column in Table II.

It is in this "analytic continuation" sense that we include the "meaningless" column for $N \leq 2$. They are needed to construct the correct entries for $N \geq 3$. Indeed, the reader will note that if one continues Table II to negative N, one finds, amusingly enough, the same entries up to some signs as in Table I! More precisely, one finds

$$A([m]_N) = (-)^{m+1} A(\{m\}_{-N}) \quad . \tag{12}$$

It is perhaps not surprising that such reflection relations exist since the general formula involves the factorial which when defined as gamma function has a definite reflection property. We have, however, no easy "intuitive" way to see why Eq. (12) must be true. The reflection formula is useful in that it is easier to calculate anomalies for antisymmetric representations. For example, knowing that $A([3]_N) = \frac{1}{2}(N-3)(N-6)$ we can immediately obtain $A(\{3\}_N) = \frac{1}{2}(N+3)(N+6)$.

One could go on to construct more tables. For example, consider the traceless tensor antisymmetric in m upper indices and antisymmetric in k lower indices which we denote by $\begin{bmatrix} m \\ k \end{bmatrix}$. The recursion equation reads

$$A\left(\begin{bmatrix} m \\ k \end{bmatrix}_N\right) = A\left(\begin{bmatrix} m \\ k \end{bmatrix}_{N-1}\right) + A\left(\begin{bmatrix} m \\ k-1 \end{bmatrix}_{N-1}\right) + A\left(\begin{bmatrix} m-1 \\ k \end{bmatrix}_{N-1}\right) + A\left(\begin{bmatrix} m-1 \\ k-1 \end{bmatrix}_{N-1}\right) \quad . \tag{9}$$

This is manageable in simple cases. For instance

$$A\left(\begin{bmatrix}2\\1\end{bmatrix}_N\right) = A\left(\begin{bmatrix}2\\1\end{bmatrix}_{N-1}\right) + A([2]_{N-1}) + A([1]_{N-1}) \qquad (10)$$

which yields

$$A\left(\begin{bmatrix}2\\1\end{bmatrix}_N\right) = \frac{1}{2}\,(N-3)(N-4) - 7 \quad . \qquad (11)$$

We could also have derived this by using Eq. (4) and considering $A([2] \times [\bar{1}])$.

These techniques were developed in discussion with S. Barr, to whom we are indebted. Table I has already been given in the first paper cited in Ref. (1) of Section 2.

TABLE I. Anomaly and Dimension of $[m]_N$.

Here, as in Table II, the first number in each entry gives the anomaly, the second the dimension.

m \ N	0	1	2	3	4	5	6	7	8	9	10	11
0				0/1	0/1	0/1	0/1	0/1	0/1	0/1	0/1	0/1
1				1/3	1/4	1/5	1/6	1/7	1/8	1/9	1/10	
2				-1/3	0/6	1/10	2/15	3/21	4/28	5/36	6/45	
3					-1/4	-1/10	0/20	2/35	5/56	9/84	14/120	
4						-1/5	-2/15	-2/35	0/70			
5							-1/6	-3/21	-5/56			
6												
7												
8												
9												

TABLE II. Anomaly and Dimension of $\{m\}_N$.

m \ N	0	1	2	3	4	5	6	7	8	9	10
0	0/1	0/1	0/1	0/1	0/1	0/1	0/1	0/1	0/1	0/1	0/1
1	1/0	1/1	1/2	1/3	1/4	1/5	1/6	1/7	1/8	1/9	
2	4/0	5/1	6/3	7/6	8/10	9/15	10/21	11/28	12/36	13/45	
3	9/0	14/1	20/4	27/10	35/20	44/35	54/56	65/84			
4	16/0	30/1	50/5	77/15	112/35	156/70	210/126				
5	25/0	55/1	105/6	182/21							
6	36/0	91/1	196/7	378/28							
7		140/1	336/8	714/36							
8											
9											

References: Appendix III-1

1. J. Banks and H. Georgi, Phys. Rev. D14 (1976).
2. S. Okubo, Phys. Rev. D16 (1977).
3. Indeed, the authors in Ref. (1) used recursion techniques to derive a general formula.

4. Operator Analysis of New Physics

So far I have lectured on rather speculative topics. You have heard many exciting lectures at this Summer School but I think that most of the lecturers will agree with me that the likelihood that the material presented will all turn out to be correct is substantially less than unity. I will shift gears here and discuss an analysis much more likely to be correct. I wish to discuss how the standard SU(3) x SU(2) x U(1) symmetry may be used to constrain new physics. By new physics we mean any hitherto unknown physical phenomenon involving the known particles. The discovery of neutrino oscillation, or of $\mu \to e\gamma$, for example, would be new physics. This definition excludes the detection of new quark flavor. Also, by the term "mass scale of the new physics" we mean the mass scale underlying the physics responsible for a phenomenon such as $\mu \to e\gamma$.

A point which appears not as well appreciated as it should be is that the standard SU(3) x SU(2) x U(1) symmetry can be used not only to fit the known phenomena but also to constrain new physics. This point can be illustrated by precedents in the history of physics. When Fermi was confronted with the new phenomenon of weak decay, he recognized that any phenomenological thoery of the decay must be constrained by the symmetry already known to him in 1934, namely the U(1) of electromagnetism. To be sure, this only led to the requirement that weak decays conserve electric charge. As we will see, the larger symmetry SU(3) x SU(2) x U(1) imposes somewhat stronger constraints.

Let us now formulate our ideas more precisely. We believe that the standard theory describes physics up to a mass scale of at least M_W. Suppose new physics first occurs at a mass scale of

$M^* > M_W$. To the extent that M_W/M^* is a small number it is a good approximation to set $M_W = 0$ and to treat SU(3) x SU(2) x U(1) as an exact symmetry. A phenomenological theory constructed to describe the new physics would consist of sums of local operators. If the approximation $M_W/M^* << 1$ is valid, then these operators must transform as singlets under SU(3) x SU(2) x U(1).

The operators are to be constructed out of the fields already present in the standard theory, namely an octet of color gluons, W, Z, γ, the usual quarks and leptons, and some Higgs doublets with the standard SU(2) x U(1) transformation properties. At this stage we choose to be conservative and will not introduce any exotic fields, such as a right-handed neutrino field, or Higgs fields not transforming as the standard doublets. If one should so desire or if experimental observation should warrant, one could easily incorporate any such additional fields into the analysis below. We also assign quarks and leptons in the standard fashion, but one could easily modify the analysis to suit any unorthodox fermion assignment.

While we restrict ourselves to the standard Higgs doublets, we do not necessarily suppose that only one Higgs doublet exists. There is no experimental evidence, at the moment, against the existence of more than one Higgs doublet.

Since the Lagrangian of the standard theory contains, by construction, all possible SU(3) x SU(2) x U(1) invariant operators of dimension four or less, composed of the fields admitted above, the operators describing new physics must have dimensions greater than four. The effective Lagrangian describing new physics must be non-renormalizable, just as Fermi's theory of weak decay. Indeed, the whole point of favoring a renormalizable theory is that with a renormalizable theory phenomena at low energies are not controlled by physics at much higher mass scale. Here we want precisely the opposite; we want to know if physics at a mass scale much higher than M_W induces phenomena such as $\mu \to e\gamma$, ν oscillation, and proton decay.

It is logical to order the operators by their dimensions since operators of successively higher dimensions, when appearing in the Lagrangian, have to be multiplied by successively higher powers of $1/M^*$. (We are always assuming M_W/M^* to be small.) The contributions of these higher dimensional operators to physical amplitudes will be suppressed by powers of (hadron mass/M^*) or (lepton mass/M^*).

It is one of the great virtues of the standard theory that, once the Lagrangian is constructed out of all possible SU(3) x SU(2) x U(1) invariant operators of dimension four or less composed of the assumed fields, then conservation of baryon number B, electron number L_e, muon number L_μ, and tauon number L_τ all follow automatically. Thus, for instance, the separate conservation of muon and electron numbers, which forbids $\mu \to e\gamma$, is correlated in the standard theory with the absence of right-handed neutrinos. Thus it is most interesting, both theoretically and experimentally, to look for phenomena violating one or more of these "sacred" numbers: B, L_e, L_μ, L_τ, and $L \equiv L_e + L_\mu + L_\tau$. We can restrict ourselves to operators violating these numbers. Any processes which conserve these quantities could in principle be generated in the standard theory by radiative effects and does not therefore signal any new physics. (An example is the operator $\phi^\dagger \tau^n \phi W^n_{\mu\nu} B^{\mu\nu}$ which induces a violation of the relation $M_W = M_Z \cos\theta$.)

This program was analyzed in some details by Weinberg[1)], Wilczek and Zee[2)], Weinberg[3)], and Weldon and Zee[4)]. We will content ourselves with giving a few examples. The reader should consult the references cited (and in particular ref. 4) for further discussion.

The most interesting example involves proton decay. The effective Lagrangian must involve three quarks because of color symmetry. To form a Lorentz scalar we must have also a fermion field which a priori could be either a lepton or an anti-lepton field. It turns out that with SU(3) x SU(2) x U(1) symmetry it is impossible to construct a dimension six four-fermion operator involving three quark fields and an anti-lepton field. Thus effective B-L conservation is actually a consequence of SU(3) x SU(2) x U(1) symmetry and does not depend on the specific model of grand unification. The allowed operators are in fact quite few in number and model-independent predictions for proton decay may be made.

It is worth remarking that back in the days of a pion-nucleon theory of strong interaction nothing would have prevented us from writing a dimension four term such as

$$\bar{p} e^+ \pi^0$$

to describe proton decay (or even a dimension three term such as $\bar{p}e^+$). We forbid such a term only be decreeing baryon and lepton number conservation. Otherwise there is no reason why the coefficient in front of such a term should be extremely small.

The possibility of neutrino masses offers another example. Since we do not allow ν_R fields, the simplest $\Delta L = 2$ operators are of dimension five:

$$D_{ab}^{(1)} = (\tilde{\ell}_{aL}\, C\, \ell_{bL})(\tilde{\phi}\phi')$$

$$D_{ab}^{(2)} = (\tilde{\ell}_{aL}\, C\, \tau_n\, \ell_{bL})(\tilde{\phi}\tau_n\phi) \quad ,$$

where a,b label the lepton choices ee, eμ, μμ. Note that $D^{(1)}$ exists only for $a \neq b$ and only if there are two distinct Higgs doublets. With the Higgs fields replaced by their vacuum expectation values, only $D^{(2)}$ survives and it generates Majorana mass terms for the neutrino. Since $D^{(2)}$ has dimension five it appears in the effective Lagrangian multiplied by 1/M where M is the mass scale of the physics responsible for neutrino masses. Replacing ϕ by its vacuum expectation value in $D^{(2)}$ we see that m_ν is of the form

$$\langle\phi\rangle^2/M \quad ,$$

where $\langle\phi\rangle$ measures the breaking of SU(2) x U(1). It is well known by now that in specific grand unified theories such as SO(10) the neutrino mass is given by the above form with the replacement $\langle\phi\rangle \rightarrow$ typical mass of some quarks and $M \rightarrow$ typical mass of some right-handed neutrinos. The operator analysis, while not providing the numerical value for m_ν, states that the form $\langle\phi\rangle^2/M$ is quite general.

One could clearly go on to analyze a number of other processes, to explore what additional symmetries might imply, and so forth. These and other questions are discussed in ref. 4, for example.

References: Section 4

1. S. Weinberg, Phys. Rev. Lett 43, 1566 (1979).
2. F. Wilczek and A. Zee, Phys. Rev. Lett. 43, 1571 (1979).
3. S. Weinberg, "Variaties of Baryon and Lepton Non-Conservation."
4. H.A. Weldon and A. Zee, Nucl. Phys. B (1980-81).

5. Dynamically Generated Gravity

Of the fundamental interactions in the world, the two more feeble interactions are characterized by dimensional coupling constants: one observes Fermi's coupling constant $G_F \simeq (300 \text{ GeV})^{-2}$ and Newton's coupling constant $G_N \simeq (10^{19}\text{GeV})^{-2}$. As was noted long ago by Heisenberg, dimensional analysis implies that interactions with coupling constants with dimensions of inverse mass to some positive power are highly divergent and non-renormalizable. Over the last two decades or so, we have come to realize that the weak interaction at a more fundamental level is actually characterized by a dimensionless coupling constant and that the dimensional nature of G_F results from a spontaneous symmetry breaking. Indeed, $G_F \simeq 1/V_w^2$, where $V_w \simeq 300$ GeV is the vacuum expectation value of some scalar field (which may be elementary or composite).

It is tempting to suggest, in light of the above, that gravity is also in fact characterized by a dimensionless coupling constant and that the weakness of gravity is associated with symmetry breaking at a high mass scale. We imagine that G_N, similar to G_F, is given by the inverse square of the vacuum expectation value V of some scalar field ϕ (which may be elementary or composite). Clearly, we should not push the analogy too far since the physics of the weak interaction and of gravity is quite different. Gravitation is long ranged and so V cannot be associated with the mass of a mediating particle. Indeed, $1/G_N$ appears in the Einstein-Hilbert action, $S_E = \int d^4x\sqrt{-g}R(16\pi G_N)^{-1}$, multiplying what is essentially the kinetic energy term for the graviton.

Nowadays, the weakness of the weak interaction is understood in terms of the fact that 300 GeV is large (compared to 1 GeV). If we believe that the breaking at 300 GeV is due to technicolor, in analogy to the breaking of chiral symmetry breaking by color at 1 GeV, then we are led, rather amusingly, to explain the weakness of the weak interaction in terms of the weakness of the strange interaction compared to the techni-strong interaction. In contrast, we have no reasonable understanding of why gravity is weak. One is tempted to suggest that ϕ, the field whose vacuum expectation value determines G_N, is also responsible for the breaking of grand unified theory into strong, weak and electromagnetic interactions. In that case, gravity is weak because the other three coupling constants move so slowly (logarithmically) under the renormalization group. This "modern" view of why gravity is weak was in fact known to Landau. Unfortunately, it is now believed that the relevant symmetry scale of grand unification is lower than what was originally calculated and is only about $10^{-4}\ M_{Pl}$. (M_{Pl} denotes the Planck mass $\sim 10^{19}$ GeV).

If we expand the metric in terms of the graviton field $h_{\mu\nu}$ by writing $g_{\mu\nu} = \eta_{\mu\nu} + h_{\mu\nu}$ we can expand $\Gamma \sim \partial h(1 + h + \ldots)$ and $\sqrt{-g}R \sim \partial\Gamma + \Gamma\Gamma \sim \partial^2 h + \partial h + h\ \partial h \partial h + \ldots$ (in a self-evident schematic notation). The term linear in h is a total divergence and can be dropped in S_E (but not in the action described below). The Einstein-Hilbert action is reminiscent of the Yang-Mills action $S_{YM} = \int d^4x(-\frac{1}{4}F^2)(g^2)^{-1}$, where F^2 has the schematic (terminating) expansion $\partial A \partial A + AA\partial A + AAAA$. The couplings G_N and g^2 measure the "stiffness" of the graviton and the gluon field against excitation. Confinement is essentially equivalent to having a large effective g^2 outside and a small effective g^2 inside hadrons. This may be described phenomenologically by an effective action $S = \int d^4x\phi^2(-\frac{1}{4}F^2)$, where ϕ^2 varies from a large value inside hadrons to a small value outside hadrons. This effective phenomenological action is similar in spirit to the action described below.

Actually, the gauge potential A corresponds to Γ and the Yang-Mills Lagrangian F^2 corresponds to gravitation theories involving the square of curvature tensors R^2 and $R^2_{\mu\nu}$.

It was proposed by the present author[1], and independently by Smolin[2], that the action S_E be replaced by the action[3]

$$S = \int d^4x\sqrt{-g}\left[\frac{1}{2}\,\varepsilon\,\phi^2 R + \frac{1}{2}\,\partial_\mu\phi\partial_\nu\phi g^{\mu\nu} - V(\phi)\right] \tag{1}$$

The coupling constant ε is dimensionless. The potential $V(\phi)$ is assumed to attain its minimum value when $\phi = V$. Then

$$G_N = \frac{1}{16\pi}\left(\frac{1}{\frac{1}{2}\,\varepsilon\,V^2}\right) \tag{2}$$

The introduction of scalar fields into gravity has a long history. Here, the crucial feature is the incorporation of spontaneous symmetry breaking. As a consequence the scalar field is "anchored" in a deep potential well $V(\phi)$ and thus the physical consequences of the present theory are indistinguishable from Einstein's theory except under extreme conditions of space-time curvature. This is in sharp contrast to earlier work such as that of Brans and Dicke.

It would be very nice indeed if ϕ is associated with the scale of a grand unified theory. But unfortunately, as is already remarked, this is numerically the case with the presently known grand unified theories. These theories are however incomplete in a number of ways and we can always hope that eventually the present grand unified theory may be extended to a theory set at the Planck mass.

Starting from the action in Eq. (1) one can easily derive the equations of motion (see Ref. (1)). Note that these equations are not derivable from the usual equations of motion by merely replacing G_N by $1/\varepsilon\phi^2$. The term linear in h in the action in Eq. (1) reads $\varepsilon\phi^2\partial^2 h$. Unlike the standard theory, in which $\varepsilon\phi^2$ is a constant, this term cannot be integrated away. Instead, setting $\phi = V + \zeta$, we find that this term leads to a mixing between ζ and h. At the graviton pole $k^2 = 0$ this mixing vanishes. Notice that the particle associated with ζ would decay into gravitons rapidly due to a term like $V\varepsilon\zeta\ \partial h\partial h(1 + h + \ldots)$ in the Lagrangian.

As has already been remarked, the present theory is indistinguishable from Einstein's theory except under extreme conditions such as may exist in the early Universe. Newton's gravitational "constant" may then vary with temperature. It is not inconceivable that this phenomenon of a varying gravitational "constant" may be relevant for the horizon problem. Any serious discussion is necessarily highly speculative, however.

The attractiveness of scale-invariant theories has been much discussed in recent years. Such theories are apparently renormalizable by power counting. The renormalization procedure introduces a mass scale and as a result of this so-called "dimensional transmutation" all dimensionless physical quantities are calculable. It was suggested by the present author in Ref. (1) that the underlying interactions of the world, including gravity, are scale invariant. (For simplicity, we do not insist on local scale invariance here since we do not need it to make our point. While it is most appealing to impose local scale invariance, we could do so only at the price of introducing additional structures not relevant to our discussion.) In that case, the Einstein-Hilbert action for gravity is not admissible. Instead, a scale-invariant action for the world would read

$$S = \int d^4x\sqrt{-g}\Big[\varepsilon\phi^2 R + \beta R^2 + \gamma R_{\mu\nu}R^{\mu\nu}$$

$$+ \frac{1}{2} g^{\mu\nu}\partial_\mu\phi\partial_\nu\phi - \lambda\phi^4$$

$$+ \mathcal{L}(\psi, A, \phi, \eta)\Big] \quad . \qquad (3)$$

Here $\mathcal{L}$ is a scale-invariant gauge-invariant Lagrangian describing the interactions between quarks and leptons ψ, gauge fields A, the scalar field ϕ, and possibly other scalar fields η. Note the presence in general of the dimension-4 terms R^2 and $R_{\mu\nu}R^{\mu\nu}$.

We note in passing that the presence of the R^2 and $R_{\mu\nu}R^{\mu\nu}$ terms, which involve four derivatives, implies that the graviton propagator would go as k^4 for large momentum k thus rendering the theory, at least formally, renormalizable. As is well known, the price to be paid for this improved convergence is the occurrence of a ghost pole in the propagator. This ghost pole, however, occurs at momentum so large, of order $\beta^{-1/2}M_{Pl}$ or $\gamma^{-1/2}M_{Pl}$, that it is not even clear whether local quantum field theory would continue to hold.

Coleman and Weinberg[4)] had shown that in the flat-space

limit the action S, with suitable choice of coupling constants, is such that ϕ has a nonzero VEV. We expect the same would hold in curved space, at least for spaces of small curvature. Thus, the $\varepsilon\phi^2 R$ term would then lead to effectively Einstein's theory of gravity. (In general, the VEV of ϕ will be a functional of the metric $g_{\mu\nu}$, and thus for spaces of high curvature the effective gravitational action may be quite complicated and nonlocal.)

Recently, Adler[5)] has gone one step further; he proposed in a very interesting paper that elementary scalar fields should not be present in the fundamental action in Eq. (3). Elementary scalar fields are generally regarded with repugnance by the particle physics community. It is generally believed that the elementary scalar fields needed in present-day theories are merely the phenomenological manifestations of some composite scalar field such as $\bar{\psi}\psi$. In the present context, Adler made the important observation that in that case scale and gauge invariance combine to forbid terms proportional to R. The terms R and $\bar{\psi}\psi R$ have mass dimension 2 and 5, respectively, while $A_\mu A^\mu R$, although of dimension 4, is not gauge invariant. However, if dynamical symmetry breaking occurs, such that $\bar{\psi}\psi$ has some nonzero VEV, then, in general, we expect that a term such as $\langle\bar{\psi}\psi\rangle^{2/3}R$ would be effectively induced in the action. Thus, we are led to the rather amusing view that gravity may be, at least in some sense, an inevitable consequence[6)] of the dynamical breaking of a grand unified symmetry describing the strong, weak, and electromagnetic interactions.

Thus, the roots of gravity may well lie in Lorentz invariance. To write down Lorentz-invariant interactions between fields we have to introduce the Minkowski metric $\eta_{\mu\nu}$. Once we admit the possibility of $\eta_{\mu\nu}$ depending on space-time, we promote the metric to a field $g_{\mu\nu}(x) = \eta_{\mu\nu} + h_{\mu\nu}(x)$. In a field-theory language, $h_{\mu\nu}$ is then a field without a proper kinetic energy term. In the view discussed here, the appropriate kinetic energy term arises as a consequence of dynamical scale-symmetry breaking.

We suggest that this view leads to a severe constraint on dynamical symmetry breaking. Scale invariance forbids the appearance of a cosmological constant term $\sim\langle\bar{\psi}\psi\rangle^{4/3}\int d^4x\sqrt{-g}$ in the action which would in general appear[7)] upon symmetry breaking. At the moment, no one knows how to avoid generating this undesirable term. This is perhaps the weakest point in the program to generate gravity spontaneously, and indeed, this problem afflicts all current theories in particle physics which utilize the notion of spontaneous symmetry breaking. We imagine that the ultimate correct theory of dynamical symmetry breaking will not produce a cosmological term. (In a soluble two-dimensional model of dynamical symmetry breaking a "cosmological" term does appear. However, it is not clear how this may be relevant since gravity does not exist in two dimensions.)

As a crucial test of this idea of gravity as a consequence of dynamical symmetry breaking, we have to perform a calculation to see whether the sign of the induced R term is positive, leading to attractive gravity.

It may be illuminating to consider a simpler analog problem. It turns out that the problem of "spontaneously" generating a graviton in four dimensions shares some common features with the problem of generating a photon in six dimensions. Consider a six-dimensional world with local U(1) invariance. Thus, the fermion kinetic energy term in the Lagrangian would read $\bar{\psi} i(\not\partial - i\not{A})\psi$. We note that ψ has dimension 5/2 while the gauge field A has dimension +1 (as is the case in any space-time dimension). Thus, if we insist on a scale-invariant Lagrangian, the "photon" kinetic energy term $F_{\mu\nu}F^{\mu\nu}$ with dimension 4 would not be allowed. Dimension-6 terms, such as $\partial_\lambda F_{\mu\nu}\partial^\lambda F^{\mu\nu}$ or F^3, are allowed. (The F^3 term may be eliminated by invoking charge conjugation invariance.) The $(\partial F)^2$ terms, involving four derivatives, are analogous to the R^2 and $R_{\mu\nu}R^{\mu\nu}$ terms in the gravity problem. Also note that, were an elementary scalar field ϕ present, the dimension-6 term $\phi^2 F_{\mu\nu}F^{\mu\nu}$ would be allowed. To summarize, scale invariance combined with gauge invariance forbids, in the absence of elementary scalar fields, a photon kinetic energy term in six dimensions and a graviton kinetic energy term in four dimensions.

With dynamical symmetry breaking, we expect an $F_{\mu\nu}F^{\mu\nu}$ term induced by, among others, a graph such as the one in Fig. 1. The dark "blobs" on the fermion line indicate the fermion condensate $\langle\bar{\psi}\psi\rangle$. In effect, a soft fermion mass is generated, so that the chiral-noninvariant part of the fermion propagator vanishes rapidly at large momentum.

Fig. 1. Photon vacuum polarization.

By the usual discussion of renormalizable quantum field theories, the coefficient of $\sqrt{g}R$, a dimension-two term forbidden by scale invariance to appear in the Lagrangian, should be finite and calculable. However, since scale invariance is invoked, we must employ a regularization scheme such as dimensional or zeta-function regularization which respects scale invariance in that physical quantities scale by their correct dimensions. The inverse of Newton's constant $1/G_N$ is the coefficient of a dimension-two term and behaves just as a boson mass squared. In dimensional regularization there is no quadratic divergence in the calculation of $1/G_N$ and the logarithmic divergence is eliminated since scale invariance is softly (dynamically) broken.

Indeed, it is possible to give a formal representation[8,9] for Newton's gravitational constant in terms of the two point function of the stress-energy tensor $T_{\mu\nu}$. A particularly simple derivation is given in Section VI of Ref. (9). The representation reads

$$\frac{1}{8\pi G_N} = \frac{i}{96} \int d^4x x^2 \langle 0|TT(x)T(0)|0\rangle \tag{4}$$

This is strictly a formal representation. To actually evaluate it we need a theory of dynamical symmetry breaking which we do not have at present. In view of the fact that a precise understanding of dynamical symmetry breaking is lacking at present, it appears to this author that it may be better not to perform calculations tied to any specific scheme of dynamical symmetry breaking. Instead, we proposed a calculation[9] which retains some general and essential features of dynamical symmetry breaking, most notably that effective masses thus generated are expected to decrease rapidly at large momentum.

In an evaluation of G_N the magnitude is not of much interest. It is simply given by the mass scale at which scale invariance is broken. The task is rather to calculate dimensional quantities such as the proton mass in terms of G_N. The sign of G_N is however of great interest. If it comes out negative, the program described here is evidently wrong.

A very crude calculation was carried out in Ref. (9) and led us to conclude that this crucial sign depends sensitively on the infrared details of the symmetry-breaking mechanism. A calculation involving the dilute instanton gas has also been performed[10]. The cut-off on the instanton size serves to set the scale.

Newton's gravitational constant G_N is positive. Objects attract each other gravitationally. This is one of the most firmly established experimental facts in physics and undoubtedly one of the most fundamental. Field theory provides a partial understanding of this phenomenon. A fundamental theorem states that the force between two like particles generated by the exchange of a meson with spin n is attractive or repulsive according to whether n is even or odd respectively. However, in the Lagrangian, quite aside from the fact that the theory involves a spin two particle, we have a choice on the sign of the coefficient $1/G_N$ in front of $\sqrt{g}R$. We choose G_N to be positive so that the kinetic energy of the graviton is positive. This is the same stability argument which forces us to choose the coefficient in front of $(-F_{\mu\nu}^2)$ in a gauge theory to be positive. In fact that coefficient is conventionally written as $1/g^2$ to remind us that it is positive. In the same way, we can write $1/G_N$ as $1/\kappa^2$. Of the four fundamental couplings the sign of Fermi's constant used to be arbitrary but is now also fixed with the advent of the gauge theory of electroweak interactions. In this sense, we understand the sign of all four fundamental couplings. Nevertheless, we feel that the stability argument is somewhat less satisfactory in the case of gravity in that we may add the dimension four terms R^2 and $R_{\mu\nu}^2$. A theory with these terms may well make sense even with G_N negative; of course, it would not describe gravity as we know it. An attractive feature of the present program is that it opens the door to possibly an improved understanding of this fundamental sign.

Is it possible, by just looking at Eq. (4), to give a general argument for the sign of G_N? That would be deliriously exciting.

In conclusion we have suggested that gravity may be generated as a result of quantum fluctuations. A heuristic understanding of this is not difficult to find and is readily suggested by Fig. (1). In the absence of a $(\partial h)^2$ term in the action, a spatially varying gravitational field contains no energy. While this may be true classically it cannot be true quantum mechanically. With quantum fluctuation a pair, say an e^-e^+ pair, could always be created. With $\vec{\nabla} h_{\mu\nu} \neq 0$ this pair will then fall, gaining kinetic energy. By bringing the e^-e^+ pair together and annihilating them we can always extract energy out of the gravitational field. Another version given by Adler involves the legendary Einstein's elevator. The usual statement is that for a small enough elevator one cannot say whether the elevator is uniformly accelerating or whether it is falling in a gravitational field (say that of the earth). But with quantum fluctuations an e^-e^+ pair may be created and the e^- could tunnel out of the elevator to make a grand tour sampling the curvature tensor before coming back to annihilate the positron.

References: Section 5

1. A. Zee, Phys. Rev. Lett. 42, 417 (1979); 44, 703 (1980).
2. L. Smolin, Nucl. Phys. B160, 253 (1979).
3. Earlier work which discusses spontaneous symmetry breaking includes Y. Fujii, Phys. Rev. D9, 874 (1974); P. Minkowski, Phys. Lett. 71B, 419 (1977); T. Matsuki, Prog. Theor. Phys. 59, 235 (1978); A.D. Linde, Pis'ma Zh. Eksp. Teor. Fiz. 30, 479 (1979) [JETP Lett. 30, 447 (1979)].
4. S. Coleman and E. Weinberg, Phys. Rev. D7, 1888 (1973).
5. S. Adler, Phys. Rev. Lett. 44, 1567 (1980). [Very similar ideas have been expressed by K. Akama, Y. Chikashige, T. Matsuki, and H. Terazawa, Prog. Theo. Phys. 60, 1900 (1980).
6. This represents, in some sense, the modern realization of ideas of A.D. Sakharov, Dokl. Akad. Nauk. SSSR 177, 70 (1967) [Sov. Phys.--Dokl. 12, 1040 (1968)] and of O. Klein, Phys. Scr. 9, 69 (1974).
7. A.D. Linde, Pis'ma Zh. Eksp. Teor. Fiz. 30, 479 (1979) [JETP Lett. 30, 447 (1979)]; J. Dreitlein, Phys. Rev. Lett. 33, 1243 (1974); M. Veltman, ibid. 34, 777 (1975).
8. S. Adler
9. A. Zee, Phys. Rev. D23, 858 (1981).
10. B. Hasslacher and E. Mottola, Phys. Lett. (1980).

6. Kaluza Theory and Grand Unification

A. Two major concepts lie at the foundation of modern physics: local coordinate invariance and local internal symmetry. The former leads to the theory of gravity while the latter leads to the gauge theory of strong, weak and electromagnetic interactions. The remarkable discovery of Kaluza is that if we suppose that space-time is embedded in a space with dimension higher than four these two concepts may not be independent--the latter may be derived from the former. The physics is so astonishing and the mathematics is so elegant that we might be tempted to rephrase G. Marx in remarking that Nature is not worth studying if She does not in fact use the Kaluza mechanism at some level. Whether or not this level is already accessible to present-day physics is of course a so-called "crucial question."

The Kaluza mechanism, while profound, is easy to explain. Imagine, as in the original version, that the world is actually five-dimensional but the fifth dimension is curled up in a tiny circle. In other words, each space-time point is actually a tiny circle, with a radius so much smaller than any length scale explored experimentally that we have all been fooled and have mistaken the circle for a point. Denote the coordinates by

$$x^{\hat{\mu}} = \{x^{\mu}, x^5\} \quad , \quad \hat{\mu} = 1\cdots 5 \quad , \quad \mu = 1\cdots 4 \quad .$$

Here $x^5 \equiv \theta$ runs from 0 to 2π. Consider an infinitesimal coordinate transformation

$$\begin{aligned} x^{\mu} &= x'^{\mu} \\ x^5 &= x'^5 + \lambda(x') \quad . \end{aligned} \tag{1}$$

In words, we rotate the circles by an infinitesimal space-time dependent amount. The metric $g_{\hat{\mu}\hat{\nu}}$ contains a piece $g_{\mu 5}$ which transforms as a four-vector under the usual coordinate transformation. Under the transformation (1) we find that

$$g_{\mu 5} \to g_{\hat{\sigma}\hat{\rho}} \frac{\partial x^{\hat{\sigma}}}{\partial x'^{\mu}} \frac{\partial x^{\hat{\rho}}}{\partial x'^{5}}$$

$$= g_{\sigma 5}\delta^{\sigma}_{\mu} + g_{55}\partial_{\mu}\lambda \quad . \tag{2}$$

Setting $g_{55} = 1$ and calling $g_{\mu 5}$ A_{μ} we see that A_{μ} transforms like a gauge potential:

$$A_{\mu} \to A_{\mu} + \partial_{\mu}\lambda \quad . \tag{3}$$

Gauge transformation may be cast as a special case of coordinate transformation!

Furthermore, the gravitational action in five-dimensions $\int d^4x dx^5 \sqrt{\hat{g}}\, \hat{R}$ reduces to the action $\int d^4x \sqrt{g}(1/\ell^2\, R + F_{\mu\nu}F^{\mu\nu})$ in four dimensions. The appearance of Maxwell's action is not so surprising if we recall that $\hat{R}$ involves two derivatives and is invariant under all five-dimensional coordinate transformations. The fact that Maxwell's action appears with the correct sign is however not so evident a priori. The relative scale between Einstein's action and Maxwell's action is set by the length scale of the Kaluza circles, which clearly must have radii the order of Planck's length. It is perfectly understandable then that experimentalists all believe the world is four-dimensional.
It is also worth remarking on charge conjugation in the Kaluza framework. Of the three fundamental discrete symmetries of physics, charge conjugation stands apart from parity and time

reversal, which are in some sense more intuitively accessible. Some people find this a bit mysterious. A beautiful feature of the Kaluza theory is that charge conjugation may be interpreted as a space-time transformation as well. In what follows we hope to convince the reader of the beauty of the Kaluza theory.

B. This year marks the sixteenth anniversary of Kaluza's classic paper. Over the decades, the theory has been developed by a number of people, including Klein, Jordan, Thiery, Trautman, Cho, Freund, Cremmer, Julia, Scherk, Schwarz, Luciani, Palla, Thirring, Tanaka, Koerner, and no doubt many others. In particular, there apparently exists a body of work in Kaluza's homeland which is perhaps not as well-known in the West as it should be. Despite its elegance, the Kaluza concept has been more or less regarded as an oddity and has languished outside the mainstream of physics. Historically, this is partly due to the fact that the symmetry of the thirties, isospin, and extensions of it later, clearly does not appear geometrical in origin. Interest in Kaluza theory has been ignited once again in recent years thanks largely to the discovery that the underlying symmetry is indeed geometrical, local, and gauged and that flavor symmetries such as isospin are approximate and rather accidental. Also, supergravity has naturally led people to consider higher space-time dimensions.

The crucial question is then whether the original Kaluza concept may be generalized from Abelian Maxwell to non-Abelian Yang-Mills. Perhaps not surprisingly, the answer to this question is a resounding yes. All one has to do is to replace the circle, which Kaluza attached to each space-time point, by some general manifold M. The circle is intimately connected to U(1), the

Abelian group of electromagnetism. It turns out if the manifold M permits the action of the group G on it one obtains, instead of electromagnetism, a gauge theory based on G. We will explain this in what follows.

M could be the manifold of G itself. For example, if G = SO(3), M is a three-dimensional ball with antipodal points identified, as is well-known. However, in general M could be considerably smaller than the group manifold G. Thus, with G = SO(3), M could be S^2 = two-sphere. (Note that "sphere" is not to be confused with "ball." S^n = the surface of the (n+1)-ball = n-dimensional manifold.) More generally, M could be G/H where H is a subgroup of G.

Recall that the notation G/H indicates the coset-space of G relative to H. It is defined as follows. Let g_1 and g_2 be two elements in G. Then in G/H, g_1 is identified with g_2 if in G $g_1 = g_2 h$ for some h in H. In other words, G/H consists of the equivalence classes of G relative to H.

As an example, the coset space SO(3)/SO(2) is in fact S^2. To see this , consider a three-dimensional vector $\vec{V}$. Let g_1 and g_2 be rotations in SO(3). Let H = SO(2) be the rotations which leave $\vec{V}$ invariant. If $g_1 = g_2 h$, $h \in H$, then $g_1\vec{V} = g_2 h\vec{V} = g_2\vec{V}$ and so the actions of g_1 and g_2 on $\vec{V}$ are identical. Under the action of SO(3) on $\vec{V}$, the tip of the vector $\vec{V}$ describes S^2. The reader should be able to show that $SO(N+1)/SO(N) = S^N$.

The reader should also attempt the slightly more complicated exercise of showing that $SU(n+1)/U(n) = CP^n$. Here CP^n denotes the complex projective space of dimension 2n, which, as the reader may or may not recall, is defined as the set consisting of "points" characterized by n+1 complex numbers $(z_1, z_2, \ldots z_{n+1})$ such that (i) not all the z_k's vanish and (ii) the point $(z_1, z_2, \ldots z_{n+1})$ is to be identified with the point $(\lambda z_1, \lambda z_2, \ldots \lambda z_{n+1})$, for any complex number λ. One may note that if the z_k's and λ are restricted to be real the set just described is exactly the set of lines in (n+1) dimensional Euclidean space passing through the origin.

The assertion to be shown below is that if one applies the Kaluza concept to Minkowski 4-space + G/H one would obtain a theory consisting of Einsteinian gravity plus Yang-Mills with the gauge group G.

C. Fermions may be incorporated into the theory. Following Palla and others one considers the Dirac equation in 4+n dimensional space-time $D^{(4+n)}\psi = 0$. Schematically, the Dirac operator $D^{(4+n)}$ may be decomposed into a 4-dimensional and an n-dimensional Dirac operator, so that we have an equation of the form

$$D^{(4)}\psi = -D^{(n)}\psi \quad .$$

The observed fermion masses would then be determined by the eigenspectrum of the operator $D^{(n)}$. Since $D^{(n)}$ is defined on a compact coset manifold G/H (we consider only compact Lie groups) with size of the order of the Planck length its eigenspectrum is discrete with spacing set by the Planck mass. Thus, as far as present day physics is concerned we need only be interested in the zero eigenvalues of $D^{(n)}$.

The Kaluza program is very ambitious. Once the group and the manifold are chosen, gravity and gauge fields both follow from the geometry and the fermion spectrum is completely determined by the structure of the manifold. One might hope in this way for a "deep" solution of the family problem.

Even more ambitious is the super-Kaluza program in which one starts with supergravity. In supergravity, one is obliged to put in spin 3/2 fermions in a definite way. These spin 3/2 fermions when reduced to 4-dimensions will contain spin 1/2 fermions. In this context Witten has noted an interesting numerology. For $G = SU(3) \times SU(2) \times U(1)$ the maximal subgroup is $U(2) \times U(1)$. Thus the minimal manifold having an $SU(3) \times SU(2) \times U(1)$ gauge theory in the Kaluza framework is $SU(3) \times SU(2) \times U(1)/U(2) \times U(1) = CP^2 + S^2 + S^1$ which is seven dimensional. Together with Minkowski space-time this suggests that the basic theory should be $4 + 7 = 11$ dimensional. But workers on supergravity have long known that the maximal dimension for supergravity is precisely eleven! Whether or not this is a pure coincidence remains to be determined. Note that there is no room for SU(5) in this discussion. The manifold $SU(5)/U(4) = CP^4$ is 8 dimensional.

Unfortunately, the theory must overcome a large number of difficulties before it could be considered phenomenologically relevant. To mention but one of these difficulties, we note that we are bound to end up with equal numbers of V+A and V-A fermions, since the geometry does not distinguish between left and right. This is the same problem which plagues us in Section 1.

We will not address any of these difficulties here. My feeling is that the Kaluza scheme is so elegant that it must be relevant at some level.

The purpose of these notes is to provide a (hopefully) pedagogical introduction to the subject. Our discussion here is based on some work I did in collaboration with L. Brown and D. Boulware. It is extremely unlikely that the material presented below would contain anything new. Nevertheless, the presentation may be somewhat unusual in places.

D. We describe geometry by introducing the standard Vielbein formalism. For the sake of pedagogical completeness we give in Appendix VI-1 a brief review of some elementary notions in differential geometry. Denote the Vielbein by $e^{\hat{\alpha}\hat{\nu}}$. Here $\hat{\alpha}$ is an orthonormal index. The index $\hat{\alpha}$ is lowered by $\eta_{\hat{\alpha}\hat{\beta}}$, the Minkowski (or Euclidean) metric. We define $e_{\hat{\alpha}}^{\hat{\nu}}$ by $e_{\hat{\alpha}}^{\hat{\nu}} = \eta_{\hat{\alpha}\hat{\beta}}e^{\hat{\beta}\hat{\nu}}$. The vectors $e^{\hat{\alpha}\hat{\nu}}(x)$ span the tangent plane at the point x. Under a translation in the $\hat{\alpha}$ direction by the amount parameterized by $\delta\xi_{\hat{\alpha}}$ we have

$$\delta x^{\hat{\nu}} = \delta\xi\hat{\alpha}e^{\hat{\alpha}\hat{\nu}} \quad . \tag{1}$$

We now suppose that the index sets may be divided up

$$\hat{\mu} = (\mu, i)$$
$$\hat{\alpha} = (\alpha, a) \tag{2}$$

so that μ,α corresponds to the space-time we know and i,a to the internal space. Our notation is such that the hat symbol ^ always refers to the whole space. The Kaluza picture is such that x^{μ} should not change under translation in the "a" direction. In other words,

$$e^{a\mu} = 0 = e_a^{\mu} \quad . \tag{3}$$

Equation (3) defines spaces of the Kaluza type. The inverse

Vielbein $e^{\hat{\beta}}_{\hat{\nu}}$ is determined by the orthonormality condition

$$e^{\hat{\beta}}_{\hat{\nu}} e^{\hat{\alpha}\hat{\nu}} = \eta^{\hat{\beta}\hat{\alpha}} \quad . \tag{4}$$

We find that Eq. (3) then implies

$$e^{\alpha}_{j} = 0 = e_{\alpha j} \quad . \tag{5}$$

The crucial equations (3) and (5) determine the form of the metric, which is given as usual by

$$\hat{g}_{\hat{\mu}\hat{\nu}} = \eta_{\hat{\alpha}\hat{\beta}} e^{\hat{\alpha}}_{\hat{\mu}} e^{\hat{\beta}}_{\hat{\nu}} \quad . \tag{6}$$

Plugging Eqs. (3) and (5) into (6) we find

$$\hat{g}_{\hat{\mu}\hat{\nu}} = \left(\begin{array}{c|c} g_{\mu\nu} + N_{i\mu} g^{ij} N_{j\nu} & N_{i\nu} \\ \hline N_{j\mu} & g_{ij} \end{array}\right) \quad . \tag{7}$$

We have introduced a number of definitions as follows:

$$g_{\mu\nu} \equiv \eta_{\alpha\beta} e^{\alpha}_{\mu} e^{\beta}_{\nu} \tag{8}$$

$$g_{ij} \equiv \eta_{ab} e^{a}_{i} e^{b}_{j} \tag{9}$$

$$N_{\mu j} \equiv e_{a\mu} e^{a}_{j} \quad . \tag{10}$$

Note that our definitions are such that $g_{ij} = \hat{g}_{ij}$ but $g_{\mu\nu} \neq \hat{g}_{\mu\nu}$. Also $N_{\mu j} = \hat{g}_{\mu j}$. In Eq. (7) g^{ij} is the inverse of g_{ij} (it is not $\hat{g}^{ij}$).

The form of the metric given in Eq. (7) generalizes an expression found in the textbook by Misner, Thorne and Wheeler, for instance (see p. 506; described there as "pushing forward the many fingers of Time"). The metric describes a space with a "layered" or "foliated" structure. A more geometrical discussion is given in Appendix VI-2.

Vielbein transforms under general coordinate transformation according to

$$e_{\hat{\nu}}^{\hat{\alpha}} \to e'^{\hat{\alpha}}_{\hat{\nu}} = \frac{\partial x^{\hat{\mu}}}{\partial x'^{\hat{\nu}}} e_{\hat{\mu}}^{\hat{\alpha}} \quad . \tag{11}$$

Now we wish to preserve the fundamental structural condition $e_j^{\alpha} = 0$. So we must have

$$0 = e'^{\alpha}_{j} = \frac{\partial x^{\mu}}{\partial x'^{j}} e_{\mu}^{\alpha} \tag{12}$$

and thus we only allow those transformations in which x^{μ} does not depend on x'^{j}.

We now ask the crucial question: when does the geometry decouple? More precisely, under what conditions can we find a coordinate transformation so that $\hat{g}'_{\mu j} = 0$? In that case, the transformed metric has the decoupled form

$$\hat{g}'_{\hat{\mu}\hat{\nu}} = \left(\begin{array}{c|c} X & O \\ \hline O & X \end{array}\right) \quad . \tag{13}$$

Space-time is totally decoupled from internal space. Note that we cannot ask $\hat{g}_{\mu j} = 0$, of course.

Starting with the transformation law

$$\hat{g}'_{\hat{\mu}\hat{\nu}} = \frac{\partial x^{\hat{\sigma}}}{\partial x'^{\hat{\mu}}} \frac{\partial x^{\hat{\tau}}}{\partial x'^{\hat{\nu}}} \hat{g}_{\hat{\sigma}\hat{\tau}} \tag{14}$$

we can easily work out the decoupling condition:

$$0 = \hat{g}'_{\mu j} = \frac{\partial x^{\hat{\sigma}}}{\partial x'^{\mu}} \frac{\partial x^{k}}{\partial x'^{j}} \hat{g}_{\hat{\sigma}k}$$

$$= \frac{\partial x^{k}}{\partial x'^{j}} \left(\frac{\partial x^{\nu}}{\partial x'^{\mu}} \hat{g}_{\nu k} + \frac{\partial x^{\ell}}{\partial x'^{\mu}} \hat{g}_{\ell k} \right) \quad . \tag{15}$$

We have used the fact that x^{μ} does not depend on x'^{j}. In further simplifying Eq. (15) it is extremely important to keep track of what variables are held fixed (as when doing thermodynamics) in the various partial derivatives. Writing

$$\frac{\partial x^{\ell}}{\partial x'^{\mu}} = \left.\frac{\partial x^{\ell}}{\partial x^{\nu}}\right|_{x'^{k}} \frac{\partial x^{\nu}}{\partial x'^{\mu}} \tag{16}$$

and supposing that $\partial x^{k}/\partial x'^{j}$ and $\partial x^{\nu}/\partial x'^{\mu}$ correspond to non-singular transformations we find from Eq. (15)

$$\hat{g}_{\nu k} + \left.\frac{\partial x^{\ell}}{\partial x^{\nu}}\right|_{x'^{k}} \hat{g}_{\ell k} = 0 \quad . \tag{17}$$

Remembering the definition of $N_{\nu k}$ we rewrite Eq. (17) somewhat more compactly as

$$\left.\frac{\partial x^{\ell}}{\partial x^{\nu}}\right|_{x'^{k}} = -g^{\ell k} N_{\nu k} \equiv -N^{\ell}_{\nu} \quad . \tag{18}$$

The last step is a definition.

Differentiating once more we find

$$- \left.\frac{\partial^{2} x^{\ell}}{\partial x^{\mu} \partial x^{\nu}}\right|_{x'^{k}} = \left.\frac{\partial}{\partial x^{\mu}}\right|_{x'^{k}} N^{\ell}_{\nu}$$

$$= \frac{\partial}{\partial x^\mu} N^\ell_\nu + \left.\frac{\partial x^j}{\partial x^\mu}\right|_{x',k} \frac{\partial}{\partial x^j} N^\ell_\nu$$

$$= \left[\frac{\partial}{\partial x^\mu} N^\ell_\nu - N^j_\mu \frac{\partial}{\partial x^j} N^\ell_\nu\right] \quad . \tag{19}$$

In this final form N^ℓ_μ is to be thought of as a function of x^μ and x^j. Let us invite ourselves to introduce an object $F^\ell_{\mu\nu}$ defined to be

$$F^\ell_{\mu\nu} \equiv \frac{\partial}{\partial x^\mu} N^\ell_\nu - N^j_\mu \frac{\partial}{\partial x^j} N^\ell_\nu - (\mu \leftrightarrow \nu) \quad . \tag{20}$$

The condition for the geometry to decouple is then simply

$$F^\ell_{\mu\nu} = 0 \quad . \tag{21}$$

If the internal space were not to be linked to space-time the tensor $F^\ell_{\mu\nu}$ vanishes.

The notation is provocative and the form in Eq. (20) is suggestive. Nevertheless, it is clear that we should not yet identify $F^\ell_{\mu\nu}$ as the gauge field. After all, $F^\ell_{\mu\nu}$ carries only geometrical indices. So far, only geometry has entered the game. There has been no algebraic concept.

E. We now introduce algebra by specifying that the internal space is not just any old space but is a coset manifold M associated with a simple compact Lie group G. Let M be = G/H.

In this subsection we concentrate entirely on the internal manifold. We will forget for the moment the Minkowski space-time. Also, the notation in this subsection will be (unfortunately) independent of that in the previous section.

Given M = G/H. The group transformations in G move the points in M around. Let us lay down a system of coordinates x^μ on M so that we can describe the movement of the points under the action of G. Denote the generators of the Lie algebra of G by X^a so that

$$[X^a, X^b] = if^{abc}X^c \quad . \tag{1}$$

Under an infinitesimal transformation $I + i\theta^a X^a$ the points move by an infinitesimal amount

$$x^\nu \to x^\nu + \theta^a A^{a\nu} \quad . \tag{2}$$

This equation defines a set of vectors $A^{a\nu}$ labelled by the index a and which in general will depend on x. The generator iX^a is represented on the manifold by the differential operator $A^{a\mu}(x)\partial_\mu$. The structure equation (1) translates into

$$A^{\mu a}\partial_\mu A^{\nu b} - A^{\mu b}\partial_\mu A^{\nu a} = f^{abc}A^{\nu c} \quad . \tag{3}$$

This condition on the A vectors is known as Lie's equation. As explained in Appendix VI-1 it can be written in the more compact form involving the commutator of two vectors:

$$[\vec{A}^a, \vec{A}^b] = f^{abc}\vec{A}^c \quad . \tag{3'}$$

Here we have used a notation explained in the Appendix: $\vec{A}^a \equiv A^{\mu a}\vec{e}_\mu$. Not surprisingly, Lie's equation could also be written very compactly in terms of the Lie derivative. (Again, for pedagogical completeness, we remind the reader what a Lie derivative is in Appendix VI-3.) Lie's equation now reads

$$L_{A^a} A^{\nu b} = f^{abc} A^{\nu c} \quad . \tag{3"}$$

It tells us that the Lie derivative along the vector $\vec{A}^a$ acting on the vector $A^{\nu b}$ amounts to a "rotation" on the group index b.

Thus far, no metrical concept has been introduced on G/H yet. While we are free to introduce any metric we please, it would be foolish to do so. There is clearly a "natural" metric induced by the action of G. Intuitively, we want to have distance relationships between neighboring points preserved when the group moves the points around on the manifold. More formally, the metric is to be such that group operations on the manifold correspond to isometries (see Appendix VI-3). Those readers who recall the definition of Killing vectors will recognize that this means that $A^{a\mu}$ are precisely Killing vectors with components indexed by μ. (In Appendix VI-4 we work out explicitly the Killing vectors for CP^n for those readers who are a bit hazy about the concept of Killing vectors.)

The desired metric may be expressed very simply in terms of the vectors $\vec{A}^a$ as follows:

$$g^{\mu\nu} = A^{\mu a} A^{\nu a} \quad . \tag{4}$$

(We have already implicitly chosen the Cartan group metric δ^{ab} for summing over group indices.) To see that the metric is in fact such that group operations generate isometries, it is easiest to note that the Lie derivative, just as an ordinary derivative, is distributive. But we have already learned that the action of the Lie derivative along the A vectors on the vector $A^{\mu a}$ amounts to a "rotation" on the group index. The group indices in Eq. (4) are summed over.

We have thus established that, with respect to this metric, the vectors $A^{\mu a}$ satisfy Killing's equation

$$A^a_{\mu;\nu} + A^a_{\nu;\mu} = 0 \quad . \tag{5}$$

The semi-colon denotes covariant derivatives.

One might wonder if the metric given in Eq. (4) is unique. We discuss this question in Appendix VI-5.

F. <u>We now digress</u> about Dirac's equation on a curved manifold:

$$\gamma^a e^\mu_a D_\mu \psi = 0 \quad . \tag{1}$$

The Dirac operator is given by

$$D_\mu \psi = (\partial_\mu + \frac{1}{4} i\sigma^{ab}\omega_{ab\mu})\psi \quad . \tag{2}$$

The connection $\omega_{ab\mu}$ is defined in the Appendix VI-1 $\sigma^{ab} \equiv (i/2)[\gamma^a,\gamma^b]$.

An extremely useful identity is

$$[D_\mu, \gamma_\nu] = 0 \quad . \tag{3}$$

In the Vielbein formalism $\gamma_\nu = \gamma^a e_{a\nu}$. The massless Dirac equation may be squared

$$\gamma^b e^\nu_b D_\nu \gamma^a e^\mu_a D_\mu \psi = 0 \tag{4}$$

to give

$$[-D^2 + R]\psi = 0 \quad . \tag{5}$$

Since $-D^2 = -D_\mu D^\mu$ is a positive definite operator Eq. (5) has no solution if the scalar curvature R is positive. This is a remarkable result stating that the Dirac operator in Eq. (1) has no zero eigenmode on a manifold with positive scalar curvature.

G. In this subsection we indicate how to calculate the Riemann curvature tensor for that special class of manifolds which interest us, namely manifolds of the form M = G/H. Since these manifolds are determined completely by the groups G and H, we expect that the Riemann curvature tensor should be calculable in terms of the structure constants of the Lie algebras. Algebra determines geometry.

The two fundamental equations at our disposal are the Lie equation and the Killing equation:

$$A^{\mu a}\partial_\mu A^{\nu b} - A^{\mu b}\partial_\mu A^{\nu a} = f^{abc}A^{\nu c} \text{ .... Lie} \tag{1}$$

$$A^a_{\mu;\nu} + A^a_{\nu;\mu} = 0 \text{ .... Killing} \tag{2}$$

By the way, note that the ordinary derivatives in Lie's equation may be replaced by covariant derivatives. We will now rely on these two fundamental equations to calculate the curvature tensor. We could not resist remarking on the brutality of this subject, inasmuch as all results are obtained by repeatedly lying and killing.

In this subsection we follow closely the treatment given by Luciani. It is useful to define

$$h^{ab} = A^{\mu a} A^{\nu b} g_{\mu\nu} \tag{3}$$

One verifies easily that

$$h^{ab} A^{b\mu} = A^{a\mu} \qquad \text{and} \qquad h^{ab} h^{bc} = h^{ac} \tag{4}$$

Thus h is a projection and acts as the identity on the Killing vectors. It follows that if H is null so that M = G, the matrix h^{ab} is just δ^{ab}. Multiplying Lie's equation (1) (the version with covariant derivatives) by $A^{b\lambda}$ and using Killing's equation and the fact that $g_{\mu\nu;\lambda} = 0$, we find

$$A^{a}_{\mu;\nu} = f^{dbc} A^{b}_{\mu} A^{c}_{\nu} \left(\delta^{ad} - \frac{1}{2} h^{ad} \right) \quad . \tag{5}$$

This tells us that the covariant derivatives of the Killing vector may be expressed in terms of products of Killing vectors. We make contact with curvature by noting that the Riemann curvature tensor is given by the commutator of two covariant derivatives. Thus

$$A^{a\lambda}{}_{\nu;\mu} - A^{a\lambda}{}_{\mu;\nu} = A^{a\sigma} R^{\lambda}{}_{\sigma\nu\mu} \quad . \tag{6}$$

Using Eq. (5) twice we find

$$\begin{aligned} R_{\rho\mu\nu\kappa} = {} & \frac{1}{2} f^{cdb} f^{efg} \left(\delta^{gb} - \frac{1}{2} h^{gb} \right) \left(A^{e}_{\mu} A^{c}_{\nu} A^{f}_{\kappa} A^{d}_{\rho} \right. \\ & \left. - A^{e}_{\nu} A^{c}_{\mu} A^{f}_{\kappa} A^{d}_{\rho} + A^{d}_{\mu} A^{c}_{\nu} A^{e}_{\rho} A^{f}_{\kappa} \right) \\ & - \frac{1}{4} f^{bcd} f^{fge} h^{de} A^{b}_{\mu} A^{c}_{\nu} A^{f}_{\kappa} A^{g}_{\rho} \\ & - (\nu \leftrightarrow \kappa) \quad . \end{aligned} \tag{7}$$

This rather messy expression may be written much more simply by the following considerations.

Since the manifold M = G/H is a homogeneous space in the sense that every point is actually equivalent to any other, let us pick one arbitrary point and call it the "North Pole" for definiteness. H is by definition the subgroup of G which leaves the "North Pole" invariant. Corresponding to the division G/H the Lie algebra $\mathcal{g}$ (of G) may be decomposed as

$$\mathcal{g} = \mathcal{H} + \mathcal{m} \; . \tag{8}$$

This is merely notation for the statement that the generators in $\mathcal{g}$ may be divided into a set which generates $\mathcal{H}$, the Lie algebra corresponding to H. The set of generators not in $\mathcal{H}$ is denoted by $\mathcal{m}$ which is of course not a Lie algebra.

It is useful to introduce the notation that the generators in $\mathcal{g}$ carry indices a,b..., those in $\mathcal{H}$ carry indices $\alpha,\beta...$, and those in $\mathcal{m}$ carry indices $\mu,\nu...$. By the very definition of H we see that the Killing vectors $A^a_\mu = 0$ if $a \in \mathcal{H}$. On the other hand $A^a_\mu = \delta^a_\mu$ if $a \in \mathcal{m}$. We have normalized or coordinates in a natural way. Recall that $g^{\mu\nu} = A^{\mu a}A^{\nu a}$. Thus, at the "North Pole" $g_{\mu\nu} = \delta_{\mu\nu}$. This merely amounts to the fact that the geometry of a manifold is locally Euclidean. Since the Killing vectors are so simple at the "North Pole" the messy expression in Eq. (7) collapses to

$$R_{\rho\mu\nu\kappa} = \frac{1}{2} f_{\mu\rho\lambda} f_{\nu\kappa\lambda} - f_{\mu\rho b} f_{\nu\kappa b} + \frac{1}{4}\left(f_{\nu\mu\lambda} f_{\kappa\rho\lambda} - f_{\kappa\mu\lambda} f_{\nu\rho\lambda}\right) \tag{9}$$

$$R_{\mu\nu} = f_{\mu c\lambda} f_{\nu c\lambda} - \frac{3}{4} f_{\mu\lambda\rho} f_{\nu\lambda\rho} \tag{10}$$

$$R = f_{\mu c\lambda} f_{\mu c\lambda} - \frac{3}{4} f_{\mu\lambda\rho} f_{\mu\lambda\rho} \quad . \tag{11}$$

This last equation (11) is remarkable. It reveals that the scalar curvature is always positive for manifold of the form M = G/H. Together with the analysis in subsection F this result leads us to conclude that the Dirac operator on this class of manifolds does not have a vanishing eigenvalue.

The expressions given above simplify further in those cases in which the coset G/H admits what is known as an "involutive automorphism." The concept is actually quite familiar to physicists as we will now explain. Since $\mathcal{H}$ is a Lie algebra the commutator of two elements of $\mathcal{H}$ is certainly in $\mathcal{H}$, a fact which we denote schematically as

$$[\mathcal{H}, \mathcal{H}] \subset \mathcal{H}. \tag{12}$$

Since structure constants are totally antisymmetric consistency with Eq. (12) demands that

$$[\mathcal{H}, \mathcal{M}] \subset \mathcal{M}. \tag{13}$$

On the other hand, the commutator of two elements of $\mathcal{M}$ may in general lie in $\mathcal{H} + \mathcal{M}$. In the special case

$$[\mathcal{M}, \mathcal{M}] \subset \mathcal{H} \tag{14}$$

we say that G/H admits an involutive automorphism. In other words, if we multiply all the generators in $\mathcal{M}$ by minus one we leave the Lie algebraic structure invariant. Gell-Mann's current algebra, based on SU(3) x SU(3)/$SU_V(3)$, offers a well-known illustration of Eq.s (12), (13) and (14). Another example is SU(3)/SU(2) x U(1) where $\mathcal{M}$ contains the strangeness changing generators. Other examples of G/H admitting involutive automorphism are SO(N+1)/SO(N), SU(N+1)/U(N), and SU(N)/SO(N). In our notation, the admission of an involutive automorphism implies that $f_{\mu\lambda\rho} = 0$. In that case

$$f_{\mu ab} f_{\mu ab} = 2 f_{\mu\nu\alpha} f_{\mu\nu\alpha} \quad . \tag{15}$$

The scalar curvature is thus given by

$$\begin{aligned} R &= \frac{1}{2} f_{\mu ab} f_{\mu ab} \\ &= \frac{1}{2} \operatorname{tr}(if^{\mu})(if^{\mu}) \\ &= \frac{1}{2} \dim M \quad . \end{aligned} \tag{16}$$

In Eq. (16) we consider $f_{\mu ab}$ as a matrix with indices a,b and we have used a definite normalization of the structure constant. Most remarkably, the scalar curvature for coset manifolds admitting an involutive automorphism is determined completely by the dimension of the manifold.

H. Now that we have discussed both geometry and algebra we could go back to subsection D and finish the discussion of how the gauge field is introduced. Recall that through the decoupling condition we introduced an object

$$F^{\ell}_{\mu\nu} = \frac{\partial}{\partial x\mu} N^{\ell}_{\nu} - N^{j}_{\mu} \frac{\partial}{\partial xj} N^{\ell}_{\nu} - (\mu \leftrightarrow \nu) \tag{1}$$

which involves only geometrical concepts. The indices μ,ν refer to space-time while j,ℓ refer to the internal Kaluza space. We now join onto this expression the algebraic notions discussed in the preceding subsections.

Unfortunately, we are faced here with a notational disaster for which the author must apologize. The notation in subsections E,F,G is not consistent with that in subsection D. Fortunately, things are not too bad. For our limited purposes here we merely make the notational substitution for the Killing vector

$$A^{\mu a} \to \xi^{aj} \quad , \quad \mu \to a \quad , \quad a \to j$$

to bring things into line with subsection D.

To see the emergence of the Yang-Mills gauge potential and field we now assume N^j_μ has the factorized form

$$N^j_\mu \equiv B^a_\mu \xi^{aj} \tag{2}$$

where B^a_μ depends only on ordinary space-time coordinates and ξ^{aj} depends only on the internal space coordinates. The Killing vectors ξ^{aj} provide the bridge between geometry and algebra. Inserting the factorized Ansatz (2) into Eq. (1) we find

$$\begin{aligned} F^\ell_{\mu\nu} = \frac{\partial B^a_\nu}{\partial x^\mu}\xi^{a\ell} - \frac{\partial B^a_\mu}{\partial x^\nu}\xi^{a\ell} \\ - B^a_\mu B^b_\nu \xi^{aj} \frac{\partial}{\partial x^j}\xi^{b\ell} \\ + B^a_\mu B^b_\nu \xi^{bj} \frac{\partial}{\partial x^j}\xi^{a\ell} \quad . \end{aligned} \tag{3}$$

Using Lie's equation (this is where algebra comes in) we see that the last two terms in Eq. (3) combine to give $-B^a_\mu B^b_\nu f^{abc}\xi^{c\ell}$. Thus

$$F^{\ell}_{\mu\nu} = F^{c}_{\mu\nu}\xi^{c\ell} \qquad (4)$$

with $F^{c}_{\mu\nu}$ given by the Yang-Mills expression

$$\partial_{\mu}B^{c}_{\nu} - \partial_{\nu}B^{c}_{\mu} - B^{a}_{\mu}B^{b}_{\nu}f^{abc} \quad . \qquad (5)$$

Appendix VI-1

For the sake of pedagogical completeness we include here a lightning review of some elementary concepts in differential geometry. We label the points on a manifold by a set of coordinates x^μ. For example, for a sphere S^2, $x^1 = \theta$ and $x^2 = \phi$. The fact that $x^\mu = (\theta,\phi)$ does not form a linear vector space (the addition $(\theta_1,\phi_1) + (\theta_2,\phi_2)$ is not particularly useful or natural) leads us immediately to the concept of a tangent space. Infinitesimal variations in x^μ do span a linear vector space; for S^2, the space consists of vectors $\delta\theta\hat{\theta} + \delta\phi \sin\theta\hat{\phi}$ where $\hat{\theta},\hat{\phi}$ are the familiar standard unit vectors in polar coordinates. In general, the tangent space is spanned by vectors of the form

$$\delta\vec{x} = \delta x^\mu \vec{e}_\mu \quad . \tag{1}$$

The set $\{\vec{e}_\mu\}$ forms a basis of this linear vector space; μ is now a label. Thus, for S^2, $\vec{e}_1 = \hat{\theta}$, $\vec{e}_2 = \sin\theta\hat{\phi}$. Attached to each point in the manifold is a tangent space with a set of basis vectors. A general vector in this tangent space will be denoted by $\vec{A} = A^\mu \vec{e}_\mu$. Note that the index μ carried by $\vec{e}_\mu$ tells us which basis vector we are considering while carried by A^μ labels the components of the vector $\vec{A}$.

As one moves from a given point on the manifold to a point infinitesimally close by the set $\{\vec{e}_\mu\}$ changes of course. We measure the infinitesimal change in $\vec{e}_\mu$ as one moves in the νth coordinate by the derivative $\nabla_\nu \vec{e}_\mu$ which lives in the tangent space and so is expressible in terms of the set $\{\vec{e}_\mu\}$. We write

$$\nabla_\nu \vec{e}_\mu = \Gamma^\lambda_{\nu\mu} \vec{e}_\lambda \quad . \tag{2}$$

This defines the Christoffel symbol $\Gamma^{\lambda}_{\nu\mu}$.

We can now define $\nabla_\nu \vec{A}$ for an arbitrary vector by invoking the chain rule:

$$\nabla_\nu \vec{A} = \nabla_\nu (A^\mu \vec{e}_\mu) \equiv A^\mu{}_{,\nu}\, \vec{e}_\mu + A^\mu \nabla_\nu \vec{e}_\mu \quad . \tag{3}$$

Here comma denotes the ordinary derivative. The change in a vector is due to the change in its components and to the change in the basis vectors. If we define the covariant derivative, denoted as usual by a semi-colon, by the equation

$$\nabla_\nu \vec{A} = A^\mu{}_{;\nu}\, \vec{e}_\mu \tag{4}$$

we get

$$A^\mu{}_{;\nu} = A^\mu{}_{,\nu} + \Gamma^\mu_{\nu\lambda} A^\lambda \quad . \tag{5}$$

The symbol ∇_ν denotes derivatives in the direction $\vec{e}_\nu$. In general, we can consider derivatives in the direction of an arbitrary vector $\vec{B}$ by defining

$$\nabla_B \vec{A} \equiv B^\nu \nabla_\nu \vec{A} = B^\nu A^\mu{}_{;\nu}\, \vec{e}_\mu \quad . \tag{6}$$

Let $\vec{A}(x)$ and $\vec{B}(x)$ be two arbitrary vectors living in the tangent space at a given point. Denote by ε and δ two infinitesimally small numbers. We now consider moving infinitesimally on the manifold, first, by $\varepsilon\vec{A}$ (i.e., in the direction $\vec{A}$ by amount $\varepsilon|\vec{A}|$) and second, by $\delta\vec{B}$. The change in x is given by

$$x^\mu \to x^\mu + \varepsilon A^\mu(x) + \delta B^\mu(x + \varepsilon A) \approx x^\mu + \varepsilon A^\mu + \delta B^\mu + \varepsilon\delta A^\nu B^\mu{}_{,\nu} \tag{7}$$

(The notation is somewhat symbolic but nevertheless clear.) We compare this change with the change if we first move by $\delta\vec{B}$ and then by $\varepsilon\vec{A}$:

$$x^\mu \to x^\mu + \delta B^\mu + \varepsilon A^\mu + \varepsilon\delta B^\nu B^\mu{}_{\nu} \quad . \tag{8}$$

The difference interests us and leads us to define, for any two vectors $\vec{A}$ and $\vec{B}$ (living in the tangent space at the same point, of course) the commutator

$$[\vec{A},\vec{B}] \equiv (A^\nu B^\mu{}_{,\nu} - B^\nu A^\mu{}_{,\nu})\vec{e}_\mu \quad . \tag{9}$$

The notation, involving the commutator of two vectors, may appear a bit strange but it expresses in fact a very useful notion (see in particular Eq. (E3'). Notice that $[\vec{e}_\mu,\vec{e}_\nu] = 0$, a fact which, if the reader thinks about it, may be trivially understood.

So far, the concept of distance is lacking. We now introduce a metric on the manifold by specifying the scalar products between the basis vectors $\{\vec{e}_\mu\}$. We define

$$g_{\mu\nu} \equiv \vec{e}_\mu \cdot \vec{e}_\nu \quad . \tag{10}$$

We characterize the manifold by specifying the lengths of the basis vectors $\vec{e}_\mu$ and the angles between them. The infinitesimal distance between x^μ and $x^\mu + \delta x^\mu$ is given by $\delta\vec{x} \cdot \delta\vec{x}$.

While the basis $\{\vec{e}_\mu\}$ is not orthonormal in general, we could always choose, in the tangent space, an orthonormal set of basis vectors which we denote by $\{\vec{e}_\alpha\}$ with the property

$$\vec{e}_\alpha \cdot \vec{e}_\beta = \delta_{\alpha\beta} \cdots \quad . \tag{11}$$

The set $\{\vec{e}_\alpha\}$ is of course expressible in terms of the set $\{\vec{e}_\mu\}$ just as for any vector $\vec{A}$ in the tangent space. Using a notation consistent with what went before, we write

$$\vec{e}_\alpha = e_\alpha^\mu \vec{e}_\mu \quad . \tag{12}$$

The numerical coefficients e_α^μ are called the Vielbien. For S^2, the set $\{\vec{e}_\alpha\}$ would just be $\{\hat{\theta},\hat{\phi}\}$ and $e_1^1 = 1$, $e_2^2 = \sin^{-1}\theta$, $e_1^2 = e_2^1 = 0$. The metric is thus related to the Vielbein via Eqs. (10) and (11)

$$\vec{e}_\alpha \vec{e}_\beta = \delta_{\alpha\beta} = e_\alpha^\mu e_\beta^\nu g_{\mu\nu} \quad . \tag{13}$$

Inspired by the definition of the Christoffel symbol (Eq. (2)) we introduce the connection symbol $\omega_{\alpha\nu}^\beta$ by considering how the set $\{\vec{e}_\alpha\}$ varies from point to point. We expand $\nabla_\nu \vec{e}_\alpha$ by writing

$$\nabla_\nu \vec{e}_\alpha = \omega_{\alpha\nu}^\beta \vec{e}_\beta \quad . \tag{14}$$

Since $\vec{e}_\alpha$ qualifies as "any vector" we may use Eq. (4) to determine the connection as

$$\omega_{\alpha\nu}^\beta = e_{\alpha;\nu}^\lambda e_\lambda^\beta \quad . \tag{15}$$

Since a Kronecker delta is a Kronecker delta everywhere $\nabla_\lambda \delta_{\alpha\beta} = 0$. Thus, we have, applying the chain rule

$$\begin{aligned} 0 = \nabla_\lambda (\vec{e}_\alpha \cdot \vec{e}_\beta) &= \omega_{\alpha\lambda}^\gamma (\vec{e}_\gamma \cdot \vec{e}_\beta) + \omega_{\beta\lambda}^\gamma (\vec{e}_\alpha \cdot \vec{e}_\gamma) \\ &\equiv \omega_{\alpha\beta\lambda} + \omega_{\beta\alpha\lambda} \quad . \end{aligned} \tag{16}$$

We raise and lower indices μ,ν,λ, etc. with $g_{\mu\nu}$ and its inverse and indices α,β,γ, etc. with $\delta_{\alpha\beta}$. Thus, $\omega_{\alpha\beta\lambda}$ is antisymmetric in its first two indices.

Having written Eq. (16), we can also consider

$$g_{\mu\nu,\lambda} = \nabla_\lambda(\vec{e}_\mu \cdot \vec{e}_\nu)$$

$$= (\nabla_\lambda\vec{e}_\mu \cdot \vec{e}_\nu) + (\vec{e}_\mu \cdot \nabla_\lambda\vec{e}_\nu)$$

$$= \Gamma^\sigma_{\lambda\mu}g_{\sigma\nu} + \Gamma^\sigma_{\lambda\nu}g_{\sigma\mu} \quad . \tag{17}$$

The first equality follows since $\vec{e}_\mu \cdot \vec{e}_\nu = g_{\mu\nu}$ is just a scalar function, the indices $\mu\nu$ are to be considered as labels. Equation (17) may be written more compactly as

$$g_{\mu\nu;\lambda} = 0 \quad . \tag{18}$$

Taking a particular linear combination of Eq. (17) with indices permitted one finds

$$g_{\lambda\mu,\nu} + g_{\lambda\nu,\mu} - g_{\mu\nu,\lambda} = \Gamma^\sigma_{\mu\nu}g_{\sigma\lambda} + (\Gamma^\sigma_{\mu\lambda} - \Gamma^\sigma_{\lambda\mu})g_{\sigma\nu} + (\mu \leftrightarrow \nu) \quad . \tag{19}$$

In general, we have no reason to think that the Christoffel symbol $\Gamma^\lambda_{\nu\mu}$ as defined in Eq. (2) should be symmetric in its two lower indices. However, as we see from Eq. (19), if the Christoffel symbol is symmetric in its two lower indices then we can determine it entirely in terms of the metric (and its first derivatives). If we insist on considering only those manifolds which are completely fixed in their geometircal properties by their metric the Christoffel symbol will be symmetric. This case is sometimes referred to as "torsion-free." We said all this in order to remark that in case there is no torsion the ordinary derivatives in the definition of the commutator of two vectors (Eq. (9)) may be replaced by covariant derivatives.

Finally, we remind the reader of the definition of the Riemann curvature tensor:

$$R^{\tau}_{\lambda\nu\mu}\vec{e}_{\tau} \equiv [\nabla_{\mu},\nabla_{\nu}]\vec{e}_{\lambda}$$

$$= \nabla_{\mu}(\Gamma^{\sigma}_{\lambda\nu}\vec{e}_{\sigma}) - (\mu \leftrightarrow \nu)$$

$$= (\Gamma^{\tau}_{\lambda\nu,\mu} + \Gamma^{\sigma}_{\lambda\nu}\Gamma^{\tau}_{\sigma\mu})\vec{e}_{\tau} - (\mu \leftrightarrow \nu) \quad . \tag{20}$$

Similarly

$$R_{\alpha}{}^{\beta}{}_{\nu\mu}\vec{e}_{\beta} \equiv [\nabla_{\mu},\nabla_{\nu}]\vec{e}_{\alpha}$$

$$= \nabla_{\mu}(\omega^{\beta}_{\alpha\nu}\vec{e}_{\beta}) - (\mu \leftrightarrow \nu)$$

$$= \left[\omega^{\beta}_{\alpha\nu,\mu} - \omega^{\gamma}_{\alpha\nu}\omega^{\beta}_{\gamma\mu} - (\mu \leftrightarrow \nu)\right]\vec{e}_{\beta} \quad . \tag{21}$$

We can also consider the operator $[\nabla_A,\nabla_B]$ with $\vec{A},\vec{B}$ any two vectors. (The operator reduces to $[\nabla_{\mu},\nabla_{\nu}]$ in the special case $\vec{A} = \vec{e}_{\mu}$, $\vec{B} = \vec{e}_{\nu}$.) We leave it to the reader to show that the difference between $[\nabla_A,\nabla_B]\vec{e}_{\lambda}$ and $\nabla_{[A,B]}\vec{e}_{\lambda}$ is just $R^{\sigma}_{\lambda\mu\nu}A^{\mu}B^{\nu}\vec{e}_{\sigma}$.

With the formalism all set up we could go on to calculate the connections and the curvature tensors for the space described in subsection D in the text. The strategy is to exploit the basic equations $e^{a\mu} = 0 = e^{\mu}_{a}$ and $e^{\alpha}_{j} = 0 = e_{\alpha j}$ by raising and lowering indices judiciously.

We limit ourselves here to sketching the calculation of a few of the connections. We start with the general formula

$$\omega_{\hat{\alpha}\hat{\beta}\hat{\mu}} = e^{\hat{\lambda}}_{\hat{\alpha};\hat{\nu}} \, e_{\hat{\beta}\hat{\lambda}} \quad . \tag{2}$$

The notation is the same as in subsection D.

For example, to calculate $\omega_{ba}{}^{i}$ we raise one index and use $e^{\mu}_{a} = 0$,

$$\omega_{bai} = e^{\hat{\lambda}}_{a} \, e_{b\hat{\lambda};i}$$

$$= e^{k}_{a}(e_{bk;i} - e_{b\hat{\mu}}\Gamma^{\hat{\mu}}_{ki}) \quad . \tag{3}$$

Repeating the same trick we write

$$e_{b\hat{\mu}}\Gamma^{\hat{\mu}}_{ki} = e^{\hat{\mu}}_{b}\{ki,\hat{\mu}\} = e^{j}_{b}\{ki,j\} \quad . \tag{4}$$

Thus

$$\omega_{bai} = e^{k}_{a}e_{bk|i} \quad . \tag{5}$$

The vertical bar denotes covariant derivative using purely the internal space metric.

Similarly, one finds that

$$\omega_{a\beta i} = -\frac{1}{2}\, e^{j}_{a}e^{\mu}_{\beta}\left[g_{jk}N^{k}_{\mu,i} + g_{ik}N^{k}_{\mu}{}_{,j} - (\partial_{\mu} - N^{k}_{\mu}\partial_{k})g_{ij}\right] \tag{6}$$

and

$$\omega_{\alpha\beta i} = \frac{1}{2}(e^{\nu}_{\alpha}e_{\beta\nu,i} - e^{\nu}_{\beta}e_{\alpha\nu,i}) + \frac{1}{2}\, e^{\mu}_{\alpha}e^{\nu}_{\beta}g_{ij}F^{j}_{\nu\mu} \quad . \tag{7}$$

Note the appearance of the object $F^{j}_{\nu\mu}$.

For completeness we list the remaining corrections:

$$\omega_{ab\mu} = \frac{1}{2}(e^{j}_{a}e_{bj,\mu} - e^{j}_{b}e_{aj,\mu}) - \frac{1}{2}\, e^{j}_{a}e^{k}_{b}(N_{k\mu,j} - N_{j\mu,k}) \tag{8}$$

$$\omega_{\alpha b\nu} = \frac{1}{2}\, e^{i}_{b}e^{\mu}_{\alpha}(-g_{ik}F^{k}_{\mu\nu} - 2N^{j}_{\nu}K_{i\mu j} + g_{\mu\nu,i}) \quad . \tag{9}$$

Here we have defined the square bracket in Eq. (6) as $-2K_{j\mu i}$. Finally,

$$\omega_{\alpha\beta\lambda} = e^{\mu}_{\alpha}e_{\beta\mu,\lambda} + \frac{1}{2}\, e^{\mu}_{\alpha}e^{\nu}_{\beta}[N_{\lambda\ell}F^{\ell}_{\nu\mu} + N^{k}_{\mu}g_{\nu\lambda,k} - N^{k}_{\nu}g_{\mu\lambda,k}] \quad . \tag{10}$$

Appendix VI-2

Here we give a geometrical and pictorial derivation of the metric given in Eq. (D7). The notation is essentially the same as in the text. We divide the coordinates $x^{\hat{\mu}}$ into two sets (x^{μ}, x^{i}). We refer to the space described by x^{μ} as space-time and that by x^{i} as internal space. For clarity, we write $\xi^{i} \equiv x^{i}$. The letter x when unspecified will refer to x^{μ} only. The metric has in general the form

$$\hat{g}_{\hat{\mu}\hat{\nu}}(x,\xi) = \begin{pmatrix} \hat{g}_{\mu\nu} & \hat{g}_{\mu j} \\ \hat{g}_{i\nu} & \hat{g}_{ij} \end{pmatrix} . \tag{1}$$

We now assume that the space has a layer structure in the following sense. We imagine that it is constructed by "piling sheets on each other." See the accompanying figure. The sheets are labelled and distinguished by x, i.e., for each value of x^{μ}

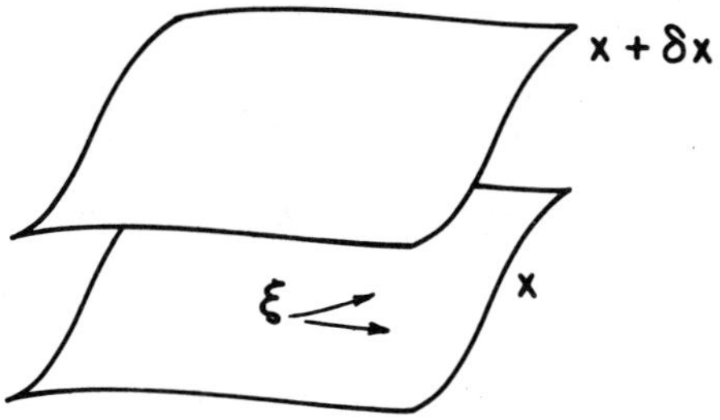

there is a sheet. Within each sheet the points are located by the coordinates ξ.

For an infinitesimal displacement characterized by $\delta\xi^i$ and lying completely within a given sheet (i.e., a displacement with $\delta x^\mu = 0$) the distance squared is

$$ds^2 = \delta\xi^i \hat{g}_{ij}(x,\xi)\delta\xi^j \quad . \tag{2}$$

Thus, the metric in internal space g_{ij} is then simply

$$g_{ij}(x,\xi) = \hat{g}_{ij}(x,\xi) \quad . \tag{3}$$

Next, we would like to consider a displacement perpendicular to a given sheet. The physical motivation for considering this is that in the Kaluza context if we want to make a translation purely in space-time we had better be sure that the displacement is "perpendicular," in an invariant sense, to the internal space. We have to be a bit careful. A displacement perpendicular to the surface in general will involve displacing in ξ as well. Roughly speaking, the ξ coordinate markings in the sheet labelled by $x + \delta x$ will not in general be "lined up" with the coordinate markings in the sheet labelled by x. We will presently determine how much ξ has to be changed by.

Let us write the desired "perpendicular" displacement as $(\delta x^\mu, \delta\xi^i)$ and impose the condition that it has to be perpendicular (as defined by the metric $\hat{g}_{\hat{\mu}\hat{\nu}}$ of course) to a displacement lying purely in the given sheet. Thus, we demand (in a self-explanatory notation)

$$(\delta x^\mu, \delta\xi^i)\hat{g}_{\hat{\mu}\hat{\nu}}(0, \delta'\xi^j) = 0. \tag{4}$$

i.e.,

$$\delta x^{\mu}\hat{g}_{\mu j} + \delta\xi^{i}\hat{g}_{ij} = 0 \quad . \tag{5}$$

This fixes the necessary displacement in ξ^{i} so as to maintain perpendicularity:

$$\delta\xi^{i} = -g^{ij}\hat{g}_{j\mu}\delta x^{\mu} \quad . \tag{6}$$

We introduce the notation g^{ij} as the inverse of $g_{ij} \equiv \hat{g}_{ij}$. It is necessary to emphasize that g^{ij} is not the i-j component of $\hat{g}^{\hat{\mu}\hat{\nu}}$.

Now that we have determined a perpendicular displacement we can compute its length to be

$$\begin{aligned}(\delta x^{\mu}\hat{g}_{\mu\nu} + \delta\xi^{i}\hat{g}_{i\nu})\delta x^{\nu} &= \delta x^{\mu}(\hat{g}_{\mu\nu} - \hat{g}_{\mu j}g^{ji}\hat{g}_{i\nu})\delta x^{\nu} \\ &\equiv \delta x^{\mu}g_{\mu\nu}\delta x^{\nu} \quad . \end{aligned} \tag{7}$$

We have used Eq. (6) to reach the second line in Eq. (7). The last line in Eq. (7) is a definition of the object $g_{\mu\nu}$. The notation is however not at all capricious. $g_{\mu\nu}$ is indeed what we mean by the space-time metric. If we make a displacement purely in space-time, i.e., perpendicular to the internal space, the length of that displacement is determined by $g_{\mu\nu}$ as per the last line in Eq. (7). This is what we mean by the space-time metric.

The metric is thus of the form

$$\hat{g}_{\hat{\mu}\hat{\nu}} = \left(\begin{array}{c|c} g_{\mu\nu} + \hat{g}_{\mu j}g^{ji}\hat{g}_{i\nu} & \hat{g}_{\mu j} \\ \hline \hat{g}_{i\nu} & g_{ij} \end{array}\right) \quad . \tag{8}$$

This is the same as Eq. (D7).

By the way, the diagram is somewhat misleading. In the picture, space-time x^μ is by necessity portrayed as only one dimensional. In that case, it is obvious we can always, starting with a given sheet, readjust the ξ coordinates on the neighboring sheets so that perpendicular displacements only have $\delta\xi = 0$. Referring to Eq. (6) we see that this means $\hat{g}_{\mu j} = 0$. A physicist would say that he knew all along that there is no gauge theory in one dimensional space-time.

Appendix VI-3

We briefly explain here the notion of a Lie derivative. Consider an infinitesimal coordinate transformation $x^\mu \to x'^\mu = x^\mu - \varepsilon^\mu(x)$. Let $T^\mu(x)$ be a vector field. Under the transformation T^μ goes to a transformed vector T'^μ. The idea of the Lie derivative is to compare T'^μ and T^μ at the same coordinate point, i.e., we want to compute $T'^\mu(x) - T^\mu(x)$ and not $T'^\mu(x') - T^\mu(x)$. It is easy to calculate the Lie derivative; keeping terms to order ε we have

$$\begin{aligned} T'^\mu(x) &= T'^\mu(x' + \varepsilon) \\ &= T'^\mu(x') + \varepsilon^\lambda \partial_\lambda T^\mu(x) \\ &= \frac{\partial x'^\mu}{\partial x^\nu} T^\nu(x) + \varepsilon^\lambda \partial_\lambda T^\mu(x) \\ &= T^\mu(x) - \partial_\nu \varepsilon^\mu T^\nu + \varepsilon^\lambda \partial_\lambda T^\mu \quad . \end{aligned} \tag{1}$$

We define

$$L_\varepsilon T^\mu(x) = \varepsilon^\lambda T^\mu{}_{,\lambda} - \varepsilon^\mu{}_{,\lambda} T^\lambda \quad . \tag{2}$$

We see from Appendix VI-1 (Eq. 9) that the Lie derivative is given in terms of the commutator of two vectors:

$$[L_\varepsilon T^\mu(x)]\vec{e}_\mu = [\vec{\varepsilon},\vec{T}] \quad . \tag{3}$$

In the absence of torsion, the ordinary derivatives in Eq. (2) may be replaced by covariant derivatives. Going through manipulations similar to those in Eq. (1) we can evaluate the Lie derivative of any tensor. For example,

$$L_\varepsilon T_{\mu\nu} \equiv T'_{\mu\nu}(x) - T_{\mu\nu}(x) = \varepsilon^\lambda \partial_\lambda T_{\mu\nu} + \partial_\mu \varepsilon^\rho T_{\rho\nu} + \partial_\nu \varepsilon^\rho T\mu\rho \quad . \tag{4}$$

The Lie derivative of the metric is particularly interesting. It is given by

$$\varepsilon^\lambda g_{\mu\nu;\lambda} + \varepsilon^\rho_{;\mu} g_{\rho\nu} + \varepsilon^\rho_{;\nu} g_{\rho\mu} \quad . \tag{5}$$

We replace the ordinary derivatives by covariant derivatives. Since $g_{\mu\nu;\lambda} = 0$ we deduce that the Killing equation

$$\varepsilon_{\mu;\nu} + \varepsilon_{\nu;\mu} = 0 \tag{6}$$

if and only if

$$g'_{\mu\nu}(x) = g_{\mu\nu}(x) \quad . \tag{7}$$

A coordinate transformation whose action on the metric is such that Eq. (7) is satisfied is known as an isometry of the given metric.

Appendix VI-4

For the sake of pedagogy we give here the Killing vectors for $CP^n = SU(n+1)/U(n)$. The explicit construction should make clear the concepts of Killing vectors and induced metric discussed in the text.

Since in CP^n the point $(z_1, z_2, \ldots z_{n+1})$ is identified with the point $\lambda(z_1, z_2, \ldots z_{n+1})$ we may choose a coordinate patch to cover the region with $z_{n+1} \neq 0$ by associating with each point the (n+1)-plet

$$\vec{z} = (z_1, z_2, \ldots z_n, 1) \quad . \tag{1}$$

The n complex numbers $z_1, \ldots z_n$ certainly provide a decent coordinate system locally. We refer to the form in Eq. (1) as the canonical representation of a point. Writing $z_k = x_k + iy_k$ we list the Killing vectors below in the coordinate system $\vec{x} = (x_1, y_1, x_2, y_2, \ldots x_n, y_n)$ labelling the 2n-dimensional manifold CP^n.

We wish to find the infinitesimal effect on $\vec{x}$ of an infinitesimal transformation in SU(n+1):

$$\vec{z} \to (1 + i\vec{\theta}\vec{\lambda})\vec{z} \quad . \tag{2}$$

We normalize the (n+1) by (n+1) traceless hermitean matrices λ by $\mathrm{tr}\lambda^2 = 2$. We consider each of the λ's in turn. First, look at diagonal λ's of the form

$$\sqrt{\frac{2}{k(k+1)}} \begin{pmatrix} 1_1 & & & & & \\ & \ddots & & & 0 & \\ & & 1_k & & & \\ & & & -k & & \\ & 0 & & & 0 & \\ & & & & & \ddots \; 0 \end{pmatrix} \tag{3}$$

The change in $\vec{x}$ divided by θ gives the Killing vector

$$A^{\mathrm{diag}\,k\mu} = \sqrt{\frac{2}{k(k+1)}}\,(-y_1, x_1, -y_2, x_2, \ldots -ky_{k+1}, \underset{\substack{\uparrow \\ k+1}}{kx_{k+1}}, 0\ldots 0) \quad . \tag{4}$$

The arrow indicates the location in the vector. Here $k = 1\ldots n-1$. The case $k = n$ requires a special treatment since the corresponding $SU(n+1)$ transformation does not leave $\vec{z}$ in the canonical form and we need to do a rescaling. We find

$$A^{\mathrm{diag}\,n\mu} = \sqrt{\frac{2(n+1)}{n}}\,(-y_1, x_1, \ldots -y_n, x_n) \quad . \tag{5}$$

The rest of the λ's may be divided into two classes: those which generalize the Pauli matrix τ_1 and those which generalize τ_2. The terminology should be obvious. We distinguish the different λ's by the location of their nonzero entries in the jth row k^{th} column and in the kth row j^{th} column. If neither j nor k is equal to n+1 we find

$$A^{\tau_1(j\leftrightarrow k)\mu} = (0,\ldots,0\,\underset{\substack{\uparrow \\ j}}{-y_k}, x_k, 0\ldots 0, \underset{\substack{\uparrow \\ k}}{-y_j}, x_j, 0,\ldots 0) \tag{6}$$

$$A^{\tau_2(j\leftrightarrow k)} = (0,\ldots,0, \underset{\substack{\uparrow \\ j}}{x_k}, y_k, 0\ldots 0, \underset{\substack{\uparrow \\ k}}{-x_j}, -y_j 0\ldots 0) \quad . \tag{7}$$

Finally, corresponding to those matrices with nonzero entries in the $(n+1)^{st}$ row and column the Killing vectors are given by

$$A^{\tau_1 k\mu} = (\ldots, \underset{\substack{\uparrow \\ k}}{2x_k y_k}, 1 - x_k^2 + y_k^2, \ldots) \quad . \tag{8}$$

The dots are to be filled in as follows. In the ℓth location ($\ell \neq k$ and $\ell = 1,\ldots,n$) put in $x_k y_\ell + x_\ell y_k$ and $-x_k x_\ell + y_k y_\ell$. The "τ_2" Killing vectors are given by

$$A^{\tau_2 k\mu} = (\ldots,\ 1 + x_k^2 - y_k^2,\ 2x_k y_k,\ \ldots) \quad . \tag{9}$$

The dots mean that in the ℓth location ($\ell \neq k$, $\ell = 1 \ldots n$) the two entries are $-x_k x_\ell + y_k y_\ell$ and $x_k y_\ell + x_\ell y_k$. Actually, it turns out that to calculate the curvature tensors we do not need to know the details of these rather messy Killing vectors in (8) and (9).

The metric is constructed by $g^{\mu\nu} = A^{\mu a} A^{\nu a}$. To calculate the Riemann tensor we might as well go to the "North Pole" defined by $x_j = y_j = 0$ all j. At that point $g^{\mu\nu} = \delta^{\mu\nu}$. In a neighborhood of the "North Pole" $g_{\mu\nu}$ may be expanded as

$$g_{\mu\nu} = \delta_{\mu\nu} + x^\lambda x^\rho \xi_{\mu\nu,\lambda\rho} + \ldots \quad . \tag{10}$$

The Riemann tensor is given in terms of the expansion coefficient by

$$R_{\lambda\mu\nu\tau} = \xi_{\mu\lambda\nu\tau} + \xi_{\nu\lambda\mu\tau} - \xi_{\mu\nu\lambda\tau} \quad . \tag{11}$$

We leave it to the reader to verify for CP^n the various statements made in the text.

Appendix VI-5

A candidate for the metric might be

$$g^{\mu\nu}(x) = A^{\mu a}(x) B^{ab}(x) A^{\nu b}(x) \quad . \tag{1}$$

In order that the vectors $\vec{A}^a$ are in fact Killing vectors, namely

$$A^{\lambda c} g^{\mu\nu}_{,\lambda} = A^{\mu c}_{,\lambda} g^{\lambda\nu} + A^{\nu c}_{,\lambda} g^{\lambda\mu} \quad , \tag{2}$$

the object B^{ab} must satisfy the following easily derivable equation

$$A^{\mu a}(A^{\lambda c} B^{ab}_{,\lambda} - f^{cae} B^{eb} + B^{ae} f^{ceb}) A^{\nu b} = 0 \quad . \tag{3}$$

The vanishing of the parenthesis just expresses the fact that the Lie derivative along the Killing vectors acting on B^{ab} twists its group indices appropriately.

To solve this equation, it is easiest to go to the "North Pole" which is in fact any point on the manifold, as explained in subsection G in the text. We now use the notation of subsection G. Referring to subsection G we see that in Eq. (3) the indices a,b are restricted to be in $\mathcal{M}$. We now choose c to be in $\mathcal{H}$. Then the index e in Eq. (3) is also restricted to be in $\mathcal{M}$. The Eq. (3) simplifies to read, in an obvious notation

$$[f^c, B] = 0 \quad ; \tag{4}$$

f^c is the adjoint representation of X^c on the coset space $\mathcal{M}$. By Schur's lemma, B must be a constant times the identity provided that the action of the subalgebra $\mathcal{H}$ on the coset space $\mathcal{M}$ in the adjoint representation is irreducible in the field of real numbers.

Let us give an example in case the preceding sentence is lacking in pedagogical clarity. For $CP^2 = SU(3)/U(2)$ the coset space is four dimensional and transforms as two complex conjugate doublets under $\mathcal{H}$ = Lie algebra of U(2). Note that the representation is irreducible only if complex conjugation is included.

Acknowledgements

The author would like to take this opportunity to thank the organizers of the Summer Institute, in particular Professors M. Konuma and Z. Maki, and everyone else, students, secretaries, physicists, for making his visit to Japan such a memorable one from both the physics and the cultural point of view.

The work represented here was supported in part by the U.S. Department of Energy under contract DE-AC06-76ER01388.

COSMOLOGICAL BARYON PRODUCTION AND RELATED TOPICS

Motohiko Yoshimura
National Laboratory for High Energy Physics (KEK)
Oho-machi, Tsukuba-gun, Ibaraki-ken, 305 Japan

in

Proceedings of the Fourth Kyoto Summer Institute on Grand Unified Theories and Related Topics, Kyoto, Japan, June 29 - July 3, 1981, ed. by M. Konuma and T. Maskawa (World Science Publishing Co., Singapore, 1981).

Contents

COSMOLOGICAL BARYON PRODUCTION AND RELATED TOPICS

Motohiko Yoshimura

National Laboratory for High Energy Physics (KEK)

Abstract

This is a review on the present status of the entitled subject, starting from an elementary introduction and ending with a brief discussion on the most recent developments. This review is divided into the following sections:

1. Introduction

We have recently witnessed a sharp increase of interactions between cosmology and particle physics, and this trend will undoubtedly continue. It is now recognized that some of the most outstanding problems in cosmology has direct relevance to particle physics at extreme high energies, in fact much higher than even the most ambitious high energy experimentalist can dream of. In this lecture I shall choose among these subjects the topic on cosmological baryon production, the subject which has been studied most extensively since 1978. I believe that the basic mechanism of baryon production is well understood by now, of course modulo usual uncertainties in particle physics. It seems that one can still learn much from the cosmological baryon excess, the point I wish to stress in the later half of this talk. We still do not know the correct grand unified theory even if we admit that the general idea of grand unification is on the right track. Instead, one can take the opposite view that one should use the cosmological baryon excess as a probe of unification beyond the standard gauge theory. After all, cosmology may be the only place to test ideas of the grand unification, hopefully except the nucleon decay. I shall try to emphasize those aspects most relevant to models of grand unification, but shall mention in the end important problems in cosmology that may be directly related to the baryon production.

The plan of these lectures is as follows. In the next section I shall state the problem very briefly and mention a fundamental difficulty of the symmetric cosmology. General ideas of grand unified theories (GUTs) are only briefly touched on, and detailed discussions are relegated to extensive reviews. In sec. 3 kinematical aspects of baryon production are clarified. The standard, delayed-decay mechanism of heavy bosons is discussed at some length, followed by alternative possibilities at lower energy scales. In sec. 4 the magnitude of the cosmological baryon excess is computed in typical GUTs. A recent, new result that relates the baryon excess to the neutrino mass is also reported. We end this lecture with a list of open problems.

2. Basic Ideas

2.1. What is the problem?

The universe as we now observe seems to be dominated[1)] by matter. This is almost certain in the local group of galaxies including our own galaxy. There are observational evidences[1)] that support this, and these come from the solar wind, the cosmic ray component, the Faraday rotation of polarized photons and γ ray spectra. Recently, observation of antiprotons in the primary cosmic ray has been reported[2)]. Most of them are expected to be produced by cosmic ray collision against interstellar matter, which is estimated [1,2)] to give a primary ratio of antiproton to proton of order 10^{-4}. Buffington et al.[2)] have just recently reported detection of antiprotons that yields a relative abundance of 2×10^{-4} at energies as low as ~ 200 MeV. The flux of this size with such a low energy is somewhat puzzling at present, and one needs a further experimental study. In this regard observation of antinuclei ($A \leq -3$) would be the most dramatic evidence for existence of antimatter in our universe because heavy antinuclei are not expected to be significantly produced as secondaries.

Another important aspect of the matter content of the present universe is amount of matter or antimatter relative to the amount of radiation. A convenient measure for this is given by the number ratio of baryons to photons[3)],

$$\frac{N_B}{N_\gamma} \simeq 10^{-10\pm 1} \qquad (2.1)$$

Essentially all photons are in the form of microwave background radiation at 3°K that gives a number density $\sim 400\ cm^{-3}$. Since 3°K corresponds to $\sim 10^{-4}$eV, Eq. (2.1) means that energetically, baryons dominate over photons by a factor $\sim 10^3$. Neutrinos are also expected as abundant as photons. Thus, if the neutrino is found as massive as 10 eV in laboratory experiments, it dominates energetically over the baryon. It is then almost certain that the neutrino plays crucial roles in the missing mass problem[4)] and galaxy formation[5)].

2.2. Symmetric vs asymmetric cosmology

Observational data suggest that clusters of galaxies are made of either matter or antimatter, but not mixtures of both. As to the global structure of the whole universe there are two schools of thought, symmetric vs asymmetric cosmology. In the symmetric cosmology[6)] the universe as a whole contains equal amounts of matter and antimatter, but regions of matter and antimatter are well separated in the present universe by a distance scale larger than that of a cluster of galaxies. The observed amount of baryons or antibaryons of order (2.1) tells that the separation of matter and antimatter must have occured when the temperature was much above several tens of MeV. I shall explain this at the end of this subsection.

The asymmetric cosmology says that the present universe is made of matter alone even globally. What does the small number ratio of (2.1) imply when an average energy of abundant particles in the early universe was much above the rest energy of nucleon ~ 1 GeV? At these temperatures the pair creation and annihilation of a nucleon and an antinucleon occur frequently so that nucleons and antinucleons exist in equilibrium abundance. Assuming that the baryon number is conserved, the present number ratio (2.1) means that

$$\left(\frac{B - \bar{B}}{B + \bar{B}}\right)_{T \gg m_N} \sim O(10^{-10}) \tag{2.2}$$

Namely, there is a small asymmetry between baryons and antibaryons: For every 10^{10} pairs of baryons and antibaryons there is an excess of just one baryon. Two scenarios have been offered to explain this asymmetry. In one scenario the asymmetry itself was present initially, but to a much more modest extent; $N_B = B - \bar{B} \approx N_\gamma$. Dissipative processes later took place and the entropy generated went to equal amounts of baryons and antibaryons and also photons because of the baryon conservation. The baryon to photon ratio was thus diluted to a large extent of order 10^{10}. In another scenario[7, 8] one uses baryon nonconserving processes in grand unified theories to dynamically generate a net baryon number. Here the fundamental parameter in cosmology, N_B/N_γ, is written in terms of fundamental parameters in particle physics. The main subject of this lecture is to discuss this GUT scenario with some depth. The troubles with the first scenario are that the initial asymmetry is not explained at all and that no plausible dissipative process has been found as yet.

In order to appreciate gravity of the problem let us estimate in the symmetric cosmology amount of nucleons or antinucleons left over after the annihilation process,

$$N + \bar{N} \rightarrow \text{many pions}, \tag{2.3}$$

took place. The reason why the annihilation is not complete is that the cosmological expansion pulls an antinucleon apart from nucleons and no reaction of the type (2.3) occurs later than a decoupling time. The temperature T_d when this decoupling happens can be estimated by comparing the annihilation rate and the expansion rate γ_e,

$$\langle\sigma v\rangle n_N \sim \gamma_e \quad , \tag{2.4}$$

where $\langle\sigma v\rangle \approx m_\pi^{-2}$, $\gamma_e \sim 16T^2/m_* \; (m_* \simeq 1.2 \times 10^{19}\text{GeV})$ and

$$n_N \simeq 2\left(\frac{m_N T_d}{2\pi}\right)^{3/2} \exp\left(-\frac{m_N}{T_d}\right) \text{ or } \frac{3\zeta(3)}{2\pi^2} \; T_d^3 \; , \tag{2.5}$$

depending on whether nucleons are nonrelativistic or relativistic at the decoupling. In this case it can be shown that the decoupling takes place in the nonrelativistic era, and one finds that,

$$\frac{m_N}{T_d} \sim \ln(m_N m_* \langle\sigma v\rangle/130), \tag{2.6}$$

$$\frac{n_N}{n_\gamma} = \frac{n_{\bar{N}}}{n_\gamma} \sim \frac{70}{m_N m_* \langle\sigma v\rangle}\left(\ln\frac{m_N m_* \langle\sigma v\rangle}{130}\right) \; . \tag{2.7}$$

By inserting $\langle\sigma v\rangle = 100 \; m_\pi^{-2}$, one gets

$$T_d \sim 20 \text{ MeV}, \; n_N/n_\gamma \sim 10^{-18}. \tag{2.8}$$

Thus, in the naive symmetric cosmology the amount of matter and antimatter left after the annihilation process is about eight orders of magnitude smaller[9)] than the observed amount. This spectacular failure implies either that the matter-antimatter separation must occur at a much higher temperature than T_d or that the symmetric cosmology is wrong.

Finally, we note that the separation of matter and antimatter can not be caused by a thermal fluctuation before the annihilation epoch. A region containing N photons and roughly the same number of nucleons and antinucleons might have an excess of nucleons of order $\sqrt{N}$ due to a statistical fluctuation. After the annihilation is completed, the region would be left with the baryon to photon ratio of order $N^{-1/2}$, which must be $\sim 10^{-10}$. Then the domain made of matter or antimatter alone would contain $N^{1/2} \sim 10^{10}$ nucleons, which is absurdly small. We shall return to the symmetric cosmology in the last section of this lecture.

2.3. Ideas of GUT

Motivated by the remarkable success of the electroweak gauge theory and the quantum chromodynamics, various grand unified theories have been proposed[10]. In short, GUT is a unified theory of all the fundamental forces except gravity by a single gauge principle. At present it is not fully designed to be a unified theory of matter fields as well. Thus, many ambiguities related to the Higgs sector and the generation structure are not resolved here. The most important feature of GUT is a new class of forces that cause baryon number nonconservation. Existence of such a force is easy to understand. A new class of gauge bosons, generically called X-bosons, transform a color into an electroweak flavor by virtue of grand unification and induce the following type of processes,

$$ql \leftrightarrow X \leftrightarrow \bar{q}\bar{q}, \qquad (2.9)$$

where q denotes a quark, and l a lepton. This immediately tells that baryon number is not conserved. Baryon number is not associated with any long range force and its conservation is not protected by the gauge principle.

What is striking is that one can estimate strength of this new interaction[11]. Since the grand gauge symmetry is broken by the Higgs mechanism, the X-boson acquires a mass m_x.

Roughly at this energy scale the three gauge couplings that characterize the standard gauge theory of SU(3) × SU(2) × U(1) must merge into a single coupling,

$$\alpha_3(m_x) \simeq \alpha_2(m_x) \simeq \alpha_1(m_x) \simeq \alpha_G(m_x). \tag{2.10}$$

Rate of change of the couplings when the energy scale is varied is described by the renormalization group equation and can be computed in the energy region, $m_W \le E \le m_x$, once the matter field content of the theory is specified. From this kind of analysis one finds[11] by using the couplings at m_W as inputs,

$$\ln\frac{m_x}{m_W} \simeq \frac{\pi}{11}\left[\frac{1}{\alpha(m_W)} - \frac{8}{3}\frac{1}{\alpha_3(m_W)}\right], \tag{2.11}$$

$$\sin^2\theta_W(m_W) \simeq \frac{3}{8}\left[1 - \frac{55}{9\pi}\alpha(m_W)\ln\frac{m_x}{m_W}\right], \tag{2.12}$$

$$\alpha_G(m_x) \simeq \alpha(m_W)\left[\sin^2\theta_W(m_W) - \frac{\alpha(m_W)}{3\pi}(11 - 2N_g)\right]^{-1}, \tag{2.13}$$

where α is the fine structure constant, and N_g the number of generations. More elaborate computations[12, 13] have been performed by Marciano, Goldman and Ross, and others. Assuming that there is no energy scale of new physics between 100 GeV and 10^{15} GeV, they obtain

$$m_x \simeq (4 - 6) \times 10^{14}\ \text{GeV}\left(\frac{\Lambda\ \overline{ms}}{400\ \text{MeV}}\right), \tag{2.14}$$

for three generations of fermions. The resulting mass turns out roughly proportional to the fundamental scale parameter of QCD. Once the unification scale m_x is fixed, the four-Fermi type coupling of $ql \leftrightarrow \bar{q}\bar{q}$ can be determined,

$$\alpha_G(m_x)/m_x^2 \sim 10^{-31}\ \text{GeV}^{-2} \sim 10^{-26}\ G_F. \tag{2.15}$$

In the simple SU(5) GUT this implies a proton lifetime of order[13]

$$\tau_p \simeq 8 \times 10^{30\pm 2}\ \text{years}. \tag{2.16}$$

This is tantalizingly close to the present experimental lower bound of $\sim 1 \times 10^{30}$ years[14]. If the nucleon decay is discovered in experiments now under way, this would vindicate without any doubt the idea of grand unified theories. It is fair to point out that the prediction of the electroweak mixing angle[11, 12], $\sin^2\theta_W \simeq 0.21$, already gives an impressive agreement with neutral current data.

For more detailed discussions on GUTs I recommend to look for extensive review articles recently published[10, 15]. Here we merely record particle assignments and gauge couplings of X-bosons in simple GUTs based on SU(5)[16] and SO(10)[17] groups:

(I) SU(5) model

$$\underline{5};\ \begin{pmatrix} d_R \\ d_G \\ d_B \\ e^+ \\ -\bar{\nu} \end{pmatrix}_R \qquad \underline{10};\ \begin{pmatrix} o & \bar{u}_B & -\bar{u}_G & u_R & d_R \\ & o & \bar{u}_R & u_G & d_G \\ & & o & u_B & d_B \\ -(\text{up}) & & & o & e^+ \\ & & & & o \end{pmatrix}_L \tag{2.17}$$

$$L_X = \frac{g}{2\sqrt{2}} [\bar{d}(X,Y)^\mu \gamma_\mu (1+\gamma_5)\binom{e^+}{-\bar{\nu}} - \overline{(-d,\ u)}\binom{X}{Y}^\mu \gamma_\mu (1-\gamma_5) e^+$$
$$+ \varepsilon_{ijk}\ \overline{u_i^c}\ (X,Y)_j^\mu\ \gamma_\mu (1-\gamma_5)\binom{u}{d}_k + (\text{h.c.})] \tag{2.18}$$

$$\left[{Y(-1/3) \atop -X(-4/3)}\right] \in (\underline{3},\ \underline{2}) \text{ of } SU(3)_C \times SU(2)_{EW}.$$

(II) SO(10) model

$$\underline{16};\ \begin{pmatrix} u_1 & u_2 & u_3 & \nu \\ d_1 & d_2 & d_3 & e \end{pmatrix}_L + \begin{pmatrix} \bar{u}_1 & \bar{u}_2 & \bar{u}_3 & \bar{N} \\ \bar{d}_1 & \bar{d}_2 & \bar{d}_3 & \bar{e} \end{pmatrix}_L \tag{2.19}$$
$$= (\underline{10})_L + (\underline{5}^*)_L + (\underline{1}^*)_L \text{ in terms of } SU(5).$$

$$L_X = \frac{g}{2}[\overline{(u,\ d)}\ \tilde{X}^\mu \gamma_\mu (1-\gamma_5)\binom{N^C}{-e^+} - \overline{(\nu,\ e)}\ \tilde{X}^\mu \gamma_\mu (1-\gamma_5)\binom{u^c}{-d^c}$$

$$- \varepsilon_{ijk} \overline{(u^c, d^c)_i} \tau_2 \tilde{X}^\mu_j \tau_2 \gamma_\mu (1 - \gamma_5) \binom{u}{d}_k + (\text{h.c.})] \tag{2.20}$$

$$\tilde{X} = \begin{bmatrix} X'_V(2/3) & X_V(-1/3) \\ X'_V(-1/3) & -X_V(-4/3) \end{bmatrix} \in (\underline{3}, \underline{2}, \underline{2}) \text{ of } SU(3)_C \times SU(2) \times SU(2)'.$$

3. Mechanism of Baryon Production

3.1. Delayed-decay mechanism

The baryon nonconservation is suppressed by the huge mass of the X-boson in the world we now know, but this may not be true when one goes back to the very earliest time of the big bang[18]. When the temperature is high enough to produce many X-bosons, baryon nonconserving processes are well operative. There are two aspects to the problem of dynamical generation of a net baryon number. One is how to wash out an initial baryon number which was presumably given from the pre-GUT epoch. The other is how to produce a net baryon number starting from a configuration which is symmetric between baryons and antibaryons. We call these two processes damping and production of baryon number. In a simple picture one hopes to damp first and to produce a net baryon number at some time later so that the final baryon number is independent of an initial condition. In a realistic situation these two epochs may overlap and one may need detailed numerical computation by taking account of both effects simultaneously. Even the production mechanism followed by a damping epoch has been proposed[19], which becomes a feasible possibility only if the damping process conserves a quantity like B-L. In this subsection I shall focus on the production mechanism, assuming that a baryon-neutral state has already been prepared by some other processes. In later sections I shall come back to the problem of damping.

The following are three conditions necessary to produce a net baryon number: (1) Baryon nonconservation, (2) C- and CP-noninvariance, (3) Departure from equilibrium. The first two conditions are necessary to give a microscopic arrow, and are usually satisfied by GUTs. We shall have more to say on this point. The third condition is also obviously needed because in a thermal and chemical equilibrium population of a particular type of particles in a given energy level is exactly equal to that of its antiparticle by the CPT theorem, and no asymmetry between particles and antiparticles develops. There is however a subtlety in this condition because one can find a situation which grossly deviates from equilibrium, and yet takes an equilibrium distribution. Take, for example, the energy spectrum of microwave background in the present universe. It is of a black-body form despite that these photons went out of equilibrium tens of billion years ago. The massless particles seem to keep its equilibrium form having a temperature red-shifted according to $T \propto 1/R$ once they were previously in equilibrium. If this is the case, it is difficult[20)] to generate a net baryon number in processes involving light particles alone. Here, light means that its mass is much less than the mass of X-boson m_x.

A basic framework[21)] for computing the baryon asymmetry and also for understanding the special feature of massless particles just mentioned is to use the Boltzmann equation in a non-equilibrium situation. Rate of change of the distribution function $f_i(\vec{p}, t)$ in a homogeneous plasma is given by dynamical processes (decay, inverse decay, scattering, reaction etc.) in this medium,

$$L(f_i) = \Gamma_i[f]. \qquad (3.1)$$

In the left side one includes the effect of cosmological expansion and for the isotropic expansion

$$L = p^\alpha \frac{\partial}{\partial x^\alpha} - \Gamma^\alpha_{\beta\gamma}\, p^\beta p^\gamma \frac{\partial}{\partial p^\alpha}$$

$$= E\frac{\partial}{\partial t} - \frac{\dot{R}}{R}\, \vec{p}^{\,2} \frac{\partial}{\partial E}, \qquad (3.2)$$

with R(t) the Robertson-Walker scale factor. An example of the right side when only two-body reactions are considered is given by

$$\Gamma_i[f] = \frac{1}{2} \sum_{jkl} \int dP_j dP_k dP_l\, [(1 \pm f_i)(1 \pm f_j) f_k f_l \Gamma_{kl \to ij} - (1 \pm f_k)(1 \pm f_l) f_i f_j \Gamma_{ij \to kl}], \qquad (3.3)$$

where + (-) should be taken for bosons (fermions), and Γ is the invariant rate essentially given by the cross section times the relative velocity σv. It can be shown [21, 22] by using the unitarity, but without assuming the time reversal invariance that the entropy in a comoving volume ($\propto R^3$) never decreases and the state of maximal entropy is characterized by the familiar distribution

$$f_i = [\exp(E_i/kT) \mp 1]^{-1},$$

when there exists no finite conserved quantity.

The equations for the distribution function can be integrated over the momentum space to obtain equations for number densities, $n_i(t)$,

$$\left(\frac{d}{dt} + \gamma_e\right) n_i = \int \frac{d^3p_i}{(2\pi)^3} \frac{\Gamma_i[f]}{E_i}, \qquad (3.4)$$

with $\gamma_e = 3\dot{R}/R$ the expansion rate. In an average sense the right hand side of (3.4) can be split into an average production (or destruction in some case) rate times the number density,

$$\left(\frac{d}{dt} + \gamma_e\right) n_i = \langle\Gamma\rangle_i n_i \quad \text{(no sum)} \qquad (3.5)$$

The expansion rate when the majority of particles is relativistic and is in equilibrium with a temperature T is given by

$$\gamma_e = 3\sqrt{\frac{8\pi}{3} G\rho} = (\frac{8\pi^3}{5})^{1/2} \sqrt{N} \frac{T^2}{m_*} , \tag{3.6}$$

$$m_* = \frac{1}{\sqrt{G}} \simeq 1.2 \times 10^{19} \text{ GeV},$$

$$N = \sum_{boson} \frac{1}{2} + \sum_{fermion} \frac{7}{16}.$$

N counts the relativistic degree of freedom at T, and for example,

$$N = \frac{43}{8} \text{ for } \gamma,\ e^{\pm},\ \nu_e,\ \nu_\mu,\ \nu_\tau, \tag{3.7}$$

$$N = 80.4 \text{ for 3-family SU(5) particles.} \tag{3.7'}$$

Both of T and N are dependent on the time, and in particular, when N is almost independent of time,

$$t \simeq [\sqrt{\frac{32\pi^3}{45}} \sqrt{N} \frac{T^2}{m_*}]^{-1} = \frac{1.7 \times 10^{-6} \text{ sec}}{\sqrt{N}\ T^2/\ \text{GeV}^2} = \frac{3}{2} \gamma_e^{-1}. \tag{3.8}$$

Typically, at higher temperatures the particle distribution takes the equilibrium form if reactions occur frequently compared to the expansion time scale, namely $\gamma_e \lesssim \langle\Gamma\rangle$. As the universe expands, momentum of abundant particles becomes red-shifted, $|\vec{p}| \propto 1/R$, and reactions occur less frequently. At some average temperature T_d the expansion takes over the reaction and this reaction is said to decouple. After this decoupling the particular reaction in question effectively ceases, and particles involved in this reaction expand freely unless they spontaneously decay or take part in other reactions. The equation for the particle distribution after interactions are switched off is given by,

$$(E\frac{\partial}{\partial t} - \frac{\dot{R}}{R} \vec{p}^2 \frac{\partial}{\partial E}) f(\vec{p}, t) = 0. \tag{3.9}$$

This equation has solution of a general form,

$$f(|\vec{p}|R),$$

with an arbitrary function f. The important point here is that the equilibrium distribution for massless particles,

$$f = [\exp(\frac{|\vec{p}|}{kT_o} \frac{R(t)}{R(t_o)}) \mp 1]^{-1} , \tag{3.10}$$

has the above form. Hence, the massless particles keep its

quasi-equilibrium form with a temperature,

$$T = T_o \frac{R(t_o)}{R(t)},$$

almost at all times once these particles were in equilibrium. On the other hand, massive particles can deviate from equilibrium because

$$\sqrt{m^2 + \vec{p}^2 [\frac{R(t)}{R(t_o)}]^2} \neq E.$$

As mentioned above, this distinction between massless and massive particles necessitates[20)] baryon nonconserving processes having a threshold around the energy scale of baryon nonconservation in order to generate a net baryon number.

The simplest and the most promising candidates[20), 23)] of processes with a threshold are decay and inverse decay of heavy particles, for example those of X-bosons;

$$X \leftrightarrow ql, \overline{q}\overline{q}, \overline{X} \leftrightarrow \overline{q}\overline{l}, qq. \qquad (3.11)$$

A crucial question is at what temperatures these X-bosons decouple and when the decay of X-bosons occurs frequently. If these X-bosons decouple after they become nonrelativistic, their number after the decoupling is presumably too small by the Boltzmann factor, $\exp(-\frac{m_X}{T_d})$. Since the final amount of baryons depends on this number of X-bosons, this is not a kind of situation one would like to get. Instead, if they decouple while they are still relativistic, their frozen number is essentially comparable to the number of photons present at the decoupling. For this to happen, the decay lifetime must catch up the cosmological time after X-bosons become nonrelativistic, so that the inverse decay is blocked energetically. Otherwise, both the decay and the inverse decay would occur frequently when they are still relativistic, and their number would decrease according to the equilibrium abundance. Qualitatively, this condition is given by

$$T_D \lesssim m_X \text{ with } \alpha m_X \sim \frac{T_D^{\,2}}{m_*}, \tag{3.12}$$

with α the coupling strength for the X-decay. Here, T_D is a temperature below which the decay proceeds faster than the expansion. From Eq. (3.12) one gets a crude lower bound for the mass of X-boson,

$$m_X \gtrsim O(\alpha\, m_*). \tag{3.13}$$

Thus, the X-boson can not be arbitrarily light in order to produce a net baryon number[23].

A more careful way[24] to compute the lower bound is to impose the condition that the inverse decay rate never exceeds the expansion rate, namely, the inverse decay does not take place substantially at any time;

$$\gamma_{id} \lesssim \gamma_e . \tag{3.14}$$

The effective production rate of a X-boson via the inverse decay is given by

$$\gamma_{id} = \frac{1}{3} g^2 N_d m_X^2 \frac{1}{(2\pi)^2} \iiint \frac{d^3p\, d^3q_1\, d^3q_2}{2E_p \cdot 2|q_1| \cdot 2|q_2|} \delta^{(4)}(p - q_1 - q_2)$$
$$\times [\exp(\frac{|q_1|}{T}) + 1]^{-1} [\exp(\frac{|q_2|}{T}) + 1]^{-1}$$
$$\times [1 - \exp(-\frac{E_p}{T})]^{-1} [8\pi\zeta(3)T^3]^{-1}, \tag{3.15}$$

with N_d the number of X-decay channels. Result of an explicit computation may be summarized as follows,

$$\frac{\gamma_{id}}{\gamma_e} = K\, F(\frac{T}{m_X}), \tag{3.16}$$

where K is ratio of the total decay rate to the expansion rate at the temperature m_X

$$K = \frac{\alpha_G N_{d}}{12} \left(\frac{5}{8\pi^3}\right)^{1/2} \frac{m_*}{\sqrt{N}\, m_X}, \tag{3.17}$$

and

$$F(x) = \frac{1}{2\zeta(3)} \int_0^\infty du \int_0^\infty dv\, \theta(uv - \frac{1}{4x^2})\, (e^u + 1)^{-1} (e^v + 1)^{-1} \times (1 - e^{-u-v})^{-1} \tag{3.18}$$

This function has a maximum of ~ 1.1 at $x \sim 0.3$ as plotted below. The condition of no inverse decay (3.14) then implies that

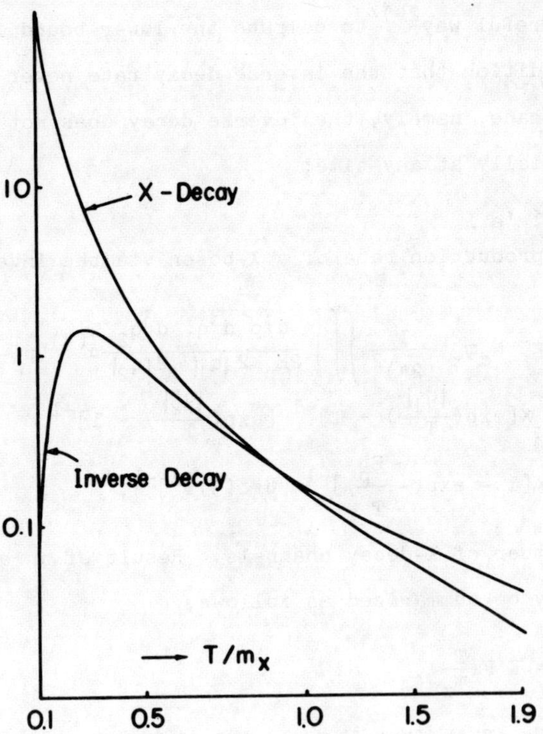

Fig. 1

$$K \lesssim (1.1)^{-1}$$

Namely,

$$m_X \gtrsim 0.013\, \alpha_G (N_d/\sqrt{N})\, m_* . \tag{3.19}$$

For the simplest SU(5) model with three generations, $N_d = 12$, $\alpha_G \simeq 0.024$, $N \simeq 80$ so that[24)]

$$m_X \gtrsim 5 \times 10^{15} \text{ GeV}. \tag{3.20}$$

This lower bound is about one order of magnitude larger than the mass of X-boson estimated from evolution of the low energy gauge couplings, namely Eq. (2.14). This argument indicates that the gauge boson is unlikely to produce a net baryon number.

At the same level of semiquantitative discussion I shall explain how the present baryon to photon ratio, N_B/N_γ, is related[24)] to the baryon asymmetry produced by a pair decay of X and $\bar{X}$. Assuming that the inverse decays can be ignored, one sets up an equation for the baryon number density n_B,

$$\dot{n}_B + \gamma_e\, n_B = \varepsilon\, \gamma_t\, n_X, \tag{3.21}$$

with ε the baryon number produced by the pair decay and γ_t the total rate of X. For instance, main decay modes of the X-boson are

$$X \to ql \text{ with rate } \gamma_l,\ \bar{q}\bar{q} \text{ with } \gamma_q, \tag{3.22}$$

$$\bar{X} \to \bar{q}\bar{l} \text{ with } \bar{\gamma}_l,\ qq \text{ with } \bar{\gamma}_q. \tag{3.23}$$

In this case the net baryon number produced when a pair of X and $\bar{X}$ decay independently is

$$\varepsilon = \frac{\frac{1}{3}\gamma_l - \frac{2}{3}\gamma_q}{\gamma_l + \gamma_q} + \frac{-\frac{1}{3}\bar{\gamma}_l + \frac{2}{3}\bar{\gamma}_q}{\bar{\gamma}_l + \bar{\gamma}_q} = \frac{\bar{\gamma}_q - \gamma_q}{\gamma_l + \gamma_q} \tag{3.24}$$

because of the CPT theorem, $\gamma_l + \gamma_q = \bar{\gamma}_l + \bar{\gamma}_q$. This expression clearly shows that the asymmetry ε does not vanish in general if both C and CP are violated. The number density of X and also $\bar{X}$

in Eq. (3.21) is governed by
$$\dot{n}_x + \gamma_e n_x = -\gamma_t n_x. \qquad (3.25)$$
These equations can be simplified by using ratios to the photon number density,
$$r_i = n_i/n_\gamma \propto n_i T^{-3} \propto n_i R^3, \qquad (3.26)$$
which is proportional to the number of i in the comoving volume. Then, since
$$\dot{r}_i \propto R^3 (\dot{n}_i + 3\frac{\dot{R}}{R} n_i) = R^3 (\dot{n}_i + \gamma_e n_i),$$
one gets from Eqs. (3.21) and (3.25)
$$\dot{r}_B = -\varepsilon \dot{r}_x \qquad (3.27)$$
To integrate this differential equation we assume for simplicity that the net baryon number is zero at an intial time t_i and that X-bosons completely disappear after a final time t_f. Then,
$$(r_B)_f = \varepsilon (r_x)_i,$$
namely
$$(\frac{n_B}{n_\gamma})_f = \varepsilon (\frac{n_x}{n_\gamma})_i. \qquad (3.28)$$

Actually, the ratio to the photon number is not a good quantity because at much later stages of cosmological evolution many particles annihilate to heat up the photon temperature, for example via $e^+ + e^- \to \gamma + \gamma$. A better ratio is n_B divided by the entropy density s/k, which is conserved if the expansion is adiabatic, thus,
$$(\frac{k n_B}{s})_{present} = \varepsilon (\frac{k n_x}{s})_i \qquad (3.29)$$
The entropy density s is given by
$$s = \frac{4\pi^2}{45} T^3 N, \qquad (3.30)$$
with N the same effective degree of freedom of massless particles as in Eq. (3.7). In the present universe $s \simeq (4\pi^2/45) T_\gamma^3 (1.95)$,

so that

$$n \simeq \frac{s}{7} \text{ at present.}$$

Thus, the present baryon to photon ratio is given by

$$\left(\frac{N_B}{N_\gamma}\right)_{present} \simeq 3.7 \times 10^{-2} \ \varepsilon \left(\frac{n_x}{n_x^o}\right)_i , \qquad (3.31)$$

where we assumed the initial value of N for the SU(5) model, Eq. (3.7)'. The last factor $(n_x/n_x^o)_i$ represents a dilution effect at the onset of X-decay. For this estimate one needs a more detailed analysis[24], but it is of order 0.1 for the $m_x \simeq 5 \times 10^{15}$ GeV of Eq. (3.20). Thus, our analysis shows that

$$\left(\frac{N_B}{N_\gamma}\right)_{present} \simeq 4 \times 10^{-3} \cdot \varepsilon. \qquad (3.32)$$

Fry, Olive and Turner[25] made a more extensive numerical analysis. They integrated a truncated version of the Boltzmann equations by taking into account of decay, inverse decay and scatterings mediated by gauge X-bosons. Their computation is very complicated, but they express the final baryon to entropy ratio by an approximate numerical formula,

$$\frac{kn_B}{s} \simeq \frac{1.5 \times 10^{-2}\varepsilon}{1 + (24\ K)^{1.3}} , \quad K = 1.4 \times 10^{17} \text{ GeV } \frac{\alpha_G N_d}{\sqrt{N} m_x} . \qquad (3.33)$$

This tells that for a small enough value of m_x there is a suppression of the produced baryon asymmetry due to a partial thermalization. This suppression factor is

$$\left(\frac{m_x}{1.1 \times 10^{17} \text{ GeV}}\right)^{1.3} ,$$

for the canonical SU(5) model with three generations. With $m_x = 5 \times 10^{15}$ GeV the result of Fry et al. gives

$$\frac{N_B}{N_\gamma} \simeq 2 \times 10^{-3} \cdot \varepsilon ,$$

which is roughly consistent with our simple analysis, Eq. (3.32).

For the popular SU(5) value of $m_x \sim 6 \times 10^{14}$ GeV Eq. (3.33) gives

$$\frac{kn_B}{s} \simeq 1.7 \times 10^{-5} \cdot \varepsilon \left(\frac{m_x}{6 \times 10^{14}\ \text{GeV}}\right)^{1.3} . \qquad (3.34)$$

As will be explained in the next section, the asymmetry ε is too small in this simple SU(5) model to be compatible with the observed size of the baryon excess, Eq. (2.1). Thus, the conclusion of this long subsection is that the decay of the SU(5) gauge boson X_V is not a viable mechansim of baryon production.

3.2. A possible scenario

An alternative mechanism of baryon production is to use the Higgs boson that causes the baryon nonconservation. For example, the color triplet Higgs boson $X_s(-1/3)$ of charge -1/3 in the SU(5) model has the following Yukawa coupling,

$$L = f\, [-\overline{(u,\, d)}_k \tfrac{1}{2}(1+\gamma_5) \binom{e^+}{-\bar{\nu}} + \varepsilon_{ijk}\, \overline{u^c_i}\, \tfrac{1}{2}(1+\gamma_5)\, d_j]\, X^k_s + h\, [-\bar{u}_k \tfrac{1}{2}(1-\gamma_5)\, e^+ + \varepsilon_{ijk}\, \overline{u^c_i}\, \tfrac{1}{2}(1-\gamma_5)\, d_j]\, X^k_s + (\text{h.c.}), \qquad (3.35)$$

with f and h coupling constants. By a similar numerical analysis as in the gauge boson case Fry et al.[26] obtain a net baryon number given by

$$\frac{kn_B}{s} \simeq \frac{0.5 \times 10^{-2}\ \varepsilon_H}{1 + (4.5\ K_H)^{1.2}}, \qquad (3.36)$$

for the Higgs boson of SU(5) model. Here ε_H is a baryon number produced by the pair decay of X_s and $\bar{X}_s$, and K_H is the ratio of the total decay to the expansion rate at $T = m(X_s)$,

$$K_H \simeq 1.9 \times 10^{17}\ \text{GeV}\ \frac{\alpha_H}{m_x}\sqrt{\frac{80}{N}} . \qquad (3.37)$$

Unlike the gauge boson the Higgs mass m_x is not well determined and the decay constant α_H tends to be smaller than the gauge coupling α_G by a factor like $(m_t/m_W)^2$. It is thus possible to have a small value for K_H so that the final asymmetry is close to the naive estimate of $\sim 0.5 \times 10^{-2}\ \varepsilon_H$.

Moreover, this opens up a new possibility[27] of combining the gauge and Higgs decays to eliminate the problem of initial condition. Although the gauge boson X_V is not effective in produc ing the asymmetry, it may be very efficient to wipe out an initial baryon number. Damping of the baryon number is a first-order process unlike the production which needs effect of CP-violation. What one needs for the damping is that baryon nonconservation occurs strongly enough compared to the expansion rate. As we discussed in the last subsection this appears to hold for decay and inverse decay of the SU(5) gauge boson. Indeed, a more extensive[25] study indicates that the damping factor for the gauge boson X_V is given by

$$\exp(-8.6\, K_g)$$

with K_g given by Eq. (3.17). For the canonical mass and coupling value of SU(5) model this is $\sim 10^{-28}$ so that the damping is practically complete. If the Higgs mass is smaller than the canonical mass $(4 - 6) \times 10^{14}$ GeV of the gauge boson, the baryon asymmetry via the Higgs decay will be developed according to Eqs. (3.36) and (3.37). This two step process[27], damping by decay and inverse decay of the gauge boson followed by production due to the Higgs decay, is called hierarchy scheme. I shall discuss the magnitude of the elementary baryon asymmetry ε_H due to the Higgs decay in the next section. The hierarchy scheme is the most promising mechanism of baryon production.

In more complicated GUTs there are many gauge bosons and many Higgs bosons that give rise to baryon nonconservation. The SU(3) × SU(2) × U(1) quantum numbers of these bosons are completely classified[28];

$$\text{gauge bosons:} \quad X_V\ (-1/3,\ -4/3),\ X_V'\ (2/3,\ -1/3), \tag{3.38}$$

$$\text{Higgs bosons:} \quad X_S\ (-1/3),\ X_S'\ (-4/3),\ X_S''\ (2/3,\ -1/3,\ -4/3), \tag{3.39}$$

where all of these bosons are color triplets and the numbers in the parentheses indicate charges in the electroweak SU(2) multiplets. With many of these bosons the time development of baryon asymmetry would be very complicated. In general, what is most important to the final baryon asymmetry is a process that occurs at the lowest mass scale, hence at the latest time. Naively, one might expect that the last process should not damp a baryon number previously produced. But some damping processes may conserve a linear combination of B and some other quantities like L, say , B + aL. In this case the process does not actually wipe out the baryon number given from a previous epoch , but it rather redistributes[29] B and L. By writing the chemical potential associated with the conserved B + aL as μ , one can relate[30] equilibrium number densities, n_B and n_L, to those of initial densities, n_B^o and n_L^o, by

$$\frac{n_B}{n_L} = \frac{\sum_i B_i \; e^{\mu(B_i + aL_i)}}{\sum_i L_i \; e^{\mu(B_i + aL_i)}} \simeq \frac{\sum_i B_i^2}{\sum_i aL_i^2} ,$$

$$n_B + an_L = (n_B^o + an_L^o)\left(\frac{T}{T^o}\right)^3 , \qquad (3.40)$$

to $O(\mu)$. The sum over the fermion species i gives

$$\frac{n_B}{n_L} = \frac{3 \cdot 2 \cdot 2 \cdot (1/3)^2}{(2+1) \cdot 1^2} = \frac{1}{a} \cdot \frac{4}{9} , \qquad (3.41)$$

for the familiar set of fermions. Thus, together with Eq. (3.40)

$$n_B = (1 + 9/4 . a^2)^{-1} \; (n_B^o + a\, n_L^o) \; \left(\frac{T}{T^o}\right)^3 . \qquad (3.42)$$

Namely, if there exists a nonthermalizing mode that contains the baryon number, one should not expect the baryon damping. Existence of such a nonthermalizing mode would complicate the scenario of baryon production with many mass scales[31-33].

3.3. Baryon production at lower temperatures

The lower bound (3.13) for the temperature scale of baryon production was derived by comparing decay rate with the expansion rate. Since the decay rate depends on the mass of the parent particle, what would happen if one uses a decay vertex different from that of X-boson? This is a question that one has to face when considering a heavy fermion as a candidate of baryon production[34-36]. For example, take a heavy fermion F decaying into modes with different baryon numbers,

$$F \to qqq \text{ and } \bar{q}\bar{q}\bar{q}. \tag{3.43}$$

A difference in rates of the two modes results in a net baryon number, which we denote by ε_F. The point here is that the decay vertex for (3.43) is of the four-Fermi type and nonrenormalizable unlike that of $X \to ql$ and $\bar{q}\bar{q}$. Hence the delayed-decay condition similar to Eq. (3.12) leads to

$$T_D \lesssim m_F \text{ with } \tilde{G}^2 m_F^5 \sim \frac{T_D^2}{m_*}$$

with $\tilde{G}$ the four-Fermi coupling. This gives

$$m_F \lesssim (\tilde{G}^{-2}/m_*)^{1/3} \sim (\alpha^{-2} m_X^4/m_*)^{1/3}, \tag{3.44}$$

since in gauge theories $\tilde{G} \sim \alpha/m_X^2$ with m_X mass of either a gauge or Higgs boson. The important point[34] is that one gets an upper bound for the F mass due to the dimensional coupling $\tilde{G}$. Thus, one is relieved from the tight lower bound and one may hope to produce the baryon asymmetry at a much lower energy scale. How low one can push down this energy scale are now being studied in a general context[37], and will be discussed in the next section for the SO(10) model. This question is most relevant to the partial unification scheme[38] in which baryon nonconservation with $\Delta B = 2$ occurs at an energy scale between 100 GeV and 10^{15} GeV. As will be shown in the next section it seems difficult to accomodate both the intermediate baryon nonconservation and the cosmological baryon production in the left-right symmetric GUT.

4. Magnitude of Baryon Asymmetry

4.1 General remarks

The asymmetry parameter ε that determines the final amount of baryon to photon ratio, N_B/N_γ, should ultimately be expressed in terms of particle physics parameters. It is a parameter that reflects an intricate interplay of baryon nonconservation and CP violation as will be made clear. To eliminate unimportant complications that exist in any realistic theory, let us take again the illustrative example of two-body decays of X-boson as given by Eq. (3.22) $\sim$ (3.24). The asymmetry ε is given by a difference between partial rates of X and $\bar{X}$;

$$\varepsilon = \frac{\bar{\gamma}_q - \gamma_q}{\gamma_l + \gamma_q} . \tag{4.1}$$

Suppose for simplicity that these partial rates consist of two amplitudes,

$$\gamma_q = |g_1 f_1 + g_2 f_2|^2 , \tag{4.2}$$

$$\bar{\gamma}_q = |g_1^* f_1 + g_2^* f_2|^2 . \tag{4.3}$$

Here we decompose the individual amplitudes into products of a coupling factor g_i and a genuine, dynamical part f_i. Note that only coupling factors get complex conjugated when one goes to the antiparticle. The difference between the two rates is relevant to the asymmetry (4.1) and

$$\gamma_q - \bar{\gamma}_q = -4 \text{ Im } (g_1 g_2^*) \times \text{Im } (f_1 f_2^*). \tag{4.4}$$

Besides complex coupling factors provided by CP violation, one needs different dynamical phases to get a nonvanishing asymmetry ε. This immediately tells that a summation of an arbitrary number of Born terms can never produce a finite ε.

An example of such a dynamical phase is a rescattering phase due to some sort of final state interaction. Necessity of a dynamical phase implies that an extra coupling factor of higher order effects is necessary for the difference (4.4) compared to the rate itself. Thus,

$$|\varepsilon| \lesssim O(\alpha). \qquad (4.5)$$

This coupling factor α usually involves a Higgs boson because CP violation is needed. Typically,

$$\alpha \sim \alpha_H \sim \alpha_G \left(\frac{m_t}{m_W}\right)^2,$$

where m_t is the largest quark mass. It is important to realize that there is a natural mechanism of suppressing the size of the baryon asymmetry. At which order of the coupling factors a finite asymmetry occurs depends on details of a specific model, most importantly how CP violation is incorporated in GUT. As will be shown in the rest of this section, the cosmological baryon asymmetry is a sensitive probe into the structure of unification beyond the standard gauge theory.

I shall now compute the baryon asymmetry by taking an example shown in Fig. 2. This is an interference between a Born term and a one-loop diagram

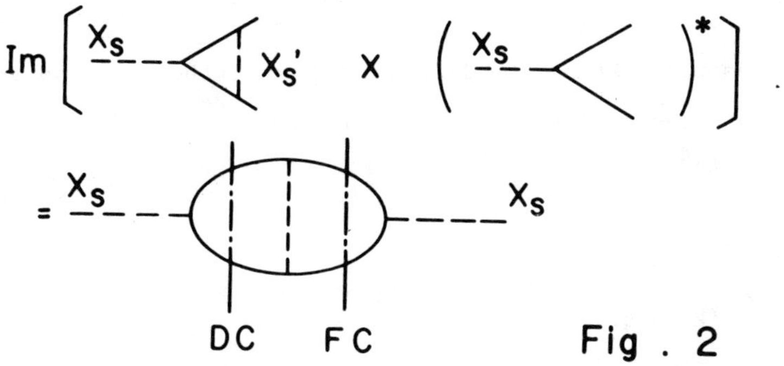

Fig . 2

for a process $X_s(-1/3) \to \bar{u}_L \bar{d}_L$. It is a typical example of computations that one encounter in the asymmetry due to the Higgs decay, and we shall later use this result when we discuss in left-right symmetric GUT a relation between the neutrino mass and the baryon excess. Since the purpose of this calculation is to illustrate a general method of computation rather than to discuss a particular result, I shall focus on the imaginary part of dynamical amplitudes and ignore the coupling factors. Since Born terms can be made real by a judicious choice of phase factors,

$$\mathrm{Im}\,(F_L F_B^*) = F_B \,\mathrm{Im}\, F_L ,$$

where $F_B(f_L)$ is a Born (one-loop) diagram. An imaginary part of Feynman diagrams can always be obtained after a direct integration, but a more efficient way to get the imaginary part, or rather the discontinuity, 2i Im, is to use the Landau-Cutkosky rule[39]. Namely, one cuts a loop diagram as illustrated by the line DC in Fig. 2 between an initial state and a final state (denoted by FC in Fig. 2), and replaces propagators cut this way by on-shell factors,

$$-2\pi i\, \delta(k^2 - m^2)\, \theta(k_o) \equiv -2\pi i\, \delta^{(+)}(k^2 - m^2) \text{ for a spinless particle,}$$

$$-2\pi i\, \delta^{(+)}(k^2 - m^2)(k \cdot \gamma + m) \text{ for a spinor particle.} \tag{4.6}$$

In the example at hand the imaginary part of the one-loop integral is given by (we take all fermions massless)

$$-\frac{1}{2}\int \frac{d^4k_1\, d^4k_2}{(2\pi)^4}\, \delta^{(4)}(q - k_1 - k_2)(-2\pi i)^2\, \delta^{(+)}(k_1^2)\, \delta^{(+)}(k_2^2)$$

$$\times \frac{1}{(k_1 - q_1)^2 - m'^2} \cdot \overline{v^c(q_2)}\, k_2 \cdot \gamma\, k_1 \cdot \gamma\, \tfrac{1}{2}(1 + \gamma_5)\, v(q_1), \tag{4.7}$$

where q and q_1 are 4-momenta of the initial Higgs and final fermions and m' is the mass of the exchanged Higgs boson.

We are finally interested in the rates so that we may take the decaying Higgs boson at rest;

$q = (m, \vec{o})$, $k_1 = \frac{m}{2}(1, \pm\hat{k})$, $q_1 = \frac{m}{2}(1, \pm e_z)$, $k_1 \cdot q_1 = \frac{m^2}{4}(1 - \hat{K}_z)$,

where the momentum of a final fermion is along the z-axis. Thus, Eq. (4.7) is reduced to (note $k_2 = q - k_1$, $k_1^2 = 0$)

$$\overline{v^c(q_2)}\ F'_{L\mu\nu}\ \gamma^\mu \gamma^\nu \frac{1}{2}(1 + \gamma_5)\ v(q_1), \tag{4.8}$$

$$F'_{L\mu\nu} = -(2\pi)^{-2}\frac{1}{8}\int d\Omega_{\hat{k}} \frac{(1, \vec{0})_\mu\ (1, \hat{k})_\nu}{1 - \hat{k}_z + \frac{2m'^2}{m^2}} .$$

By an explicit integration one finds that

$$F'_{Loo} = -\frac{1}{16\pi} f_1(\frac{m'^2}{m^2}),\quad F'_{Loz} = -\frac{1}{16\pi} f_2(\frac{m'^2}{m^2}), \tag{4.9}$$

$$\text{other } F'_{L\mu\nu} = 0,$$

$$f_1(x) = \int_{-1}^{1} dz \frac{1}{1 + 2x - z} = \ln(\frac{1 + x}{x}),$$

$$f_2(x) = \int_{-1}^{1} dz \frac{z}{1 + 2x - z} = -2 + (1 + 2x)\ln(\frac{1 + x}{x}).$$

To get the difference of rates one multiplies Eq. (4.8) by the Born term $-\bar{v}(q_1)\ 1/2(1 - \gamma_5)\ v^c(q_2)$, and sums over final momenta. Thus,

$$\gamma - \bar{\gamma} = -(2^6\pi^2)^{-1}\ m\ f\ (\frac{m'^2}{m^2})\ \text{Im}(g_L g_B^*), \tag{4.10}$$

$$f(x) \equiv f_1(x) - f_2(x) = 2 - 2x\ln(\frac{1 + x}{x}). \tag{4.11}$$

We have put coupling factors according to (4.4). On the other hand, the total rate is given by $m/16\pi$ without coupling factors, and the asymmetry ε is found as follows;

$$\varepsilon = (\gamma - \bar{\gamma})/\gamma = -\frac{1}{4\pi} f\ (\frac{m'^2}{m^2}) \frac{\text{Im}\ (g_L g_B^*)}{|g_B|^2} \tag{4.12}$$

where f is a function of the mass ratio of the exchanged Higgs to the parent Higgs boson in Fig. 2, and is explicitly given by

Eq. (4.11). Similarly, when one of the intermediate fermions is massive, the asymmetry is computed to give

$$= - \frac{1}{2\pi} F\left(\frac{m'^2}{m^2}, \frac{\mu^2}{m^2}\right) \frac{\mathrm{Im}\,(g_L g_B^*)}{|g_B|^2} , \qquad (4.13)$$

$$F(x, y) = [1 - y - x \ln(\frac{1 + x - y}{x})] \; \theta(1 - y). \qquad (4.14)$$

The step function tells that the fermion mass $\mu < m$ in order to get a finite discontinuity; the decay process to intermediate states must be energetically possible as a real process. This is a general feature common in all cases, and enables one to interprete the extra dynamical phase due to some sort of final state interaction. As seen from this example, computation of the discontinuity is much easier than that of the amplitude itself, and in lower orders it does not involve renormalization.

4.2. SU(5) model

Fermions in the SU(5) model are of the familiar type; u, d, e, ν_e and their family repetition. Their masses are generated[16] by the Higgs mechanism that breaks the ordinary electroweak gauge symmetry. Thus, one can effectively take all fermions massless in comparison to 10^{15} GeV of various X-bosons (gauge as well as Higgs). CP violating complex parameters appear in Yukawa coupling constant, f_{ij} and h_{ij}, as follows;

$$L_Y = (f_{ij}\, \overline{10}_f^i \cdot 5_f^j + h_{ij} \cdot 10_f^i \cdot 10_f^j)\, 5_H + (\text{h.c.}). \qquad (4.15)$$

This formula generalizes Eq. (3.35) for the colored-Higgs coupling to a case of many generations distinguished by indices, i and j. One may introduce coupling matrices,

$$\tilde{f} = (f_{ij}),\ \tilde{h} = (h_{ij}),\ \tilde{g} = (g\,\delta_{ij}). \qquad (4.16)$$

The gauge coupling matrix $\tilde{g}$ is generation-diagonal and real, and commutes with other coupling matrices. A contribution to the baryon asymmetry contains an imaginary part of a product of coupling constants a la Eq. (4.4). Various contributions from different fermion flavors must be summed and one ends up with a sum over generations, namely a trace of a product of the coupling matrices of (4.16);

$$\sum_{\substack{\text{fermion}\\ \text{flavors}}} (\ldots) = \mathrm{tr} \sum_{\substack{\text{EW}\\ \text{flavors}}} (\tilde{f}, \tilde{h})\tilde{g}^n .$$

One is thus led to investigate imaginary parts of these traces, which can be classified by the number of Yukawa coupling matrices. Actually, one can make $\tilde{f}$ real and diagonal by appropriate, unitary transformations among generation indices of 10_f and 5_f, and $\tilde{h}$ is a symmetric matrix by the Fermi statistics. Up to 6-th orders allowed types of traces are

$$\mathrm{tr}(\tilde{h}\tilde{h}^\dagger)^2,\ \mathrm{tr}(\tilde{h}\tilde{h}^\dagger\tilde{f}^\dagger\tilde{f}),\ \mathrm{tr}(\tilde{f}^\dagger\tilde{f})^2,$$
$$\mathrm{tr}(\tilde{h}\tilde{h}^\dagger)^3,\ \mathrm{tr}(\tilde{h}\tilde{h}^\dagger)^2\tilde{f}^\dagger\tilde{f},\ \mathrm{tr}\ \tilde{h}\tilde{h}^\dagger(\tilde{f}^\dagger\tilde{f})^2,$$
$$\mathrm{tr}\ (\tilde{h}\tilde{f}\tilde{f}^\dagger\tilde{h}^\dagger\tilde{f}^\dagger\tilde{f}).$$

Note that for instance $\mathrm{tr}(\tilde{h}^3\tilde{h}^\dagger)$ is not allowed by a mismatch of SU(5) quantum numbers. It is not difficult to see that all the traces listed above are real. What happens here is that contributions from different generations cancel in the sum although individual terms give finite asymmetry. The first, complex coupling appears at 8-th order[32] and is given by

$$\mathrm{tr}(\tilde{h}\tilde{h}^\dagger\tilde{h}\tilde{f}\tilde{f}^\dagger\tilde{h}^\dagger\tilde{f}^\dagger\tilde{f}), \tag{4.17}$$

which has an imaginary part unless $\tilde{h}$ commutes with $\tilde{f}\tilde{f}^\dagger$. The asymmetry ε is thus of order

$$\varepsilon_H = O(\alpha_H^3).$$

A more careful estimate[32, 40] gives for three generations of fermions

$$\varepsilon_H \simeq \frac{\sqrt{2}\, G_F^3\, m_b^4\, m_t\, m_c}{3^6 \cdot 2^8 \cdot \pi^3} \times (\text{mixings, phases}). \qquad (4.18)$$

The CP-violating phase that appears is independent of that of the quark mixing. The numerical factor in front can be estimated by taking a maximal value of order 200 GeV for the top mass allowed by vacuum stability[55]. Then,

$$|\varepsilon_H| \lesssim 5 \times 10^{-17} \frac{m_t}{200\ \text{GeV}}\ . \qquad (4.19)$$

This is too small to explain the observed size of the baryon asymmetry. If a fourth generation of fermions exists,

$$|\varepsilon_H^{max}| \sim \frac{\sqrt{2}\, G_F^3\, m_{b'}^4\, m_{t'}\, m_t}{3^6 \cdot 2^8 \cdot \pi^3}\ . \qquad (4.20)$$

This may become $\sim 10^{-8}$ to be marginally consistent with $N_B/N_\gamma \sim 10^{-10}$ only if all undiscovered quarks, t, t' and b', are as massive as 200 GeV. Segrè and Turner[40] derived a larger asymmetry because they ignored the factor 3^6 due to the difference of Yukawa couplings at m_x and m_w. We have so far discussed the Higgs boson decay, but from a similar argument one can conclude that the gauge boson X_V can produce an asymmetry even smaller, of order α_H^4. Athough not completely ruled out, the simplest SU(5) model has a serious difficulty in getting a sizable baryon excess[41-44].

A way out of this difficulty[41-44] is to add more 5 of Higgs bosons within the SU(5) scheme. In this case one has to distinguish two couplings, one coupled to the parent Higgs and the other coupled to the exchanged Higgs boson. Thus, the asymmetry is found even at $\sim \alpha_H$

$$\varepsilon_H = 1/4\pi\ [4\ \text{tr}(\tilde{f}^\dagger \tilde{f}) + 3\ \text{tr}(\tilde{h}^\dagger \tilde{h})]^{-1}\ \text{Im tr}(\tilde{f}\tilde{h}'^\dagger \tilde{h}^\dagger \tilde{f}'^\dagger) \cdot f(\frac{m'^2}{m^2}), \qquad (4.21)$$

with f(x) given by Eq. (4.11). The important point here is that the imaginary part does not vanish since

$$f(\frac{m'^2}{m^2}) \neq f(\frac{m^2}{m'^2}),$$

as seen from Eq. (4.11). Thus, there is no essential difficulty in getting a sizable baryon asymmetry once the Higgs system is extended beyond the minimal set, namely one 5 and one 24. In passing we should point out that there is a partial cancellation between the two contributions as shown in Fig. 3. But in this

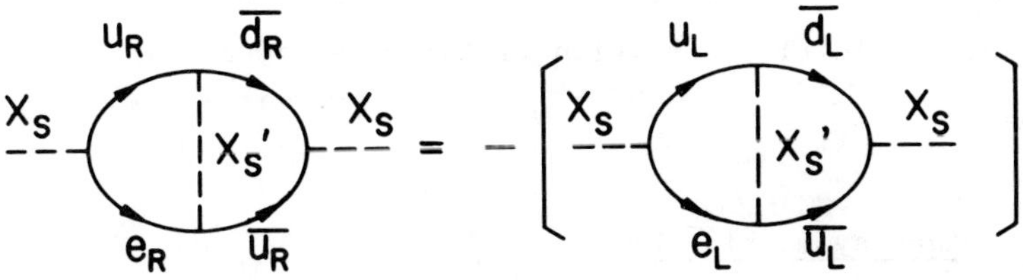

Fig. 3

SU(5) model there is an additional contribution via the diagram of Fig. 4.

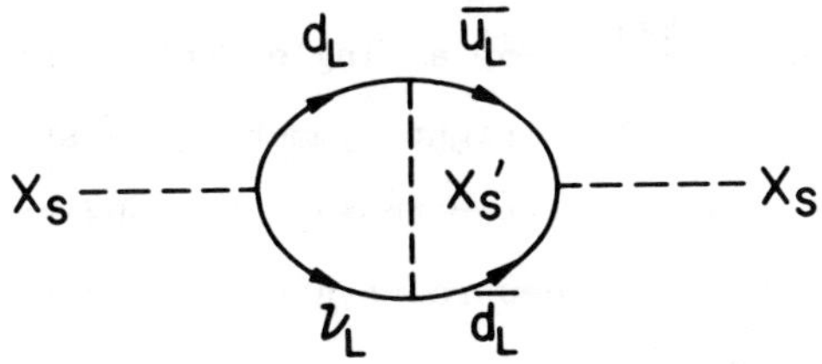

Fig. 4

This problem of cancellation becomes more acute when we discuss the baryon asymmetry in left-right symmetric GUT as will be shown in the next section.

Another fundamental problem of SU(5) models, at least in its simple version, is that B-L is conserved in any order of perturbation, and the cosmological B-L must be given by the initial condition. Even an SU(5) invariant, initial condition can give an arbitrary B-L. It would be nicer either if one had a model without any global conservation law or if one had a gauge invariant, initial condition giving rise to the vanishing global charge.

4.3. Left-right symmetric GUT

The SU(5) model is the most economical GUT and in this regard it should perhaps be compared to the standard SU(2) × U(1) model in the electroweak gauge theory. But it has some distasteful features such as B-L conservation just mentioned and reducibility of one family representation, $(5^*)_L + 10_L$. There are many extended GUTs[10] that overcome these problems. Among these, left-right symmetric models are particularly interesting because in these schemes there is a good chance of understanding why the neutrino is very light[45-47]. For a single family the argument goes as follows. Due to a left-right symmetry one starts with two Weyl spinors, ν_L and N_R. After the spontaneous symmetry breakdown mass terms for this neutral system may look like

$$m \bar{\nu}_L N_R + \frac{M}{2} N_R^T C N_R + (h.c.). \qquad (4.22)$$

Since N_R does not carry SU(2) × U(1) quantum numbers, the Majorana mass M may be much larger than a typical mass scale of the electroweak gauge symmetry, for example m. In this case two mass eigenstates of Majorana type emerge;

$$N'_R \simeq N_R + \frac{m}{M}(\nu_L)^c \text{ with mass} \sim M,$$

$$\nu'_L \simeq \nu_L - \frac{m}{M}(N_R)^c \text{ with mass} \sim \frac{m^2}{M}. \qquad (4.23)$$

The neutrino mass is suppressed by m/M in comparison to any typical fermion. The important question is then whether there is any useful constraint on the size of M. We have recently shown[48) that the cosmological baryon excess indeed gives such a constraint, and that there is a general link between the neutrino mass and the baryon excess. I shall explain this result in the following discussion.

As we discussed in the previous subsection, there is a cancellation mechanism working between the two contributions related by the parity operation. An example of this kind is shown in Fig. 3. The way how this comes about is easy to understand in a left-right symmetric GUT. Here a single family of fermions is contained in an irreducible representation ψ_L which decomposes into

$$\psi_L = (\phi_L) + (\phi_R)^*. \qquad (4.24)$$

Two Weyl fields, ϕ_L and ϕ_R, are mirror reflections of each other and * means the charge conjugation. An example of such decomposition is shown for the SO(10) spinor in Eq. (2.19). The Yukawa coupling of Higgs multiplets H_a to a number of families is expressed symbolically as

$$L_Y = f^{(a)}_{ij}\, \psi^i_L\, \psi^j_L\, H_a + (\text{h.c.})$$
$$= f_{ij} \cdot H\, \phi^i_L\, \phi^j_L + f^*_{ij} \cdot H\, \phi^i_R\, \phi^j_R + (\text{h.c.}) + (\ldots), \qquad (4.25)$$

where only terms relevant to X_S bosons are displayed and we assumed H real. We thus see that there is a symmetry under the transformations,

$$\phi_L \leftrightarrow \phi_R, \; f \leftrightarrow f^*, \; X_S \leftrightarrow X_S. \qquad (4.26)$$

Namely, the coupling factors related by the parity operation are a pair of a complex number and its conjugate. On the other hand, the baryon asymmetry is proportional to the imaginary part of coupling factors. This then means that in the Higgs decay there is no baryon asymmetry unless the cancellation is somewhat avoided by mass differences between chiral partners[49, 50]. Since charged fermions form massive, four-component Dirac spinors, this disparity of masses can only occur for neutral leptons. In a simple left-right symmetric scheme such as the SO(10) model this would imply[50] that the Majorana lepton N_R must be roughly as massive as the Higgs boson X_S to produce a sizable asymmetry. This in turn gives a constraint on the neutrino mass through the formula of Eq. (4.23).

Let us now consider an SO(10) model[17] both to illustrate the general idea outlined above and to estimate the size of neutrino mass consistent with the baryon excess. For definiteness take the baryon asymmetry due to the decay of Higgs boson $X_S(-1/3)$ (see Eq. (3.39) for notation) that belongs to the vector representation <u>10</u> of SO(10). The main contribution comes from the diagram of Fig. 2 in which the ordinary neutrino ν_L in one case and its chiral partner N_R in the other case appear in one of the intermediate states. The Yukawa coupling of X_S to a family of fermions is given by

$$L = \sqrt{2}\, f\, [-\overline{(u,\, d)}_k \tfrac{1}{2}(1 + \gamma_5) \binom{e^+}{-\nu} + \varepsilon_{ijk}\, \overline{u^c_i}\, \tfrac{1}{2}(1 + \gamma_5) d_j]\, X^k_s$$

$$+ \sqrt{2}\, f^*\, [-\overline{(u,\, d)}_k \tfrac{1}{2}(1 - \gamma_5) \binom{e^+}{-\bar{N}} + \varepsilon_{ijk}\, \overline{u^c_i}\, \tfrac{1}{2}(1 - \gamma_5) d_j]\, X^k_s$$

$$+ \text{(h.c.)} \qquad (4.27)$$

Note a similarity between this formula and that for the SU(5) model, Eq. (3.35); the symmetry (4.26) is explicitly realized. The total decay rate of X_S is given by a sum of two-body fermionic modes,

$$\gamma = (1/8\pi)\, m\, [7\, \mathrm{tr}(\tilde{f}^\dagger \tilde{f}) + \sum_i (1 - \frac{\mu_i^2}{m^2})^2\, \mathrm{tr}(\tilde{f}^\dagger \tilde{f} \rho_i^\dagger \rho_i)], \quad (4.28)$$

where $\tilde{f}$ is the Yukawa coupling matrix with generation indices and ρ_i is a projection matrix onto a Majorana mass eigenstate N_R^i of mass μ_i. Using this rate and the previous formula (4.13) one finds for the baryon asymmetry due to the $X_s^{(1)}$ decay,

$$\varepsilon_H^{(1)} = -1/8\pi\, [\frac{7}{8}\, \mathrm{tr}(\tilde{f}_1^\dagger \tilde{f}_1) + \sum_i \frac{1}{8}(1 - \frac{\mu_i^2}{m^2})^2\, \mathrm{tr}(\tilde{f}_1 \tilde{f}_1^\dagger \rho_i \rho_i^\dagger)]^{-1}$$

$$\sum_i \mathrm{Im}(\tilde{f}_1^\dagger \tilde{f}_2 \tilde{f}_1^\dagger \tilde{f}_2 \rho_i^\dagger \rho_i)[F(\frac{m_2^2}{m_1^2}, 0) - F(\frac{m_2^2}{m_1^2}, \frac{\mu_i^2}{m_1^2})], \quad (4.29)$$

where F(x,y) is given by Eq. (4.14), and the exchanged Higgs boson in Fig. 2 is $X_S^{(2)}$ with the Yukawa coupling $\tilde{f}_2$. Very crudely the asymmetry is (we assume CP violating phases of order unity)

$$\varepsilon_H \sim \frac{f^2}{8\pi} \sim \frac{1}{8} \cdot \frac{1}{9}\, \alpha_G\, (\frac{m_t}{m_W})^2 \quad \text{for } \mu_i > m,$$

$$\varepsilon_H \sim \frac{f^2}{8\pi} \frac{\langle\mu^2\rangle}{m_1^2 + m_2^2} \sim \frac{1}{72}\, \alpha_G (\frac{m_t}{m_W})^2 \frac{\langle\mu^2\rangle}{m_1^2 + m_2^2} \quad \text{for } \mu_i \ll m_i, \quad (4.30)$$

since for a small y, $F(x,y) \sim 1 - x \ln[(1 + x)/x)] - y/(1 + x)$. The factor 9 here is due to the renormalization effect of Yukawa couplings. With $\alpha_G \sim 1/40$

$$|\varepsilon_H| \sim 2 \times 10^{-5} (\frac{m_t}{18\ \mathrm{GeV}})^2 \quad \text{for } \mu_i > m, \quad (4.31)$$

which is too large. To get ε_H of order 10^{-8} we need[50)]

$$(\frac{2\langle\mu^2\rangle}{m_1^2 + m_2^2})^{1/2} \sim 0.02\ \frac{30\ \mathrm{GeV}}{m_t}\ . \quad (4.32)$$

It is interesting to see that a slightly smaller value than $m(X_S)$ comes out for the Majorana mass of N_R.

To make a direct connection with the neutrino mass one needs an explicit formula for the Dirac mixing term m between ν_L and N_R as defined by (4.22) This Dirac mass term depends on the Higgs Yukawa coupling to fermions which would be read from

$$\underline{16} \times \underline{16} = \underline{10} + \underline{120} + \underline{126}. \tag{4.33}$$

The simplest Higgs system that one can adopt in SO(10) models is to use representations of $\underline{16}$, $\underline{45}$, $\underline{10}$. In this case only $\underline{10}$ can directly couple to fermions, and one has to utilize the Witten mechanism[51, 52] of generating a large Majorana mass for N_R. Dirac mass matrices that are derived when vacuum expectation values of $\underline{10}$ are put in have symmetry relations,

$$\tilde{m}_e = \tilde{m}_d = \sqrt{2} \sum_i \tilde{f}_i v_i^*, \tag{4.34}$$

$$\tilde{m}_{N\nu} = \tilde{m}_u = \sqrt{2} \sum_i \tilde{f}_i v_i, \tag{4.35}$$

where $v_j = \langle H_j^9 + i H_j^{10} \rangle / \sqrt{2}$ with H^9 and H^{10} being two neutral components in $\underline{10}$. These relations are easy to understand since neutral components in $\underline{10}$ are invariant under the subgroup SO(6) ≃ SU(4), and mass terms look like

$$\mathrm{tr}\overline{\begin{pmatrix} u_1 & u_3 & u_3 & N \\ d_1 & d_2 & d_3 & e \end{pmatrix}_R^T} \begin{pmatrix} v & o \\ o & v^* \end{pmatrix} \begin{pmatrix} u_1 & u_2 & u_3 & \nu \\ d_1 & d_2 & d_3 & e \end{pmatrix}_L .$$

If only one $\underline{10}$ is introduced, one gets a mass relation like $m_b = m_t$ even after renormalization effects being taken into account of[52]. This can be avoided with many $\underline{10}$'s because relative phases of v_i then make a crucial difference in Eqs. (4.34) and (4.35). One troublesome relation is $m_d \simeq 3\, m_e$ after renomalization. But, unlike the SU(5) model there are radiative corrections[53] to mass terms which come from Witten-like diagrams (Fig. 5). These may

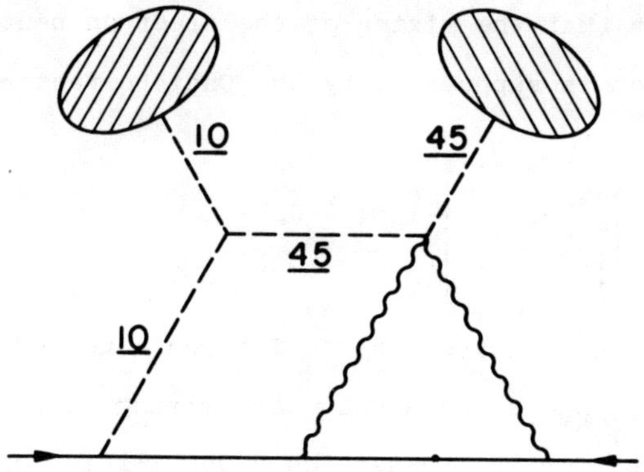

Fig. 5

easily give mass terms of order MeV, hence substantially modify mass relations in the first generation without much affecting those in higher generations. Due to this possibility we do not regard the mass relations of Eq. (4.34) and (4,35) as a failure, but they may well be roughly consistent with experimental data.

What about the neutrino mass? With the Dirac mixing given by (4.35) the neutrino mass matrix is found to be

$$\tilde{m}_\nu = \tilde{m}_u \, \tilde{m}_N^{-1} \, \tilde{m}_u, \qquad (4.36)$$

where $\tilde{m}_u^T = \tilde{m}_u$ was used, and the initial mass terms are written as,

$$1/2 \; \bar{N}_R \, \tilde{m}_N \, (N_R)^c + \bar{N}_R \, \tilde{m}_u \, \nu_L + (\text{h.c.}).$$

The form of the neutrino mass matrix, Eq. (4.36), is obtained if all Dirac masses of electroweak scale are derived from the vector representation 10 and this form is independent of the Witten mechanism. One important consequence[52)] of (4.36) combined with

(4.34), $\tilde{m}_e = \tilde{m}_d$, is that the mixing of the electron neutrino ν_e with other neutrinos is suppressed by the Cabibbo factor. This can be seen from

$$\tilde{m} \simeq U_{KM}^T \begin{pmatrix} o & o & o & \cdots \\ o & & & \\ o & & * & \\ \vdots & & & \end{pmatrix} U_{KM}, \qquad (4.37)$$

which follows when the u-quark mass is ignored compared to m_c, m_t, ... Here, U_{KM} is the usual quark mixing matrix[54]. From (4.37) one gets for the transition probabilities of the electron neutrino

$$w(\nu_e \to \nu_\mu + \nu_\tau + \ldots) \lesssim 4 \sin^2\theta_1 \sim 0.21. \qquad (4.38)$$

A stronger result follows if one assumes a mimimal set of Higgs bosons, one <u>16</u>, one <u>45</u> and two <u>10</u>'s. As shown in Ref. (52) this case gives for three generations

$$U_\nu \simeq U_{KM} \text{ for } m_s \ll m_c,\ m_b,\ m_t, \qquad (4.39)$$

$$m(\nu_1)/m(\nu_2)/m(\nu_3) = 0/m_c/m_t. \qquad (4.40)$$

Namely, the neutrino spectrum imitates that of up-type quarks. Crudely speaking, what happens is that the Witten mechanism forces a linear relation among mass matrices,

$$\tilde{m}_N = \mu_1 \tilde{f}_1 + \mu_2 \tilde{f}_2 = \xi_1 \tilde{m}_u + \xi_2 \tilde{m}_d \sim \xi_1 \tilde{m}_u,$$

thus

$$\tilde{m}_\nu = \tilde{m}_u \tilde{m}_N^{-1} \tilde{m}_u \sim \xi_1^{-1} \tilde{m}_u,$$

ignoring less dominant term. Off course, one has to be more careful in treating the small up-quark mass, which makes the derivation[52] of Eqs. (4.39) and (4.40) a little more complicated.

Let us put together these formulas into the expression of the baryon asymmetry, Eq. (4.29). To a good approximation (see (4.14))

$$\sum_i \rho_i^\dagger \rho_i \ [(\frac{m_2^2}{m_1^2}, 0) - F(\frac{m_2^2}{m_1^2}, \frac{\mu_1^2}{m_1^2})]$$

$$\simeq \frac{\tilde{m}_N^\dagger \, \tilde{m}_N}{m_1^2 + m_2^2} = \frac{\tilde{m}_u^\dagger \, \tilde{m}_\nu^{\dagger -1} \, \tilde{m}_u^\dagger \, \tilde{m}_u \, \tilde{m}_\nu^{-1} \, \tilde{m}_u}{m_1^2 + m_2^2} \qquad (4.41)$$

The relations between the mass matrices and coupling matrices, Eqs. (4.34) and (4.35), can be converted when there are only two <u>10</u>'s,

$$\tilde{f}_1 = (4\sqrt{2}\, G_f)^{1/2} \, (\eta_1 \tilde{m}_u + \eta_1^* \tilde{m}_d) \sim (4\sqrt{2}\, G_F)1/2 \, \eta_1 \tilde{m}_u, \quad (4.42)$$

$$\tilde{f}_2 = (4\sqrt{2}\, G_f)^{1/2} \, (\eta_2 \tilde{m}_u + \eta_2^* \tilde{m}_d) \sim (4\sqrt{2}\, G_f)^{1/2} \, \eta_2 \tilde{m}_u, \quad (4.43)$$

with $\eta_1/\eta_2 = -(v_2/v_1)^* = O(1)$. We have assumed that couplings to quarks of charge 2/3 dominate due to larger masses. Thus, by keeping dominant coupling terms proportional to $\tilde{m}_u$, one finds from Eq. (4.29), (4,41) - (4,43) that the $X_S^{(1)}$ decay gives

$$\varepsilon_H^{(1)} \simeq (9^3\sqrt{2}\pi)^{-1} \, G_F \, |\eta_1|^{-2} \, \mathrm{Im\ tr} \, [\eta_1^2 \, \eta_2^{*2} (\tilde{m}_u \tilde{m}_u^\dagger)^3 \, \tilde{m}_\nu^{\dagger -1} \cdot (\tilde{m}_u^\dagger \tilde{m}_u) \, \tilde{m}_\nu^{-1}] / [\mathrm{tr}(\tilde{m}_u^\dagger \tilde{m}_u) \cdot (m_1^2 + m_2^2)]. \quad (4.44)$$

The factor 9 here is due to the renormalization effect. Finally, we assume, as it is the case for the simplest model, that both $\tilde{m}_u$ and $\tilde{m}_\nu$ can be diagonalized by U_{KM} a la U_{KM}^T (diagonal) U_{KM}. Then,

$$\varepsilon_H^{(1)} \simeq (9^3\sqrt{2}\pi)^{-1} \, G_F \, |\eta_2|^2 \, \sin[2\ \mathrm{Arg}(\frac{\eta_1}{\eta_2})] \, \frac{m_t^6}{m_\nu^2 \, (m_1^2 + m_2^2)}, \quad (4.45)$$

where m_t and m_ν are the largest quark and the largest neutrino masses, respectively. A similar contribution due to the $X_S^{(2)}$ decay adds up unless the two masses are very much different,

$$\varepsilon_H = \varepsilon_H^{(1)} + \varepsilon_H^{(2)} = (9^3\sqrt{2}\pi)^{-1} \, G_F \, \frac{m_t^6}{m_\nu^2 \, (m_1^2 + m_2^2)} \cdot C, \quad (4.46)$$

$$C \simeq (|\eta_2|^2 - |\eta_1|^2) \, \sin[2\ \mathrm{Arg}\ (\frac{\eta_1}{\eta_2})] \sim O\ (1). \quad (4.47)$$

Note that this result is independent of U_{KM}, and in particular the baryon asymmetry has no direct relation to the CP violation observed in K-$\bar{K}$ system.

The formula for the asymmetry (4.46) should be combined with the baryon to entropy ratio (3.36),

$$\frac{kn_B}{s} \simeq \frac{0.5 \times 10^{-2}\ \varepsilon_H}{1 + (4.5\ K_H)^{1.2}}\ , \tag{4.48}$$

where K_H is the ratio of total rate to the expansion rate at $T = m(X_s)$ and in this case

$$K_H \simeq 3.8 \times 10^{11} \cdot \sqrt{\frac{80}{N}} \cdot \frac{m_t^2}{m_x}\ \mathrm{GeV}^{-1}\ , \tag{4.49}$$

from (3.37), (4.28), (4.42) and (4.43). Combining (4.46), (4.48) and (4.49) with $N = 80$ and $C = 1$, one finds[48] that

$$\frac{m_t^6 / (m_\nu^2\, m_x^2)}{1 + 5 \times 10^{14}\ (m_t^2 / m_x)^{1.2}} \simeq 2 \times 10^{\pm 1} \quad \text{for } \frac{kn_B}{s} = 10^{-10.8 \pm 1} \tag{4.50}$$

All masses are measured in unit of GeV.

To obtain a bound on the neutrino mass m_ν, one needs an estimate for the magnitudes of m_t and m_x. For the heaviest quark we use

$$18\ \mathrm{GeV} < m_t < 200\ \mathrm{GeV}. \tag{4.51}$$

The lower bound is the present experimental limit and the upper bound is derived from the condition of vacuum stability in perturbation[55]. The unitarity bound[56] gives a similar mass bound, and the following result is not sensitive to a precise value of this bound. For the mass of the Higgs X-boson we use

$$3 \times 10^{10}\ \mathrm{GeV} < m_x < 6 \times 10^{14}\ \mathrm{GeV}\ (1.2 \times 10^{19}\ \mathrm{GeV}). \tag{4.52}$$

The lower bound is due to stability of nucleon. Without this bound nucleon would decay via Higgs exchange at a rate above the present experimental bound.[14] The upper bound of 6×10^{14} GeV is what we take as the best guess for the mass of X_V' $(\frac{2}{3}, -\frac{1}{3})$. If $m(X_S) > m(X_V')$, the baryon asymmetry produced by the Higgs X_S decay would be wiped out by the decay and inverse decay of gauge X_V' boson, which we do not like. Actually, there are two types of gauge bosons, X_V and X_V' in the SO(10) model, and one of them, X_V, does not thermalize B-L as discussed previously. Thus, the case of $m(X_V) < m(X_S) < m(X_V')$ might still be allowed. At the present stage of development we only know $m(X_V) \simeq 6 \times 10^{14}$ GeV, but the mass of X_V' is unknown except that $m(X_V) \lesssim m(X_V') \lesssim m_*$ (Planck mass). We have chosen two extreme values for the upper bound of $m(X_S)$ as written in (4.52), and shall indicate how the result depends on this choice. The bound of neutrino mass that we have found[48] is as follows;

$$0.5 \times 10^{-2}\ \text{eV}\ (3 \times 10^{-7}\ \text{eV}) \times 3^{\pm 1} < m_\nu^{max} < 30\ \text{eV} \times 3^{\pm 1}. \tag{4.53}$$

The uncertainty of factor ~ 3 is due to the observational uncertainty in kn_B / s. If the unknown factor due to CP violation, C in (4.46), is arbitrarily put, we would get the upper bound replaced by

$$m_\nu^{max} < 30\ \text{eV}\ |C|^{1/2} \times 3^{\pm 1}.$$

Since $|C| \leq 1$, this upper bound is more stringent than that of (4.53). Furthermore, if m_t is experimentally known, we can replace the upper bound of (4.53) by

$$m_\nu^{max} < 2.4\ \text{eV}\ \left(\frac{m_t}{50\ \text{GeV}}\right)^{1.8}, \tag{4.54}$$

which may become quite a strong bound. The mass bound we have found is in the range of immediate interest both to laboratory experiments and to cosmology. For more details on this bound I shall refer to our recent paper[48].

Before closing this subsection I should stress that although particular values of the mass bounds are not trustable due to specific features of this SO(10) model, the general link between the neutrino mass and the baryon excess is a common feature in any left-right symmetric GUT. All that is needed to get some numbers for the neutrino mass is to specify the Dirac mixing term between ν_L and N_R. A possible exception to this general link would arise when baryon production goes through N_R decay. Here the symmetry of couplings, (4.26), which was the essential bridge between the neutrino mass and the baryon excess, would be irrelevant because only one of the chiral partners can produce the baryon asymmetry. The problem of baryon production by N_R decay will be addressed in the following subsection.

4.4. <u>Heavy fermion decay</u>

Heavy neutral leptons can contribute to the baryon asymmetry via decay processes of either

$$F \to X_s\bar{q},\ \bar{X}_s q, \quad \hookrightarrow \bar{q}\bar{q},\ \hookrightarrow qq \text{ etc.} \qquad (4.55)$$

or

$$F \to qqq,\ \bar{q}\bar{q}\bar{q}. \qquad (4.56)$$

Kinematical constraints of delayed-decay mechanism are quite different in the two cases as discussed in the previous section. A general discussion on the magnitude of baryon asymmetry[37] is beyond the scope of this short lecture, and I shall make a few comments in the case of SO(10) model. In the SO(10) model $F = N_R$. The two-body decays (4.55) might give a reasonable asymmetry, but the subsequent decay of X_s either damps this asymmetry or produces too much an asymmetry. Since at least one of N_R's is heavier than X_s, the partial cancellation factor $m_N^2 / m^2(X_s)$ does not exist and the produced asymmetry is as large as 10^{-5}. See (4.31). It is thus unlikely for the two-body (4.55) to give the correct magnitude of baryon excess. The two-body decays do not occur if $m(N) < m(X_s)$.

The baryon asymmetry due to the three-body decays, (4.56), was also computed[50]. Here the main problems are; (1) a large total decay rate via $N_R \to \ell^- \phi^+$ ($\ell = e, \mu, \ldots, \phi$ = ordinary Higgs doublet), (2) suppression factor due to a propagator. Symbolically, the magnitude of asymmetry is written as

$$\varepsilon_H \sim \alpha_H^2 \cdot \left(\frac{m_N}{m_X}\right)^2 \cdot \left(\frac{\text{3-body phase space}}{\text{2-body phase space}}\right).$$

A more careful calculation in the SO(10) model shows that ε_H is at most $\sim 10^{-9}\ (m_t / m_W)^4$, which we regard too small. I shall refer to our paper[50] for a more detailed discussion. Our conclusion in the SO(10) model is that the extra heavy lepton N_R is not a viable candidate of baryon production. We believe that the mechamism of baryon production at an intermediate energy scale, proposed by Masiero and Mohapatra[57], has also a serious difficulty although their numerical estimate is optimistic. Off course, the intermediate baryon nonconservation with $\Delta B = 2$ would be compatible with the cosmological baryon production if the intermediate unification is not related to the left-right symmetry[58].

5. Summary and Open Problems

In summarizing long discussions of sec. 3 and 4, I would say that a basic principle of baryon production is now well understood, and that a few processes of baryon production have been identified, at least as promising candidates. The simplest possibility is to use baryon nonconserving decay of the Higgs boson X_s. Decay and inverse decay of a gauge boson, X_V or X_V', may wipe out an arbitrary initial baryon number given from a pre-GUT epoch so that the final baryon to photon ratio becomes independent of an initial condition. What deserves further investigation is heavy fermion decay with a dimensional conpling. In this case there is no tight constraint on the fermion mass, and one might be able to produce the baryon asymmetry at a temperature much below 10^{15} GeV. Even the decay of a gauge boson may produce a reasonable asymmetry in a complicated GUT scheme[44), 58)]. The magnitude of baryon excess is sensitive to detailed structures of a model. The SU(5) model with the mimimal set of Higgs bosons seems almost ruled out. More complicated schemes with a complex Higgs system would work, but there are too many adjustable parameters in these models. In a left-right symmetric GUT the situation is different and there is always a partial cancellation of baryon asymmetry between parity partners. In particular, the size of the neutrino mass is related to the cosmological baryon excess and tends to be $\sim$ 1 eV in typical cases.

There are several open problems that remain to be studied on the present subject. I shall only mention a few of them. In our discussion we have assumed that the GUT phase transition is of a second order or weakly first order. If it is of a strongly first order, it may affect the scenario of baryon production in a drastic way. The first order phase transition has recently been much discussed[59)] as a solution to outstanding problems in cosmology, homogeneity and flatness, and furthermore it may suppress[60)] overproduction of magnetic monopoles. The problem that has to be solved before accepting the first order GUT phase transition is whether the latent heat released after termination of the phase transition can actually reheat the bulk of abundant particles so that the universe goes back to a smooth Friedmann model. If this really happens and the reheated temperature is close enough to m_x, then the scenario of baryon production will follow what have been discussed here. On the other hand, a strongly first order phase transition of the electroweak gauge symmetry may cause a trouble. If the entropy generated by the electroweak transition is substantial, it would dilute the baryon to entropy ratio because the baryon number is well conserved at this low temperature and the generated entropy goes to baryons and antibaryons equally and also to radiation. Indeed, Witten and others[61] computed amount of the generated entropy and found it large (of order 10^6) for the Coleman-Weinberg model[62]. This seems to exclude the Coleman-Weinberg mechanism for the electroweak symmetry breakdown unless one produces a maximally allowed asymmetry of $\epsilon \sim 10^{-3}$ at the GUT epoch.

Another important problem has to do with a nature of CP violation. So far no one has succeeded in relating the CP violation in the baryon asymmetry to that observed in K-$\bar{K}$ system. It may well be that the two sources of CP violation are completely different and independent[63]. Then, we will not be able to tell that the matter defined by the baryon excess is the same as that defined by K-$\bar{K}$ system. One might also prefer a soft or spontaneously broken CP violation to a hard CP violation that one usually assumes here. But a cosmological phase transition of a discrete symmetry such as CP generates[64] a domain structure of different discrete quantum numbers and the size of the individual domains is fixed by distance to the horizon.

$$2ct_{GUT} \sim 2\frac{m_*}{m_x^2} \sim 5 \times 10^{-25} \text{ cm } (\frac{10^{15} \text{ GeV}}{m_x})^2 \quad .$$

The domain of this size contains only a small number of particles of order

$$10^{15} \left(\frac{10^{15}\ \mathrm{GeV}}{m_x}\right)^3 .$$

This is a disaster because these small domains of matter and antimatter would be completely mixed at later times and one goes back to a symmetric cosmology. However, as K. Sato[65] discussed recently, the size of the domain is much enlarged if the phase transition is strongly of a first order and if the scale factor expands exponentially during the supercooling. Indeed, if the first order phase transition terminates at a sufficiently late time $t = t_f$, then the domain will be stretched roughly by a factor

$$\exp\ (t_f\ /\ 2t_{GUT}) ,$$

and the number of particles in the domain will be increased by a factor

$$\exp\ (3t_f\ /\ 2t_{GUT}) .$$

After the reheating the baryon production within each domain would proceed in much the same way as in the standard scenario, and this domain could grow[65] to a present cluster of galaxies if $t_f \sim 60t_{GUT}$. This new version of a symmetric cosmology has an esthetic appeal, again if the problem of reheating is really solved. Also, the amount of supercooling which easily exceeds 10^{20} in temperature ratio has been critisized[66].

Finally, the present theory of baryosynthesis restricts type of perturbation needed to form galaxies. Since in the simple scenario the baryon to entropy ratio is essentially given in terms of particle physics parameters, a baryon fluctuation would correlate with temperature fluctuation, namely the type of perturbation allowed in the simple GUT scenario is adiabatic[67]. But as recent observations[68] suggest, one may need isothermal perturbation as well. Isothermal perturbation means that only the baryon number density is spatially inhomogenious without any correlation to radiation. Recently, a number of people[69)] have discussed a possibility of generating isothermal perturbation in an anisotropic model with shear. Much remains to be studied, but it would be interesting if the GUT epoch can provide seeds of galaxy formation, at least as a viable possibility.

In conclusion, it seems almost certain to me that the cosmic matter as we now see was created because it is ultimately unstable. It would be nice if one can prove instability of matter in terrestrial experiments now under way.

Acknowledgement

I should like to thank many colleagues and collaborators who have contributed to the subject discussed here. In particular, I should like to record my sincere thanks to J. Arafune, M. Fukugita, H. Sato, H. Sugawara and T. Yanagida, who have discussed with and encouraged me with unfailing patience during various stages of this investigation.

6. References

(1) G. Steigman, Ann. Rev. Astron. Astrophys. 14 (1976) 339.

(2) A. Buffington and S. M. Schindler, Caltech preprint, to be published in Astrophys. J. Lett. (1981); and references therein.

(3) K. A. Olive, D. N. Schramm, G. Steigman, M. S. Turner and J. Yang, Astrophys. J., to be published (1981).

(4) S. Tremaine and J. Gunn, Phys. Rev. Lett. 42 (1979) 407; D. Schramm and G. Steigman, Astrophys. J. 243 (1981) 1.

(5) H. Sato and F. Takahara, Progr. Theor. Phys. 64 (1980) 2029; J. Bond, G. Efstathiou and J. Silk, Phys. Rev. Lett. 45 (1980) 45.

(6) R. Omnès, Phys. Rept. C3 (1972) 1 and references therein.

(7) A. D. Sakharov, Zh. Exsp. Theor. Fiz. Pis'ma 5 (1967) 32 (English translation, JETP Lett. 5, 24); V.A. Kuzmin Zh. Exsp. Theor, Fiz. Pis'ma 12 (1970) 335 (English translation, JETP Lett. 12, 228)

(8) M. Yoshimura, Phys. Rev. Lett. 41 (1978) 281; 42 (1979) 746 (E).

(9) Ya. B. Zeldovich, Zh. Eksp. Theor. Fiz. 48 (1965) 986.

(10) For a review, see P. Langacker, Phys. Rept. C to be published (1981).

(11) H. Georgi, H. R. Quinn and S. Weinberg, Phys. Rev. Lett. 33 (1974) 451.

(12) W. J. Marciano, Phys. Rev. D20 (1979) 274; T. J. Goldman and D. A. Ross, Phys. Lett. 84B (1979) 208.

(13) J. Ellis, M. K. Gaillard, D. V. Nanopoulos and S. Rudaz, Nucl. Phys. B176 (1981) 61 and references therein.

(14) J. Learned, F. Reines and A. Soni, Phys. Rev. Lett. 43 (1979) 907.

(15) J. Ellis, Lectures at the 21st Scottish Universities Summer School in Physics (1980); CERN preprint, TH. 2942.

(16) H. Georgi and S. L. Glashow, Phys. Rev. Lett. 32 (1974) 438; A. J. Buras, J. Ellis, M. K. Gaillard and D. V. Nanopoulos, Nucl. Phys, B135 (1978) 66.

(17) H. Georgi, in Particles and Fields ed. by C. E. Carlson, AIP Conf. No. 23, New York (1975); H. Fritzsch and P. Minkowski Ann, Phys. (N. Y.) 93 (1975) 193; M. S. Chanowitz, J. Ellis and M. K. Gaillard, Nucl. Phys. B128 (1977) 506; H. Georgi and

D. V. Nanopoulos, Nucl. Phys. B155 (1979) 52.

(18) For a review of the big bang cosmology, see S. Weinberg, Gravitation and Cosmology: Principles and Applications of the General Theory of Relativity, Chap. 15 (John Wiley and Sons, Inc., New York, 1972).

(19) J. A. Harvey, E. W. Kolb, D. B. Reiss and S. Wolfram, Caltech preprint (1981).

(20) D. Toussaint, S. B. Treiman, F. Wilczek and A. Zee, Phys. Rev. D19 (1979) 1036.

(21) This part of discussion follows the auther's review article; M. Yoshimura in Proc. Workshop on the Unified Theory and the Baryon Number in the Universe, ed. by O. Sawada and A. Sugamoto, KEK-79-18 (1979) pp. 1-18.

(22) A. Aharony, in Modern Developments in Thermodynamics (Wiley, New York, 1974) pp. 95-114.

(23) S. Weinberg, Phys. Rev. Lett. 42 (1979) 850.

(24) M. Yoshimura, Phys. Lett. 88B (1979) 294.

(25) J. N. Fry, K. A. Olive and M. S. Turner, Phys. Rev. D22 (1980) 2953. See also E. W. Kolb and S. Wolfram, Phys. Lett. 91B (1980) 217; Nucl. Phys, B172 (1980) 224. Note that our definition of K is different from that of Fry et al. by 2/3.

(26) J. N. Fry, K. A. Olive and M. S. Turner, Phys. Rev. D22 (1980) 2977.

(27) J. N. Fry, K. A. Olive and M. S. Turner, Phys. Rev. Lett. 45 (1980) 2074.

(28) S. Weinberg, Phys. Rev. Lett. 43 (1979) 1566; F. Wilczek and A. Zee, Phys. Rev. Lett. 43 (1979) 1571.

(29) S. B. Treiman and F. Wilczek, Phys. Lett. 95B (1980) 222.

(30) S. Weinberg, Phys. Rev. D22 (1980) 1694.

(31) The following references (31)-(33) discuss nonstandard scenarios of baryon production. S. Dimopoulos and L. Susskind, Phys. Rev. D18 (1979) 4500; Phys. Lett. 81B (1979) 416.

(32) J. Ellis, M. K. Gaillard and D. V. Nanopoulos, Phys. Lett. 80B (1979) 360; 82B (1979) 464 (E).

(33) A. Yu Ignatiev, N. V. Krasmikov, V. A. Kuzmin and A. N. Tavkhelidze, Phys. Lett. 76B (1978) 436.

(34) T. Yanagida and M. Yoshimura, Phys. Rev. Lett. 45 (1980) 71.

(35) J. A. Harvey, E. W. Kolb, D. B. Reiss and S. Wolfram, Nucl. Phys. B177 (1981) 456.

(36) R. Barbieri, D. V. Nanopoulos and A. Masiero, Phys. Lett. 98B (1981) 191.

(37) M. Fukugita, T. Yanagida and M. Yoshimura, in preparation.

(38) R. N. Mohapatra and R. E. Marshak, Phys. Rev. Lett. 44 (1980) 1316.

(39) For example, R. J. Eden et al., The Analytic S-Matrix, Chap. 2 (Cambridge University Press, London, 1966).

(40) G. Segrè and M. S. Turner, Phys. Lett. 99B (1981) 399.

(41) S. M. Barr, G. Segrè and H. A. Weldon, Phys. Rev. D20 (1979) 2494.

(42) D. V. Nanopoulos and S. Weinberg, Phys. Rev. D20 (1979) 2454.

(43) A. Yildiz and P. Cox, Phys. Rev. D21 (1980) 906.

(44) T. Yanagida and M. Yoshimura, Nucl. Phys. B168 (1980) 534.

(45) M. Gell-Mann, P. Ramond and R. Slansky, in Supergravity, ed. by D. Z. Freedman and P. van Nieuwenhuizen (North Holland, 1979).

(46) T. Yanagida, in Proc. Workshop on the Unified Theory and the Baryon Number in the Universe, ed. by O. Sawada and A. Sugamoto, KEK-79-18 (1979).

(47) R. N. Mohapatra and G. Senjanovic, Phys. Rev. Lett. 44 (1980) 912.

(48) M. Fukugita, T. Yanagida and M. Yoshimura, KEK preprint, KEK-TH 27 (1981). I understand that a similar result has been obtained by P. Ramond, D. B. Reiss and their collaborators (private communications).

(49) V. A. Kuzmin and M. E. Shaposhnikov, Phys. Lett. 92B (1980) 115.

(50) T. Yanagida and M. Yoshimura, Phys. Rev. D23 (1981) 2048.

(51) E. Witten, Phys. Lett. 91B (1980) 81.

(52) T. Yanagida and M. Yoshimura, Phys. Lett. 97B (1980) 99.

(53) M. Yoshimura, Progr. Theor. Phys. 64 (1980) 1756.

(54) M. Kobayashi and T. Maskawa, Progr. Theor. Phys. 49 (1973) 652.

(55) P. Q. Hung, Phys. Rev. Lett. 42 (1979) 873; H. D. Politzer and S. Wolfram, Phys. Lett. 82B (1979) 242; 83B (1979) 421 (E); L. Maiani, G. Parisi and R. Petronzio, Nucl. Phys. B136 (1978) 115; N. Cabibbo, L. Maiani, G. Parisi and R. Petronzio, Nucl. Phys. B158 (1979) 295; H. Komatsu, Progr. Theor. Phys. 65 (1981) 779.

(56) B. W. Lee, C. Quigg and H. B. Thacker, Phys. Rev. D16 (1977) 1519.

(57) A. Masiero and R. N. Mohapatra, Max Planck preprint (1980).

(58) M. Fukugita, T. Yanagida and M. Yoshimura, in preparation.

(59) A. H. Guth, Phys. Rev D23 (1981) 347; K. Sato, Month. Not. R. Astron. Soc. 195 (1981).

(60) J. P. Preskill, Phys. Rev. Lett. 43 (1979) 1365; A. H. Guth and S. H. H. Tye, Phys. Rev. Lett. 44 (1980) 631; M. B. Einhorn, D. Stein and D. Toussaint, Phys. Rev. D21 (1980) 3295; M. B. Einhorn and K. Sato, Nucl. Phys. B180 [FS2] (1981) 385.

(61) E. Witten, Nucl. Phys. B177 (1981) 477; A. H. Guth and E. J. Weinberg. Phys. Rev. Lett. 45 (1980) 1131; P. J. Steinhardt, Harvard preprint (1981).

(62) S. Coleman and E. Weinberg, Phys. Rev. D7 (1973) 1888.

(63) A. Masiero, R. N. Mohapatra and R. D. Peccei, Max Planck preprint 27/81 (1981).

(64) I. Yu Kobsarev, L. B. Okun and Ya. B. Zeldovich, Phys. Lett. 50B (1974) 340.

(65) K. Sato, Phys. Lett. 99B (1981) 66.

(66) M. Horibe and A. Hosoya, Osaka preprint HET 41 (1981); Y. Fujii, Phys. Lett. 103B (1981) 29.

(67) W. H. Press, Physica Scripta 21 (1980) 702; M. S. Turner and D. N. Schramm, Nature 279 (1979) 1979.

(68) R. Fabbri et al., Phys. Rev. Lett. 44 (1980) 1563; S. P. Boughn et al., Astrophys. J. Lett. 243 (1981) L113.

(69) J. D. Barrow and M. S. Turner, Chicago preprint (1981); J. R. Bond, E. W. Kolb and J. Silk, Berkeley preprint (1981).

COSMOLOGICAL CONSEQUENCES OF FIRST-ORDER PHASE TRANSITION

Katsuhiko Sato

Department of Physics, Kyoto University, Kyoto 606, Japan

in

Proceedings of the Fourth Kyoto Summer Institute on Grand Unified Theories and Related Topics, Kyoto, Japan, June 29 - July 3, 1981, ed. by M. Konuma and T. Maskawa (World Science Publishing Co., Singapore, 1981).

Contents

COSMOLOGICAL CONSEQUENCES OF FIRST-ORDER PHASE TRANSITION

Katsuhiko Sato

Department of Physics, Kyoto University, Kyoto 606

Abstract

Recently, a first-order phase transition of a vacuum has become a topic of great interest, because it provides a suppression mechanism against overproduction of monopoles, a reason for homogeneity and isotropy of the universe, and a possibility that the universe has a baryon-number domain structure. This is a review on the cosmological consequences of first-order phase transition, starting from an elementary introduction and ending with the surprising consequence "multiproduction of the universes".

1. Introduction

Recently it has been shown that gauge theories with a spontaneously broken symmetry not only provide a possibility of unifying the four basic interactions in nature, but also imply interesting cosmological predictions.[1)] One of the most interesting predictions is a cosmological evolution of interactions; namely, that there may have been solely a unitary interaction between particles when the universe was created. This unitary interaction, however, may have shot branches by the successive cosmological phase transitons of vacua with the expansion of the universe, and thus the present four basic interactions may have been evolved in a similar way to the evolution of creatures (see Fig. 1-1). It is very natural to question whether these phase transitions of vacua necessarily affected the cosmic evolution or not.

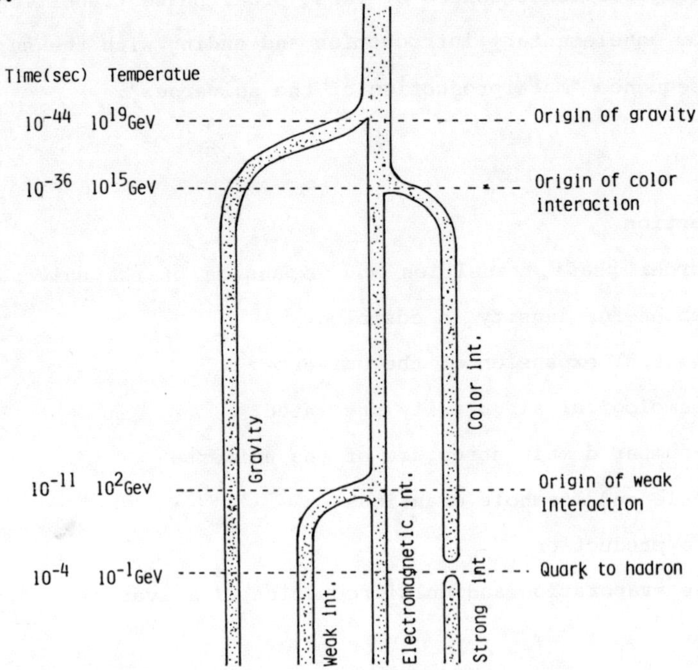

Fig. 1-1 Evolution of interactions

In a first half of the present lecture (sections 2 and 3), I will show that the standard big bang model is greatly modified and some difficulties in this model may be removed, if the phase transition of vacuum is of first order. In a last half (sections 4 ~ 6), I will discuss the first-order phase transition taking into account a general relativistic effect and will show that black holes and wormholes are produced copiously by the phase transition.

2. First order phase transition and expansion of the universe

2.1 Vacuum energy density in cosmology

If we take a simple ϕ^4 model for a model of phase transition, the energy density of the vacuum $V(\phi,T)$ is given by

$$V(\phi,T) = -\frac{1}{2}\mu^2\{1-(T/T_c)^2\}\phi^2 + \frac{\lambda}{4}\phi^4 + \rho_v \qquad (2.1)$$

where ϕ is a Higgs field or an order parameter of the vacuum. The critical temperature is of an order of $T_c=\mu/\sqrt{\lambda}$ if the fourth power of gauge coupling constant is much less than λ, i.e., $g^4<\lambda$.[2] Usually the last constant term ρ_v is neglected in particle physics, but it plays an important role in cosmology. As is well known, the energy density of the present stable vacuum with $\phi=\mu/\sqrt{\lambda}$ and $T<<T_c$ would be extremely small from the upper limit of the cosmological constant $|\Lambda|< 10^{-56}$ cm^{-2}, because the cosmological constant is given by $\Lambda=8\pi G\, V(\mu/\sqrt{\lambda},\, 0)$,[3] i.e.

$$|V(\mu/\sqrt{\lambda},0)|<10^{-122}\, m_p^4 \approx 10^{-106}\, m_x^4 \qquad (2.2)$$

where m_p is the planck mass $G^{-1/2} \approx 10^{19}$ GeV and m_x is the X-boson mass which is of an order of 10^{15} GeV. This value is extremely small if the natural scale of the energy density in super GUTs or GUTs is of an order of m_p^4 or m_x^4. In order for the condition

(2.2) to be satisfied, very fine tuning is necessary when we adjust the value ρ_V. This may be one of deep mysteries in physics.[1]

The observational fact that the present vacuum energy density $V(\phi, 0)$ is almost vanishing implies that it was huge and positive in the very early hot ($T \gg T_C$) universe. Linde[4] and Bludman and Ruderman[5] investigated whether the energy of the vacuum $V(0, T>>T_C) = \rho_V$ in the early universe would affect the expansion of the universe before phase transition or not. They showed that the energy density of a vacuum is always less than the radiation energy density T^4 if the phase transition is of a second order described by a simple ϕ^4 model (2.1). (The detailed estimates of the ratio ρ_V/T^4 in a more realistic Higgs model are shown by Linde[2] and Kolb and Wolfram.[6])

On the contrary, Sato,[7],[8] Guth[9] and Kazanas[10] showed that the vacuum energy density ρ_V becomes dominant and changes the expansion law of the universe greatly if the phase transition is of a first order. As shown in Fig. 2-1 the vacuum stays at the

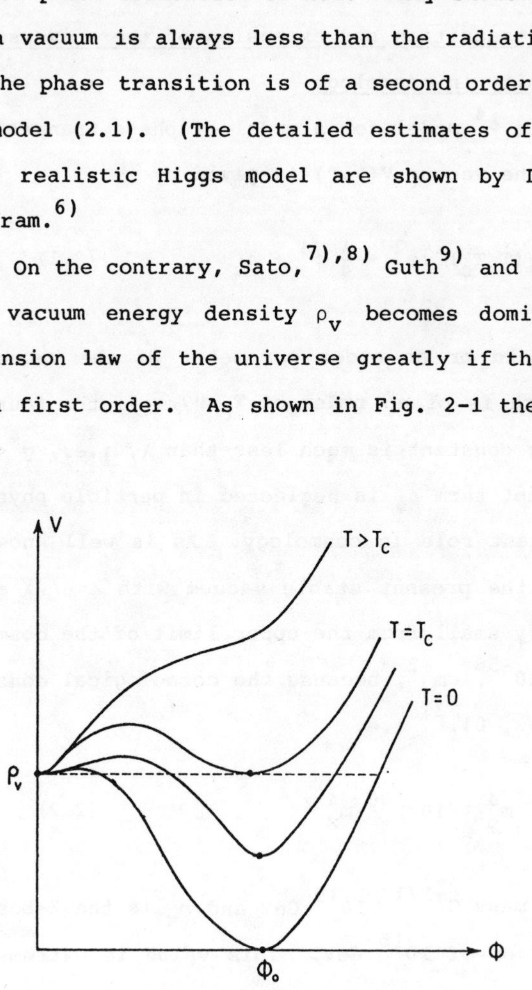

Fig. 2-1. The schematic diagraom of the effective potential of vacuum energy density.

symmetric state $\phi=0$ for a while even if the cosmic temperature decreases to lower than the critical temperature. This false vacuum $\phi=0$ decays into the stable vacuum ϕ_0 by quantum fluctuations[11)] or by thermal fluctuations.[12)] Obviously if the potential barrier between two vacua $\phi=0$ and ϕ_0 is high and the transition time-scale is much longer than that of cosmic expansion, the energy density of the vacuum becomes dominant and changes the expansion law of the universe, because the radiation energy density decreases with cosmic expansion, but ρ_v is a constant. (See Fig. 2-2)

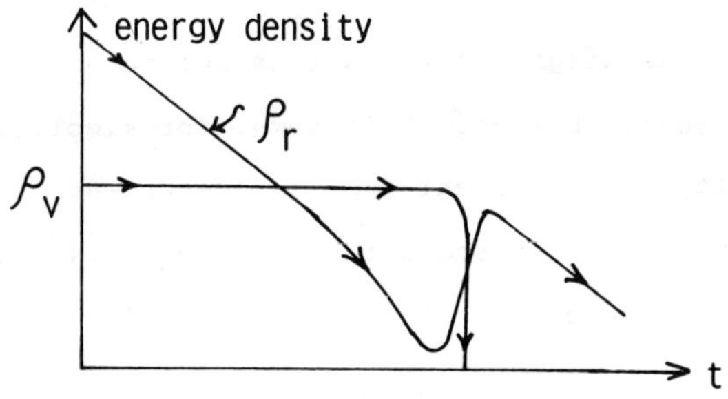

Fig. 2-2 The evolutions of the radiation energy density ρ_r and the vacuum energy density ρ_v

2.2 Exponential expansion of the universe

If we assume that our universe is homogeneous and isotropic, the metric takes the form

$$ds^2 = -dt^2 + R^2(t)\left(d\chi^2 + \begin{Bmatrix} \sin^2\chi \\ \chi^2 \\ \sinh^2\chi \end{Bmatrix} d\Omega^2\right) \tag{2.3}$$

$$\text{for } k = \begin{cases} +1; & \text{Closed universe} \\ 0; & \text{flat universe} \\ -1; & \text{open universe} \end{cases}$$

Then time evolution of a cosmic scale factor of the universe R(t) is described by the following equation

$$\dot{R}^2/R^2 + k/R^2 = 8\pi G(\rho_v + \rho_r(t))/3 , \tag{2.4}$$

where a dot denotes time differentiation and ρ_v is the vacuum energy density. The radiation energy $\rho_r(t)$ is given by

$$\rho_r(t) = N_b \pi^2 T^4(t)/30 , \tag{2.5}$$

where N_b is the statistical weight of radiations before the phase transition in units of scalar bosons. We assume, for simplicity, the vacuum energy density is a constant and is independent of time before the phase transition. Then the solution to Eq.(2.4) before the phase transition is given by

$$(R(t)/\ell)^2 = \begin{cases} b^{1/2} \sinh(2t/\ell) & \text{for } k=0 , \\ k/2 + (b-1/4)^{1/2} \sinh(2t/\ell) & \text{for } k\neq 0,\ b>1/4, \\ k/2 + \exp(2t/\ell) & \text{for } k\neq 0,\ b=1/4, \\ k/2 + (1/4-b)^{1/2} \cosh(2t/\ell) & \text{for } k\neq 0,\ b<1/4, \end{cases} \tag{2.6}$$

where ℓ is the characteristic scale factor of this model;

$$\ell \equiv (8\pi G\rho_v/3)^{-1/2} \tag{2.7}$$

and b is the non dimension numerical factor which represents the "size" of the universe;

$$b \equiv \rho_r(t)R^4(t)/\rho_v \ell^4 \, , \tag{2.8}$$

which remains constant when the universe evolves adiabatically. As easily seen from Eq.(2.6), the situation is very different according whether $b < 1/4$ or $b > 1/4$. In the limit $b >> 1/4$ or the case $k=0$, the Eq.(2.3) reduces to

$$R(t) = b^{1/4} \sinh^{1/2}(2t/\ell) = \begin{cases} b^{1/4}\ell(2t/\ell)^{1/2} & ; \quad t << \ell/2 \quad (2.9a) \\ b^{1/4}\ell/\sqrt{2}\, \exp(t/\ell) & ; \quad t >> \ell/2 \, , \quad (2.9b) \end{cases}$$

independently of the value of k. As seen from Eq.(2.9a), the universe expands as $\sqrt{t}$ in the standard big bang model when $t << \ell/2$, bacause the radiation energy density is dominant in this period. After the cosmic time $\ell/2$, however, the universe begins to expand exponentially because the vacuum energy density becomes to be dominant. This exponential expansion continues until the vacuum energy density disappears by the termination of the phase transition. Recall also that the temperature before the phase transition varies as $T \propto R^{-1}$, so the time-temperature relation can be written as

$$T = T_v \, N_b^{-1/4} \sinh^{-1/2}(2t/\ell) \, , \tag{2.10}$$

where T_v is defined by the relation

$$\pi^2 \, T_v^4/30 \equiv \rho_v \, . \tag{2.11}$$

Evidently, T_v may be interpreted as the temperature equivalent of the vacuum energy density if this were transformed into massless, spin-zero boson radiation.

As discussed by Coleman,[11] the first-order phase transition proceeds by nucleation and subsequent expansion of bubbles of true vacuum ($\phi=0$). The phase transition finishes when all the space is covered by bubbles (see sections 4 and 5). When the phase transition finishes, the energy density between two vacua (the false and the true vacua) ρ_v is released and the universe is strongly heated up if the energy is thermalized. In the case of the first order phase transition, however, the released energy goes to the kinetic energy of the bubble wall at the first step.[11] This kinetic energy will be thermalized when the bubbles collide with each other. In the present and the following sections 2 and 3, however, let us assume that the phase transition finishes instantaneously at cosmic time t_f, for simplicity. Then, the temperature immediately after te phase transition T_a is given as

$$T_a = (30\rho_v/\pi^2 N_a)^{1/4} = N_a^{-1/4} T_v \,. \qquad (2.12)$$

In this equality, we assumed that the released energy is equal to ρ_v , because the effective potential $V(\phi,T)$ is almost the same as $V(\phi,0)$, provided $T \ll T_c$.

Note that if the baryon number is conserved during the phase transition, the baryon/entropy ratio n_b/s is diluted very much by the entropy production due to the first order phase transition. The ratio after the phase transition $(n_b/s)_a$ is given by

$$(n_b/s)_a \approx (n_b/s)_b \sinh^{3/2}(2t_f/\ell) \qquad (2.13)$$

where $(n_b/s)_b$ is the ratio before the phase transition. As is well known, the ratio n_b/s at present is of an order of 10^{-9}. In order for this condition to be satisfied, t_f/ℓ should be 7.3 even if the ratio before the phase transition $(n_b/s)_b$ is of an order of unity. From this result, the phase transition must finish at least before $t_f/\ell<7.3$. Sato[7] pointed out that the constraint on the Higgs boson mass in Weinberg-Salam model is obtained with the aid of the present result, because the finishing time t_f sensitively depends on the effective potential, and the potential depends on the Higgs boson mass. The constraint obtained by precise numerical calculations,[15] $m_H \gtrsim 9$ GeV is, however, essentially the same as the value of Linde,[12] who obtained the constraint from a primitive condition for completion of the phase transition. This coincidence is because the transition rate (or the nucleation rate of bubbles of true vacuum) decreases very rapidly with decreasing Higgs boson mass.

In the case of the phase transition in GUTs which critical temperature is of an order of 10^{15} GeV, however, no constraint on the finishing time (or on the degree of the entropy production by the phase transition) exists, because the baryon-number asymmetry may be generated after the phase transition. Sato[8] estimated the temperature immediately after the phase transition T_a (Eq. (2.12)) with the aid of Abelian Higgs model and showed that T_a is a little higher than the Higgs boson mass but lower than the lepto-quark boson mass. This result suggests that baryon-number asymmetry is generated by Higgs boson decays.[14] In the following, we discuss only the case of GUT-phase transition in which no constraint on the entropy production exists.

The subsequent evolution of the universe after the phase transtion in the present model is essentially the same as that of the standard big bang model: The cosmic scale factor after the phase transition is given by

$$R(t) = R(t_f)\{2(t-t_f)/\ell + 1\}^{1/2}$$

$$\approx b^{1/4}\, e^{t_f/\ell}\, t^{1/2}\,, \quad \text{for} \quad t>t_f>\ell\,. \qquad (2.14)$$

One of the most important differences between the standard model and the present model is the size of a particle horizon, which is defined as[3)]

$$r_H(t) = R(t)\int_0^t dt'/R(t')\,. \qquad (2.15)$$

Roughly speaking, this is the maximum size of a region in which causal relations can exist. In the standard model ($R\propto t^{1/2}$), the horizon is $r_H(t)=2t$ and very small compared with the scale factor of the universe in the early universe. On the other hand, the horizon in the present model, which is calculated as

$$r_H(t) = \begin{cases} \ell\, F(\cos^{-1}(e^{-t/\ell}),1/\sqrt{2})\sinh^{1/2}(2t/\ell)\,, & \text{for} \quad t<t_f \\ \ell\, F(\pi/2,1/\sqrt{2})\exp(t_f/\ell)\,(t/\ell)^{1/2} + 2t\,, & \text{for} \quad t>>t_f \end{cases} \qquad (2.16)$$

can become very large, where $F(\cdot,\cdot)$ is the elliptic integral of the first kind and the value $F(\pi/2, 1/\sqrt{2})$ is about 1.85. As shown in Fig. 2-3, the particle horizon also increases exponentially as the scale factor increases, and becomes very larger than that of the standard model. As is well known from the observation of the cosmic back ground radiation, the universe is surprisingly homogeneous and isotropic. This fact seems to suggest that a homogenizing process had worked in the early universe. In the standard model, however, this process can never work in principle because the particle horizon is very small compared with the scale of the universe in the early universe. Obviously this difficulty is

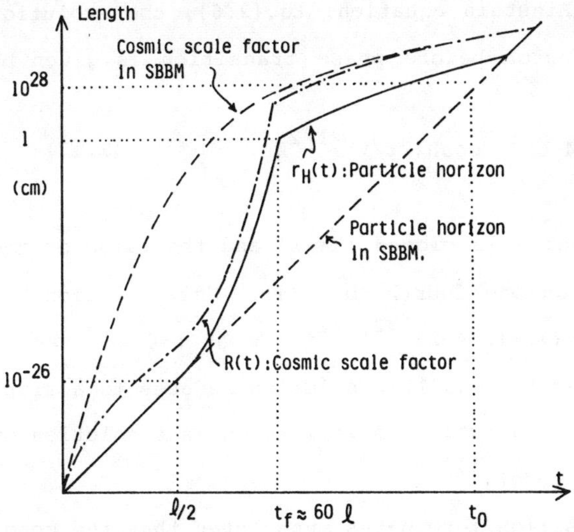

Fig. 2-3. An example of evolutions of the particle horizon and the cosmic scale factor. If $t_f \gtrsim 60$, the particle horizon at present becomes greater than $H_0^{-1} \approx 10^{28}$ cm and the homogeneity of the universe may be explained.

removed in the present model, because the horizon can become very large. If we assume that the finishing time t_f is later than the cosmic time 60ℓ, the regions which we are observing at present (the distances are comparable to or less than $H_0{}^{-1} \approx 10^{28}$ cm) had been in the particle horizon when the cosmic time is $t= \ell/2 \approx 10^{-36}$ sec as shown in Fig. 2-3. We may, therefore, presume that the universe is homogenized at this very early time 10^{-36} sec.

2-3 Is cosmological singularity prevented?[15)]

In the standard big bang model, the universe starts from a singularity, namely, the cosmic scale factor is completely vanishing at the origin of the cosmic time t=0. Now I show that this singularity is easily prevented in the model of the universe with the first-order phase transition. As we have already obtained the solutions of the Einstein equation, Eq.(2.6), the evolution of the cosmic scale factor before phase transition is given by

$$R(t) = \{1/2 + (1/4-b)^{1/2}\cosh(2t/\ell)\}^{1/2}\ell \, , \qquad (2.17)$$

provided that the universe is closed (k=+1) and the value of the parameter b is less than one fourth ($b < 1/4$). This solution has an absolute minimum $\{1/2+(1/4-b)^{1/2}\}^{1/2}\ell > 0$ at t=0 and has no singularity. In the limit $b \to 0$, this solution reduces to a simple equation $R(t) = \ell \cosh(t/\ell)$, which is well known as a solution of the de Sitter universe model.[3)]

If the phase transition terminates much later than the cosmic time ℓ, $t_f \gg \ell$ and the energy released by the phase transition goes to radiation instantaneously at this time, $\rho_r(t_f)=\rho_v$, the value of the parameter b immidiately after the phase transition is calculated from Eq.(2.17) as

$$b_a \equiv \rho_r(t_f)R^4(t_f)/\rho_v\ell^4 \approx \exp(4\, t_f/\ell)/16 \, . \qquad (2.18)$$

As discussed in the subsection 2.2, the evolution of the universe is essentially the same as that of the standard model, in which the value of b remains constant approximately because the universe expands almost adiabatically. The value b_a, therefore, must be equal to the present value of the parameter b_o, which is estimated from the observation at present t_0 as

$$b_a \approx b_0 \equiv \rho_r(t_0)R^4(t_0)/\rho_v\ell^4 > 10^{101}\ (T_v/10^{15}\ \mathrm{GeV})^4 \, , \qquad (2.19)$$

where the observational facts, $\rho_r(t_0) \approx (3K)^4$ and $R(t_0) > H_0^{-1} \approx 10^{28}$ cm were used. We obtain a constraint on the finishing time t_f from Eq.(2.18) and Inequality (2.19) as

$$t_f/\ell > 60 . \qquad (2.20)$$

This is essentially the same as the condition which is necessary to explain the homogeneity of the universe (see Fig. 2-3).

We have found that the consistent model of the universe with no singularity exists if the phase transition is strongly of first order. It should be, however, noticed that this model also implies an inevitable collapse of the universe in future. In the closed universe model, the expansion stops eventually and the universe begins to contract. If the value of b is less than 1/4, the universe bounces and the collapse will be prevented by the phase transition to the false vacuum. But the value of the parameter b_0 at present ($b_0 > 10^{101}$) indicates that the collapse of our universe is inevitable in future, even if the phase transition to the false vacuum occurs by the increase of the cosmic temperature.

3. Baryon-number domain structure [8)]

It has been suggested recently that the baryon number asymmetry of the universe is a natural consequence of the baryon- and CP-nonconserving interactions in the modern grand unified theories (Lecture given by Yoshimura). This theory is very interesting from the point of view that the gauge theory with spontaneous symmetry breaking developed recently also account for one of the most important problems of cosmology. In this theory, however, we must break the symmetry of the physical law in a special direction in order to explain this asymmetry of the universe, i.e., the sign of the CP-violating interactions would be the same over all in space. But it seems natural that the CP-conservation is broken spontaneously,[14),16)] if basic law of physics is symmetrical.

This assumption, however, predicts the baryon-number symmetric universe with local fluctuations of the baryon number. Stecker and his collaborators[17)] considered that baryon-number domains grow to the astrophysical size, such as super-clusters of galaxies, by the merging process proposed by Omnes.[18)] But the size of CP-domains is, at most, the particle horizon when the CP-conservation is broken spontaneously. It seems to be very difficult that these small domains grow sufficiently in the standard model of the universe.

Now we show that if the GUT-phase transition is of first order, this difficulty is removed and the baryon number domain structure is realized without observational conflicts.

First we assume that the CP-domain structure of the universe was created by the spontaneous breaking of the CP-conservation at the cosmic time t_{cp} and the radius of the domains is of an order of the particle horizon at this time. The CP-breaking time t_{cp} is a free parameter in the present discussion. It is, however, reasonable to presume that the time t_{cp} is equal to or earlier than the time of the phase transition in GUTs, but is later than the Planck time, i.e., $\ell/2 > t_{cp} > t_p$. Then the initial radius of CP-domains is $L(t_{cp})=2\, t_{cp}$.

Second, we assume that the GUT phase transition is of first order as discussed in the preceding section 2 and the finishing time of the phase transition is much later than the time ℓ, $t_f > \ell$. Since the size of the CP-domains is streched by the cosmic expansion, the radius of the CP-domains also increases exponentially. The radius at the finishing time $L(t_f)$ is calculated from Eq.(2.9) as

$$L(t_f) = L(t_{cp})\{\sinh(2t_f/\ell)/\sinh(2t_{cp}/\ell)\}^{1/2}$$

$$\approx (t_{cp}\ell)^{1/2} \exp(t_f/\ell) , \quad \text{for } t_{cp}<\ell<t_f . \qquad (3.1)$$

As has been discussed, if the cosmic temperature immediately after the phase transition $T_a = N_a^{-1/4} T_v$ (Eq.(2.12)) is sufficiently high, the baryon-number asymmetry is created in each CP-domain by the decays of Higgs bosons in the course of the following expansion of the universe. The subsequent evolution of the universe after the phase transition in the present model is esentially the same as that of the standard model if pair annihilations of the baryon and antibaryon domians are negligible. Now we consider the case that the domain are extremely stretched and that the radius of the domains at present is comparable to the mean separation distance between super-clusters of galaxies, d. The radius of the domain at present L_0 is estimated by the ratio of the present cosmic radiation temperature T_0 to T_a,

$$
\begin{aligned}
L_0 &= L(t_f)(N_a/N_0)^{1/3}(T_a/T_0) \\
&= (2\, t_{cp}/\ell)^{1/2}\, e^{t_f/\ell}\, (N_a/N_0)^{1/3}(T_a/T_0) \\
&\approx 100(t_{cp}/\ell)^{1/2}\, e^{t_f/\ell}\ \text{cm} \qquad (3.2)
\end{aligned}
$$

where the statistical weight of particles at present N_0 is 2+22/11 if the number of neutrino species is three, i.e., ν_e, ν_μ and ν_τ. In the third equality, we assumed, $\ell=10^{-36}$ sec$=10^{-26}$ cm, $T_a=10^{15}$ GeV, $T_0=3$ K and $N_a/N_0=10$. From the condition that the present baryon-number domains should be greater than d, i.e. $L_0 > d$, we obtain the following inequality,

$$
t_f/\ell > 57 + 1/2\, \ell n(\ell/t_{cp}) \qquad (3.3)
$$

where we assumed d=100 MPC. If the CP-breaking arises later than the planck time $t_p \approx 5\times 10^{-44}$ sec, the inequality (3.3) reduces $t_f/\ell > 58$. As Linde[2)] has shown, the finishing time of the phase transition becomes infinite in the limitting case $\lambda \to 3g^4/16\pi^2$ in the abelian Higgs model. It seems, therefore, reasonable to presume that the condition $t_f/\ell > 58$ is satisfied in GUT phase transition. We may conclude that if GUT phase transition is strongly of first order and the condition (3.3) is satisfied, the size of the stretched domains is sufficiently big to avoid pair annihilations and the baryon-number domain structure of the universe is realized without observational conflicts.

4. Black hole and wormhole creations[19)]

In preceding sections, we have assumed that the universe is homogeneous even in the period of the phase transition and the phase transition finishes instantaneously. Stricktly speaking, this is not the case. As discussed by Coleman,[11)] the first-order phase transition proceeds by nucleation and subsequent expansion of bubbles. In the period of the phase transition, the universe is never homogeneous, but is extremely bumpy because the contrast between the energy densities of the inner and the outer regions is very huge. What is then the space-time structure of the universe and how is the present homogeneous universe recovered? In order to investigate this problem, we make the following assumptions for simplicity.

1) The bubbles expand at light velocity isotropically with no energy dissipation. The energy released by the phase transition goes to the kinetic energy of bubble walls.[11)]

2) The radiation energy density is negligible compared to that of that of the false vacuum, ρ_V, which is constant with time, and that the energy density inside the true vacuum bubbles is completely vanishing.

3) The universe is flat.

Then, the metric of the universe before the nucleation of bubbles is written as

$$ds^2 = -dt^2 + R^2(t)(d\chi^2 + \chi^2 d\Omega^2) \tag{4.1}$$

with the cosmic scale factor R(t) given by

$$R(t) = \alpha_* \ell \exp((t-t_N)/\ell) \ , \tag{4.2}$$

where ℓ is the characteristic length of the present model and α_* is a non-dimensional numerical factor which represents the scale of the universe at $t=t_N$ (see Eqs.(2.6) ~ (2.8)). The cosmic time t_N denotes the ignition time of the phase transition and we presume it to satisfy the relation $t_N \gg \ell$.

At the early stage of the phase transition, overlapping of these bubbles can be ignored, because the volume of each bubble as well as the number density of them is very small. Therefore, it is sufficient to consider an isolated bubble (true vacuum) surrounded by false vacuum. By virtue of Birkoff's theorem[20] the metric in a (spherical symmetric) bubble must be Minkowskian. The embedding of the Minkowski metric into the de Sitter-like universe was discussed in Ref. (21).

On the other hand, at the late stage of the phase transition, most of the space is covered by these bubbles. According to percolation theories,[22] infinite-size networks of false vacuum disappear when a volume fraction of false vacuum regions u(t) becomes less than a critical value $u_c(\approx 0.3)$. This means all the false vacuum regions are surrounded by bubbles. In this case, the geometry of the universe becomes very complicate. In order to make clear the metric, we adopt a spherically symmetric model, in which an infinite number of bubbles are nucleated on a sphere of radius $\chi=\chi_0$ simultaneously at time $t=t_0$ ($> t_N$), and they grow with walls expanding isotropically at the speed of light, forming a spherical shell-like region (see Fig. (4-1)). From causality,

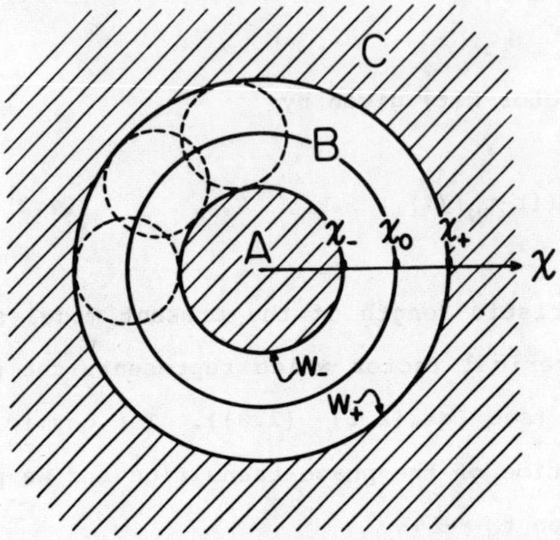

Fig.4-1 A simplified model for a false vacuum surrounded by bubbles.

the geometries of the false vacuum regions A and C are unaffected by the existence of the bubbles and still described by Eq.(4.1). Meanwhile, the shell-like true vacuum region B is found to be described by the Schwarzschild geometry from Birkoff's theorem,[20]

$$ds_B^2 = -(1-r_g/r)d\tau^2 + (1-r_g/r)^{-1}dr^2 + r^2 d\Omega^2 . \quad (4.3)$$

The corresponding Schwarzschild mass M and radius r_g are given by

$$M = \frac{4\pi}{3}\rho_v\{R(t_0)\chi_0\}^3 = (\alpha\chi_0)^3(4\pi\rho_v\ell^3/3) \quad (4.4)$$

and

$$r_g = 2GM = (\alpha\chi_0)^3\ell \ ; \ \alpha \equiv \alpha_* \exp\{(t_0-t_N)/\ell\} . \quad (4.5)$$

With the expansion of the region B, the coordinate radii $\chi_\pm$ of the outer and the inner walls $W_\pm$ change as

$$\chi_\pm(t) = \chi_0 \pm \int_{t_0}^{t} dt'/R(t') = \chi_0 \pm \alpha^{-1}\{1-\exp[-(t-t_0)/\ell]\} \quad (4.6)$$

so that their proper radii are given as

$$r_\pm(t) = R(t)\chi_\pm(t) = \ell\{(\alpha\chi_0 \pm 1)\exp[(t-t_0)/\ell] \mp 1\} \quad (4.7)$$

the paths of the walls $W_\pm$ in the Penrose diagram of the Schwarzschild geometry are shown in Fig. 4-2.

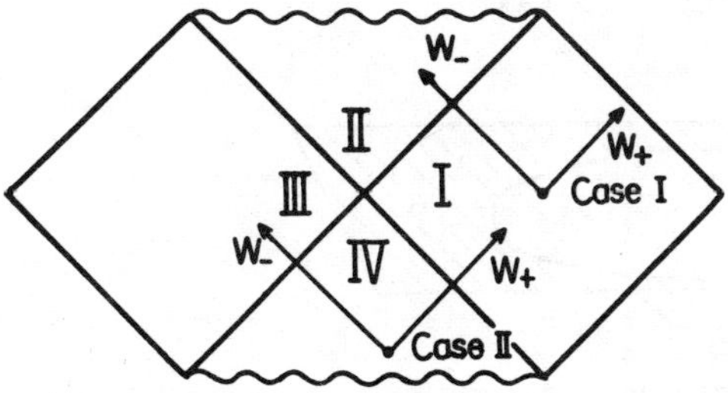

Fig. 4-2 Paths of the walls $W_\pm$ in a Penrose diagram of the Schwarzschild geometry.

If $\chi_0 < \alpha^{-1}$, r_- is initially greater than r_g, but decreases monotonically and vanishes within a finite time. Thus the inner false vacuum region eventually vanishes and a black hole of mass M remains in this case. On the other hand, if $\chi_0 > \alpha^{-1}$, r_- is initially smaller than r_g, but increases monotonically (and becomes infinite.) This means that the inner region A remains as an ever expanding de Sitter-like sub-universe connected with the outer universe C by an Einstein-Rosen (wormhole) bridge B (see Fig. 4-3 and 4-4). In order to confirm the consistency of our model we must show that we can join this geometry with the original metric Eq.(4.1) along the walls $\chi = \chi_+$ and χ_-. As an illustration we

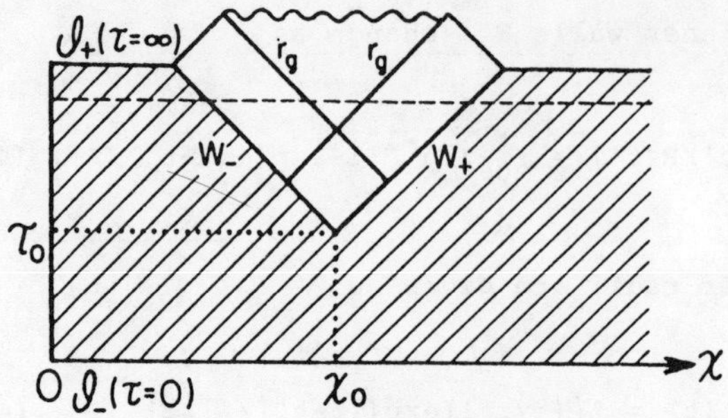

Fig. 4-3 A conformal diagram for de Sitter-like universe with the Schwarzschild wormhole. Null lines are at ±45°. A space-like hypersurface represented by the dashed line is shown schematically in Fig. 4-4.

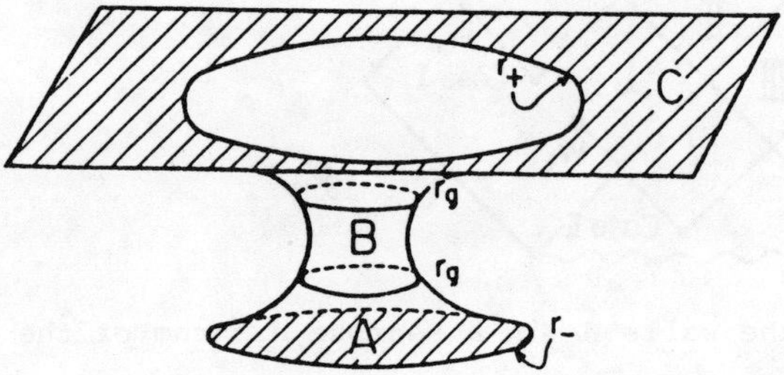

Fig. 4-4 A schematic picture of a wormhole created by the first order phase transition.

do this junction in detail for $\chi = \chi_+$. The boundary surface is expressed as

$$(1-r_g/r)d\tau^2 - (1-r_g/r)^{-1}dr^2 = dt^2 - R^2(t)d\chi_+^2 = 0 \ . \ (4.8)$$

Since this surface is null and the geometry is spherically symmetric, the junction condition reduces to just the fitting of the surface areas;

$$R(t)\chi_+(t) = r(\tau) \ . \tag{4.9}$$

This condition yeilds the relation between two time coordinates t and τ:

$$\tau = r_+(t) - r_+(t_0) + r_g \ \ell n \frac{r_+(t)-r_g}{r_+(t_0)-r_g} \tag{4.10}$$

where $r_+(t)$ is given in Eq.(4-7). In the limit $t-t_0 >> 1$ we obtain

$$\tau \approx (1+\alpha\chi_0)\ell \ \exp\{(t-t_0)/\ell\} \ . \tag{4.11}$$

In reality, the trapped false vacuum regions are highly asymmetric and new bubbles are further nucleated therein. However, we consider the above model to represent the essential feature of the actual situation. Then one naturally expects these wormholes (and black holes) to play an important role in the subsequent evolution of the universe.

5. Monopole production[23)]

Recently, it has been argued that overproduction of monopoles in the very early universe may be suppressed if the phase transition is of first order.[24)~26)] In the cosmological model in which the universe experiences a first order phase transition,[7)~9)] however, there was a serious problem to be solved as stressed by Einhorn and Sato[26)]: It is usually considered that the phase transition terminates when the volume fraction of false vacuum regions u(t) vanishes completely.

The volume fraction of false vacuum u(t) is described as[25)~26)]

$$u(t) = \exp[-\int_0^t dt'\ p_N(t')R^3(t')4\pi\chi^3(t',t)/3] \qquad (5.1)$$

where $x(\tau,t)$ is the radius (in coordinate units) of a bubble nucleated at time τ, given by $x(\tau,t)=\int_\tau^t dt'/R(t')$. The fraction u(t), however, does not vanish within a finite time unless the nucleation rate diverges, but infinite cosmic time is necessary to make u(t) vanish.[25)] Moreover, the entropy per unit comoving volume increases forever,[26)] i.e., the phase transition is never completed in the universe. This is due to the exponential expansion of the false vacuum regions. Even if the volume fraction u(t) extremely decreases, the exponential expansion never ceases. Now this defect is removed if we take into account the space-time structure of the false vacuum regions surrounded by bubbles: As discussed in the preceding section, infinite-size networks of false vacuum disappear when u(t) becomes less than a critical value $u_c(\approx 0.3)$. This means all the false vacuum regions are surrounded by bubbles. The result of the preceding section 4 predicts that false vacuum regions surrounded by bubbles become black holes or wormholes (hereafter denoted as BWHs) and drop out of our universe. We, therefore, may consider that the first-order phase transition finishes when $u(t)=u_c$.

Obviously, if the false vacuum regions surrounded by bubbles have non-vanishing winding number, the resulting BWHs have the corresponding magnetic charge (Reissner-Nordstrom type black hole or wormhole). This result implies all the monopoles are produced as magnetized black holes or wormholes in the first order phase transition, and after the evaporation of these black holes and wormholes, 't Hooft-Polyakov type monopoles remain. Now let us investigate whether overproduction of monopoles is suppressed in this scenario without conflicts with cosmological observations or not. For simplicity, we assume the nucleation rate of bubbles is given by the following simplified equation,[7),26)]

$$p_N(t) = \nu_T \ell^{-3}\delta(t-t_N) + \nu_Q \ell^{-4}\theta(t-t_N) , \quad (5.2)$$

where the first term represents the nucleation by thermal fluctuations and the second term, the nucleation by quantum fluctuations. We also assume that the ignition time of nucleation t_N is much later than the characteristic cosmic time $\ell/2$, i.e., the nucleation starts after the universe begins to expand exponentially. (see Eq.(2.9)). Then, substituting the Eqs.(5.2), (2.9a) into Eq.(5.1), we obtain

$$u(t) = u_T(t)u_Q(t) , \quad (5.3)$$

where $u_T(t)=\exp\{-(4\pi/3)\nu_T(1-e^{-(t-t_N)/\ell})^3\}$ and $u_Q(t)=\exp\{-(4\pi/3)\nu_Q(t-t_N)/\ell\}$ for $(t-t_N)\gg\ell$.

The number density of BWHs is the number density of false vacuum regions trapped by bubbles. The fraction of magnetized BWHs is given by the group theoretic factor p as shown by Kibble.[27)] We assume the number density of BWHs at the finishing time is equal to the number density of bubbles at this time $n_B(t_f)$ and the number density of magnetized BWHs $n_M(t_f)$ is given by $n_M(t_f)=pn_B(t_f)$, where

$$n_B(t_f) = \int_0^{t_f} p_N(t')u(t')R^3(t')dt'/R^3(t_f) . \quad (5.4)$$

When the phase transition finishes, the universe is strongly heated up. As discussed in section 2.2, we simply assume that the energy released by the phase transition is thermalized instantaneously at t_f. Then the temperature and the entropy after the phase transition T_a and $S(t_f)$ are easily calculated as $T_a = N^{-1/4}T_v$

and $S(t_f) = 2\pi^2 N^{1/4} T_v^3/45$, where T_v is defined by Eq.(2.11).

The volume fraction of false vacuum regions in the absence of quantum tunneling $u_T(t)$ becomes $\exp(-4\pi\nu_T/3)$ in the limit $t\to\infty$, i.e. $u_T(\infty)=\exp(-4\pi\nu_T/3)$. If the nucleation rate ν_T is less than a critical value ν_T^c, which is defined as

$$\nu_T^c \equiv -(3/4\pi)\ln u_c \approx 0.29 \qquad (5.5)$$

the value $u_T(\infty)$ is greater than u_c, i.e. the phase transition never finishes without the nucleation by quantum fluctuations. On the contrary, in the case $\nu_T > \nu_T^c$, the phase transition finishes even if nucleation by quantum fluctuations is completely negligible. In order to make the situation clear, we discuss these cases separately.

Case I: $\nu_T \geq \nu_T^c$ and $\nu_Q = 0$

In this case, the number density of bubbles at t_f is easily calculated as $n_B(t_f) = \nu_T \ell^{-3}\{1-(\nu_T^c/\nu_T)^{1/3}\}^3$. Then the magnetized BWH/entropy ratio after the phase transition is obtained as

$$(n_M/s)_{t_f} = 2\pi(8\pi^3/90)^{1/2}\, p\nu_T N^{-1/4}(T_v/m_p)^3\{1-(\nu_T^c/\nu_T)^{1/3}\}^3, \qquad (5.6)$$

where m_p is the Planck mass. If the magnetized BWHs have evaporated and simple monopoles remain at present, this ratio must be less than 10^{-25}, which is obtained from the upper limit of the energy density of the present universe.[28] From this condition, $(n_M/S)_t < 10^{-25}$, we obtain the following severe constraint on the nucleation rate ν_T, combining the condition $\nu_T^c \leq \nu_T$,

$$\nu_T^c \leq \nu_T < \nu_T^c(1+\delta), \qquad (5.7)$$

where $\delta \approx 1.0\times 10^{-4}$. In this evaluation, we assumed N=100, p=1/10,

$T_V/m_p=10^{-4}$ and $u_c=0.3$. This result shows that very unnatural fine tuning is necessary for the suppression of monopole overproduction.

If the entropy production due to BWHs evaporation is significant, the ratio (5.6) is diluted and the constraint (5.7) may be relaxed. But this entropy production must be small, because this also makes the baryon/entropy ratio reduce. According to the result of the preceding section 4, the mass of BWHs is estimated as $M=\rho_V d^3$, where d is the mean separation distance of bubbles when they are created, i.e., $d=n_B(t_N)^{-1/3}=\nu_T^{-1/3}\ell$. The ratio of the density of BWHs to that of radiations at the cosmic time t_f, which is given by

$$D(t_f) \equiv M\, n_B(t_f)/(\pi^2 N\, T_a^4/30) = \{1-(\nu_T^c/\nu_T)^{1/3}\}^3 , \qquad (5.8)$$

is very small if the value of nucleation rate ν_T is in the region (5.7). With the cooling of the universe, however, this ratio increases and the mass density of BWHs becomes dominant at the cosmic time $t_m=\ell/2\{1-(\nu_T^c/\nu_T)^{1/3}\}^{-6}$. Obviously, if the BWHs evaporate before this time t_m, the entropy production is negligible. The evaporation time of BWHs with the mass $\rho_V d^3$ is given by $\tau_{ev}=3\times N^{-1}\nu_T^{-3}(m_p/T_V)^4\ell$. From the condition $\tau_{ev}<t_m$, we obtain the condition

$$\nu_T < \nu_T^c(1+\delta') \qquad (5.9)$$

where $\delta'=5\times 10^{-3}$.

Since this condition is automatically satisfied if the constraint (5.7) is satisfied, we may conclude that the entropy production by the BWH evaporation is negligible, and the ratio (5.6) and the baryon/entropy ratio are never diluted.

Case II: $\nu_T < \nu_T^c$ and $\nu_Q << 1$

In this case, the number density of bubbles at the cosmic time t_f (Eq.(5.6)) is rewritten as

$$n_B(t_f) = (\nu_Q u_c/3)\ell^{-3} \tag{5.10}$$

provided $(t_f - t_N)/\ell >> 1$, because the number density of bubbles created by thermal fluctuation is negligible. Then, magnetized BWH/entropy ratio is obtained as

$$(n_M/s)_{t_f} = 2\pi(8\pi^3/90)^{1/2}\, p\nu_Q\, N^{-1/4}(T_v/m_p)^3(u_c/3) \ . \tag{5.11}$$

In order for this ratio to be less than 10^{-25}, the nucleation rate ν_Q must be less than 3×10^{-12} i.e.,

$$\nu_Q < 3\times10^{-12} \tag{5.12}$$

where we assumed $p=1/10$, $N=100$, $T_v/m_p=10^{-4}$, $u_c=0.3$ as in Case I.

When the nucleation rate is a constant and independent of time, BWHs would have the continuous mass spectrum. The characteristic mass is estimated as $M=4\pi\rho_v\ell^3/(-3\ell n u_c)$. Then we can easily derive the time when the mass density of BWHs become dominant, t_m and the evaporation time τ_{ev} following the discussion shown in Case I,

$$t_m=\{(-3\ell n u_c)/(4\, u_c\,\nu_Q)\}^2\,\ell/2 \tag{5.13}$$

and

$$\tau_{ev}=3\times N^{-1}\,(-3\ell n u_c/4\pi)^{-3}(m_p/T_v)^4\ell. \tag{5.14}$$

In order for the entropy production due to the evaporation to be negligible, the condition $\tau_{ev} < t_m$ must be satisfied. From this condition, we obtain

$$\nu_Q < 4\times10^{-7} . \qquad (5.15)$$

Since this condition is sufficiently satisfied if the constraint (5.12) is satisfied, the entropy production by the BWH evaporation is negligible, and the ratio (5.11) and the baryon/entropy ratio are never diluted.

In summary, we have shown that, if the nucleation rates ν_T and ν_Q are sufficiently small as shown in Fig.5-1, overproduction of monopoles is suppressed and also the baryon/entropy ratio is not diluted by the entropy prodution of BWHs.

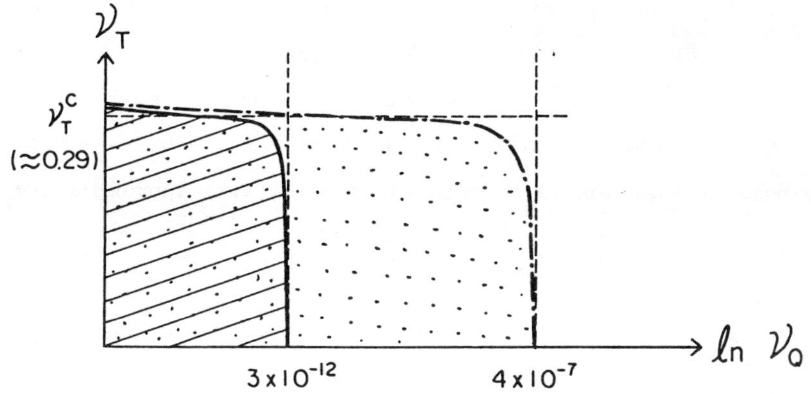

Fig. 5-1 If the values of nucleation rates ν_T and ν_Q are in the shaded region, the monopole/entropy ratio is less than 10^{-25} and the conflict with cosmology disappears. The entropy production due to the evaporation of black holes and wormholes is negligible if the values of ν_T and ν_Q are in the dotted region.

At present, the other interesting models in which overproduction of monopole is suppressed have been proposed.[29] It is, of course, very hard to judge which is the most plausible one. It should be, however, noticed that recent investigations[30] suggest strongly that the GUT phase transition is of first order.

6. Wormhole evaporation and multiproduction of universes[31]

In preceding sections 4 and 5, it was shown that wormholes are created copiously by the phase transition and the production of monoples is discussed assuming the evaporation of wormholes. But the cosmological meaning of the evaporation has not been discussed in detail. Now let us investigate the fate of wormholes created by the phase transition.

For simplicity, we assume the nucleation rate of bubbles per unit time and unit volume be independent of time and denote it by $\nu_Q \ell^{-4}$. This assumption holds when the universe cools rapidly so that the nucleation by thermal fluctuations is negligible, and the nucleation is due only to quantum fluctuations which is independent of time. Then the finishing time t_f $(u(t_f)=u_c \approx 0.3)$ of the phase transition is calculated from Eq.(5.1) approximately as

$$t_f - t_N \approx \nu_Q^{-1} \ell \ . \tag{6.1}$$

The mean separation distance of bubbles at this time is about $d = n_B(t_f)^{-1/3} \approx (\frac{10}{\nu_Q})^{1/3}$ from Eq.(5.10). We may presume that the mean diameter of trapped false vacuum regions is of the same order, i.e.,

$$r_-(t_f) \approx 0.5(10/\nu_Q)^{1/3} \ell \ . \tag{6.2}$$

By comparing this equation with Eq.(4.7), we find that, if the factor $(10/\nu_Q)^{1/3}$ is greater than unity, most of the trapped false vacuum regions become wormholes eventually. As has been discussed in the previous sections 2 ~ 3, the duration of the phase transition era must be greater than $60\,\ell$ in order to explain the homogeneity of the present universe by a GUT phase transition (as a consequence of the exponential expansion in the size of the particle horizon). If the nucleation rate ν_Q is smaller than 1/60, both this condition and the condition for the wormhole creation are satisfied.

The Einstein-Rosen bridge which connects a sub-universe (Region A in Fig. 4-4) with the original universe will disappear and the sub-universe will become entirely disconnected if wormholes evapolate as black holes do. If we take a plausible model for the phase transition, $\rho_V=(10^{15}\ \mathrm{GeV})^4$ and $\nu_Q=1/60$, the evapolation time τ_{ev} is evaluated from Eq.(5.14) as

$$\tau_{ev} \simeq 3\times10^{-22}\ \mathrm{sec}\ . \tag{6.3}$$

Note that this corresponds to a cosmic time sufficiently before the Weinberg-Salam phase transition. The time τ_{ev} in Eq.(6.3) should be considered as the lapse of the Schwarzschild time appeared in Eq.(4.3). The corresponding lapse of the de Sitter time is determined from the junction condition of the metrics (4.1) and (4.3) at the boundary, which yields a relation between the two time coordinates t and τ (Eq.(4.11)).

Phase transition proceeds also in the sub-universes. The completing time of the phase transition in these sub-universes is estimated as $t=t_0+\nu^{-1}\ell$ from Eq.(6.1), because the ignition time of the phase transition is now t_0 instead of t_N. The corresponding Schwarzschild time τ_f is given by

$$\tau_f = r_-(t_0)\exp(\nu_Q^{-1}) < 10^{26}\ell \approx 10^{-10} \text{ sec} . \quad (6.4)$$

Comparison of this equation with Eq.(6.3) shows $\tau_{ev} \gg \tau_f$, i.e., before the phase transition completes, the sub-universes connected by wormholes with the original universe become disconnected entirely.[32] We call such a sub-universe child-universe with respect to the original "mother" universe.

The radius of a child-universe immediately after the phase transition is given as

$$L \lesssim (10/\nu_Q)^{1/3} \exp(\nu_Q^{-1})\ell . \quad (6.5)$$

For $\rho_v=(10^{15}\text{ GeV})^4$ and $\nu_Q=1/60$, L is about 10 cm. This length is large enough to regard our universe as one of such a child-universe; since the universe expands 10^{28} times as large (the ratio of the GUT temperature (10^{15} GeV) to the present background radiation temperature (10^{-4} eV)) after the phase transition, the present radius of a child-universe is larger than $H_0^{-1} \approx 10^{28}$ cm (H_0: the Hubble constant). Thus there are no observational conflicts.

One easily arrives at an idea that wormholes are further created by the phase transition in the child-universes and they also evaporate to produce grandchild-universes. This sequential production of universes may continue on and on (see Fig. 6-1). Now a very peculiar consequence of a cosmological first-order phase transition is at hand; although the Creator might have made a unitary universe, the universe itself is also capable of bearing child-universes, which are again capable of bearing universes, and so on. Our universe is too old to bear more child-universes at present, though it might have born them during the period of

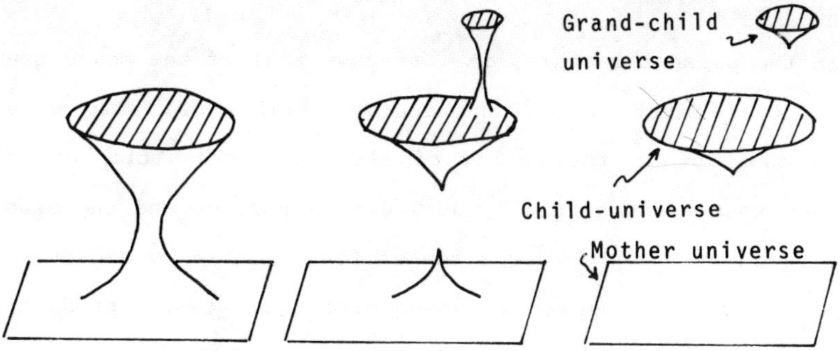

Fig. 6-1 A schematic picuture of multi-production of universes. Wormholes created by the phase transition evaporate and Einstein-Rosen bridges disappear. Wormholes are also created in the disconnected universes and they also evaporate, i.e., the disconnected universes are created infinitely.

the phase transition or/and it might have been born as one of the child-universes in this sequence.

How many child-universes can one universe bear? The number density of false vacuum domains when percolating networks of false vacua disappear is estimated from Eq.(6.2) as $n_f \approx u_c/(4\pi r_-(t_f)^3/3) = \nu/(30\ell^3)$. The volume of each disconnected universe at this instance is $4\pi L^3/3$. Thus, if all the false vacuum domains become wormholes, the production rate of child-universes per a universe is estimated as

$$p \approx n_f(4\pi L^3/3) \approx \exp(3/\nu_Q)/3 \ . \tag{6.6}$$

For ν_Q=1/60, Eq.(6.6) yields an extremely huge number 10^{77}!

7. Remarks

In the present lecture, it was shown that if the phase transition is of first order, the energy density of false vacuum becomes dominant in the period of the phase transition and the universe expands exponentially during this period, and the existence of such period provides a reason for homogeneity and isotropy of the universe, a suppression mechanism against over production of monopoles, and a possibility that the universe has a baryon-number domain structure. Furthermore it was shown that wormholes are copiously produced by the first order phase transition and the evaporation of wormholes yields a surprising prediction, namely, a sequential multi production of universes".

There remain, however, severe problems to be solved in the model of the universe with the first-order phase transition. One is on the thermalization mechanism of the kinetic energy of bubble walls at their mutual collisions. If the thermalization is ineffective and the temperature after the phase transition is lower than the mass of Higgs bosons or lepto-quark bosons, it is difficult to creat the baryon number asymmetry by the decays of these bosons. Note that, even if the kinetic energy of walls turns into the particle energy, a time longer than the characteristic time scale of the cosmic expansion is necessary in order for the radiation created at the bubble wall to go to the center of bubbles when the radius of bubbles is much greater than ℓ.

Sawyer[33)] pointed out an another possibility of the thermalization; particle creations in expanding bubbles. He showed that a bubble expanding into an infinite space filled with false vacuum can radiate the energy, released by the conversion of false vacuum to true into the interior of the bubble, in the form of particles. However, in order to make clear whether the sufficient particles are radiated or not, we must carried out more quantitative investigation, in which the effect of the reaction back on the source (wall motion) is taken into account.

Horibe and Hosoya,[34)] recently, calculated the particle creation rate in the exponentially expanding universe. They showed that the particle creations hold the temperature at the point $\ell^{-1} \approx 10^{11}$ GeV and the supercooling is prevented. Note, however, that even if the supercooling is prevented, the exponential expansion of the universe never ceases, because the radiation energy density $\ell^{-4} \approx (10^{11}\ \mathrm{GeV})^4$ is negligible compared with the energy density of the false vacuum $\rho_v \approx (10^{15}\ \mathrm{GeV})^4$. Hut and Klinkhamer[35)] discussed that the usual calculation method of the nucleation rate of bubbles[11)] breaks down in the exponentially expanding universe when the energy scale becomes less than the inverse of the event horizon $\ell^{-1} \approx 10^{11}$ GeV and speculated that the phase transition occurs at this temperature. Their guess seems to be plausible, but more rigorous proof based on the quantum field theory may be indispensable.

Abbott[36)] and Fujii[37)] also proposed a model in which the phase transition occur when the temperature becomes less than $\ell^{-1} \approx 10^{11}$ GeV. They showed that the potential barrier between false and true vacua disappears at this temperature if the coupling term $\xi R \phi^2$ exists in the Lagrangian (if $\xi = 1/12$, this gives a conformal coupling model), where R is the scalar curvature.

Barrow and Turner[38)] investigated the effects of the anisotropy of the universe on the phase transition. They showed that if the anisotropy energy density is much greater than ρ_v at $t = \ell$, the exponential expansion phase does not occur. There is no direct evidence that the early universe was extremely anisotropic, but as a reference, it was pointed out that the isothermal perturbations, which may become seeds of galaxies, can be generated in an anisotropic model with shear.[38)]

Another important problem is whether wormholes evaporate or not. The evaporation of black hole has been investigated in detail from many points of view, but the evaporarion of wormholes has not yet investigated at present. It is, of course, very reasonable to presume that wormholes also evaporate. However it seems to be not self-evident, because the global structures of black hole and wormhole are different.

It has been shown that the first order phase transition in the early universe implies many important and surprising consequences. However, in order to make clear the validity of the model of the exponentially expanding universe, more careful and precise investigations along the border line between particle physics and cosmology are necessary.

Acknowledgement

I should like to sincerely thank Prof. C. Hayashi and Prof. H. Sato for continuous encouragement and stimulating discussions. Some subjects discussed here are collaborations with M.B. Einhorn, H. Kodama, K. Maeda and M. Sasaki. I should like to thank them for important contributions and fruitful discussions.

8. References

1) See for example, A.D. Dolgov and Ya B. Zeldovich, Rev. Mod. Phys. 53 (1981) 1.

2) A.D. Linde, Rep. Prog. Phys. 42 (1979), 389 and papers cited therein.

3) See for example, C.W. Misner, K.S. Thone and J.A. Wheeler, Gravitation (W.H. Freeman and Company, 1973).

4) A.D. Linde, JETP Lett. 19 (1974), 183.

5) S.A. Bludman and M.A. Ruderman, Phys. Rev. Lett. 31 (1977), 255.

6) E.W. Kolb and S. Wolfram, Astrophys. J. 239 (1980), 428.
7) K. Sato, Mon. Not. Roy. Astron. Soc. 195 (1981), 467.
8) K. Sato, Phys. Lett. 99B (1981), 66.
9) A.H. Guth Phys. Rev. D23 (1981), 347.
10) D. Kazanas, Astrophys. J. 241 (1980), L59.
11) S. Coleman, Phys. Rev. D15 (1977), 2929.
12) A.D. Linde, Phys. Lett. 70B (1977), 306.
13) A.H. Guth and E.J. Weinberg, Phys. Rev. D23 (1981), 876.
14) See for example, Lectures given by Yoshimura at the present summer Institute.
15) S.A. Bludman, Preprint UPR-0143T (1979) - 0175T (1981).
16) Ya.B. Zeldovich, L.B. Okun and I.Yu. Kobzarev, JETP 40 (1974), 1.
17) R.W. Brown and F.W. Stecker, Phys. Rev. Lett 43 (1979), 315.
G. Senjanovic and F.W. Stecker, Phys. Lett 96B (1980), 285.
F.W. Stecker, preprint NASA TM 82083 (1980).
18) R. Omnes, Phys. Rep. 3C (1972), 1.
19) K. Sato, M. Sasaki, H. Kodama and K. Maeda, Prog. Theor. Phys. 65 (1981), 1443 and preprint KUNS 587 (1981).
20) G.D. Birkoff, Relativity and Modern Physics, (Harvard Univ. Press, Cambridge, Mass. 1923).
21) S. Coleman and F.D. Luccia, Phys. Rev. D21 (1980), 3305.
22) See for example, S. Stauffer Phys. Rep. 54 (1979), 1.
23) K. Sato, preprint KUNS 599 (1981).
24) M.B. Einhorn, D.L. Stein and D. Toussaint, Phys. Rev. D21 (1980), 3295.
G. Lazarides, M. Magg and Q. Shafi, Phys. Lett. 97B (1981), 87.
25) A.H. Guth and S.-H.H. Tye, Phys. Rev. Lett. 44 (1980) 631, 963.
26) M.B. Einhorn and K. Sato, Nucl. Phys. B180 (1981), 385.
27) T.W.B. Kibble, J. Phys. A9 (1976), 1387.

28) Ya.B. Zeldovich and M.Yu. Khlopov, Phys. Lett. 79B (1978), 239.
J.P. Preskill, Phys. Rev. Lett. 43 (1979), 1365.

29) P. Langacker and S.-Y. Pi, Phys. Rev. Lett. 45 (1980), 1.
G. Lazarides, M. Magg and Q. Shafi, preprint CERN-2856 (1980)
G. Lazarides and Q. Shafi, Phys. Lett. 94B (1980), 149.
F.A. Bais and S. Rudaz, Nucl. Phys. B170 (1980), 507.
A.D. Linde Phys. Lett. 96B (1980), 289.

30) A. Kennedy, G. Lazarides and G. Shafi, Phys. Lett. 99B (1981), 38.
M. Daniel and C.E. Vayonakos, preprint CERN-2860 (1980).
M. Daniel preprint CERN-2950 (1980).
A. Billoire and K. Tamavakis, preprint CERN-3019 (1981).

31) K. Sato, H. Kodama, M. Sasaki and K. Maeda, preprint KUNS 588 (1981).

32) Recently, Kodama et al. discussed that the fate of wormholes in detail taking into account the nucleation of bubbles in a trapped false vacuum region. The result indicates that a wormhole once formed either disappears leaving no singularity or collapses to a black hole if the phase transition in the trapped region completes before this region is disconnected by the wormhole evaporation. The condition $\tau_{ev} < \tau_f$ may be indispensable in order for these sub-universes to survive to exist. See, for detail, H. Kodama, M. Sasaki, K. Maeda and K. Sato, preprint KUNS 589 (1981).

33) R.F. Sawyer, preprint UCSB-TH-52 (1981).

34) M. Horibe and A. Hosoya, preprint OU-HET 41 (1981).

35) P. Hut and F.R. Klinkhamer, preprint (1981).

36) L.F. Abbott, preprint CERN 3018 (1981).

37) Y. Fujii, to be published in Phys. Lett. B and preprint UT-Komaba-81-11.

38) J.D. Barrow and M.S. Turner, Nature 292 (1981), 35 and preprints (1981).

On Fermion Masses and Mixings

Sandip Pakvasa

National Laboratory for High Energy Physics (KEK),

Oho-machi, Tsukuba-gun, Ibaraki-ken, 305 Japan

and

Department of Physics and Astronomy,

University of Hawaii at Manoa

Honolulu, Hawaii 96822, U.S.A.

in

Proceedings of the Fourth Kyoto Summer Institute on Grand Unified Theories and Related Topics, Kyoto, Japan, June 29 - July 3, 1981, ed. by M. Konuma and T. Maskawa (World Science Publishing Co., Singapore, 1981).

Contents

ON FERMION MASSES AND MIXINGS

S. Pakvasa* and H. Sugawara

KEK, National Laboratory for High Energy Physics
Tsukuba, Ibaraki, Japan

and

Y. Yamanaka

Department of Physics
Waseda University
Tokyo, Japan

(Presented by S. Pakvasa)

Abstract

Attempts to constrain the Fermion mass matrix by imposing permutation symmetry on the Higgs sector of $SU(2)\otimes U(1)$ are summarized. In particular, in the $SU(2)\otimes U(1)\otimes S_4$ model the KM matrix is determined completely, mass of the t-quark is predicted, and CP and flavor non-conservation arise in the neutral Higgs sector. Some of the results persist in S_n applied to n families. Restrictions on neutrino masses and mixings are also given.

*Permanent Address: Department of Physics and Astronomy
University of Hawaii at Manoa
Honolulu, Hawaii 96822 U.S.A.

INTRODUCTION

I would like to present in this talk a brief account of some attempts [1),2)] at restricting the Higgs couplings in SU(2)⊗U(1) and the resulting constraints on fermion masses and mixings. This work has been carried out by H. Sugawara, Y. Yamanaka, and I from 1977 onwards. Similar work has been done by other people.[3),4)]

The general idea of this approach is to ask whether some pattern can be recognized in the Fermion Mass Matrix already, i.e., without invoking mass scales of order $M_x \sim 10^{15}$ GeV or even 10^5 GeV. Such scales, corresponding to SU(5) or $SU(2)_R$ are presumably relevant and would give further restrictions but "can't we understand some masses here and now?" The view taken is that perhaps these higher mass scales are needed to understand m_e, m_ν but not for all masses.

To the extent that fermion masses arise from Higgs Yukawa couplings, patterns in fermion masses reflect those in Higgs Yukawa couplings. Whether the Higgs bosons are true elementary scalars or dynamical manifestations need not concern us here.

The program to be outlined below is more modest than in some GUTS[5)] in that it is possible that 1) some masses such as m_e, etc., may not be understood and 2) masses are not "calculable." All our results are at tree level and for mass ratios. However, we are more ambitious in that we present very specific proposals wherein fermion masses and mixings are <u>predicted</u> and hence are vulnerable to experiment.

It is obvious that some horizontal pattern of the form SU(2)⊗U(1)⊗G is required to constrain the fermion mass matrix. One possibility is that G is a gauge group. This has been discussed in the literature.[6)] In general, in gauging G one needs to impose more ad-hoc assumptions to make both the new gauge bosons and the new Higgs bosons sufficiently heavy. If one prefers to introduce as little new "stuff" as necessary, then it seems preferable to let G be a Global symmetry. If G is to be broken spontaneously (how else?), then to avoid zero-mass Nambu-Goldstone bosons (NGB) G should be a discrete symmetry. (Although, as we have learned recently,[7] if one is clever enough, not only can these zero-mass NGBs be made harmless, they can be even made useful!) So we are led to consider a discrete symmetry group for G to be chosen below.

It is evident that to get constraints by any of the above choices of G the Higgs sector of SU(2)⊗U(1) has to be proliferated. This is the price to be paid: either one Higgs doublet with many free parameters or several Higgs doublets with few parameters (at least in the fermion sector). With many Higgses, is general, there will be flavor-changing couplings of neutral Higgses. This need not be alarming. Since one is hoping to <u>predict</u> fermion masses and mixings, flavor-changing couplings should be allowed to occur as dictated by the model. The real constraints are <u>experimental</u>, viz., K°-$\bar{K}^\circ$, D°-$\bar{D}^\circ$ mixings, $K_L \to \mu\bar{\mu}$, $\bar{\mu}e$ rates, etc. In this connection there are a number of theorems[8] which seem to suggest that it is not possible to predict fermion mixings. However, the proofs depend upon forbidding flavor-changing couplings of neutral Higgses. Once they are allowed as just discussed, the theorems are irrelevant.

THE S_n MODELS

Next we turn to the choice of the discrete global symmetry group G. If the fermion family spectrum is sequential (as it seems to be so far), i.e., families come as LH doublets and RH singlets of SU(2) then $\mathcal{L}_{gauge}$ in

$$\mathcal{L}_{SU(2)\otimes U(1)} = \mathcal{L}_{gauge} + \mathcal{L}_{Higgs}$$

is invariant under interchange/mixing of

$$\begin{pmatrix}u\\d\end{pmatrix}_L \longleftrightarrow \begin{pmatrix}c\\s\end{pmatrix}_L \longleftrightarrow \begin{pmatrix}t\\b\end{pmatrix}_L , \begin{pmatrix}\nu\\e\end{pmatrix}_L \longleftrightarrow \begin{pmatrix}\nu\\\mu\end{pmatrix}_L \longleftrightarrow \begin{pmatrix}\nu\\\tau\end{pmatrix}_L$$

$u_R \leftrightarrow c_R \leftrightarrow t_R$, etc. In fact, $\mathcal{L}_{gauge}$ is invariant under SU(n) acting on quark LH doublets (also SU(n) on lepton doublets, SU(n) on 2/3 charge RH singlets, etc.) for n families. The discrete symmetry which suggests itself is permutation symmetry leading to a choice of S_n for G. Then we insist that $\mathcal{L}_{Higgs}$ be also symmetric under S_n and look for spontaneous breakdown of S_n by $V(\phi)$. The resulting mass-mixing pattern is the prediction of the model.

The most straightforward and obvious implementation of the above ideas would be the following. Let $\psi_L{}^i$, $\psi_R{}^i$ and ϕ^i (i = 1 to n for n families) be the objects being permuted under S_n. Here ψ_L and ψ_R are LH doublets and RH singlets respectively (both quarks and leptons). So we have

$$\begin{pmatrix} \nu_i \\ \ell_i \end{pmatrix}_L \begin{pmatrix} u_i \\ d_i \end{pmatrix}_L \qquad \phi_i = \begin{pmatrix} \phi_i^\circ \\ \phi_i^- \end{pmatrix} \qquad \ell_{iR},\ u_{iR},\ d_{iR}$$

and i = 1 to n. The Higgs Yukawa couplings are $\sum_{i,j,k} g_{ijk}\overline{(\nu_i, \ell_i)}_L\, \ell_{jR}\, \phi_k$, etc. One writes the general Higgs potential invariant under $SU(2)\otimes U(1)\otimes S_n$ and find the minimum, breaking S_n spontaneously with the corresponding vacuum expectation values for ϕ_i° : $\langle\phi_i^\circ\rangle = \eta_i$. Calculate the resulting mass matrices M_ℓ, M_u, and M_d. Find the diagonalizing matrices:

$$U^+_{uL}\ M_u\ V_{uR} = \begin{pmatrix} m_u & & \\ & m_c & \\ & & \ddots \end{pmatrix}$$

$$U^+_{dL}\ M_d\ V_{dR} = \begin{pmatrix} m_d & & \\ & m_s & \\ & & \ddots \end{pmatrix} \qquad (1)$$

Then the KM matrix is just

$$U_{KM} = U^+_{uL}\ U_{dL} \qquad (2)$$

In this class of models the form for U_{KM} turns out to be especially simple, viz., U_{KM} is block diagonal[3]: e.g.,

$$\text{for } n = 3, \qquad U_{KM} = \begin{pmatrix} \cos\theta & \sin\theta & 0 \\ -\sin\theta & \cos\theta & 0 \\ 0 & 0 & 1 \end{pmatrix}$$

$$\text{for } n = 4, \qquad U_{KM} = \begin{pmatrix} \cos\theta & \sin\theta & 0 & 0 \\ -\sin\theta & \cos\theta & 0 & 0 \\ 0 & 0 & 1 & 0 \\ 0 & 0 & 0 & 1 \end{pmatrix} \qquad (3)$$

etc.

Hence, in this class of models one quark doublet is always decoupled. This means b cannot decay by W-exchange and it can also be shown that b cannot decay via non-leptonic modes. Such properties of b seem to be ruled out experimentally. Another unsatisfactory feature is that the mixing angles are not predictable.

Let me turn to a slightly different model which is in some sense more economical. Consider n sequential families. Now if all of ψ_L^i and ψ_R^i are objects of permutation under S_n, then they transform as $\underline{(n-1)} \oplus \underline{1}$ under S_n (where $\underline{n-1}$ and $\underline{1}$ refer to irreducible representations of S_n of dimensionality n-1 and 1 respectively). In the model just discussed, the Higgses also reduced to $\underline{(n-1)} \oplus \underline{1}$. We now propose that there are only n-1 Higgs doublets transforming as $\underline{n-1}$ under S_n. Then in the Higgs Yukawa couplings of each charge sector (up, down, and lepton) there are just three terms as follows.

$$\begin{array}{ccccc} L & & R & & \text{Higgs} \\ \underline{(n-1)} & \otimes & \underline{(n-1)} & \otimes & \underline{(n-1)} \\ \underline{(n-1)} & \otimes & \underline{1} & \otimes & \underline{(n-1)} \\ \underline{1} & \otimes & \underline{(n-1)} & \otimes & \underline{(n-1)} \end{array} \tag{4}$$

Because $\underline{(n-1)} \otimes \underline{(n-1)} \otimes \underline{(n-1)}$ contains $\underline{1}$ exactly once each of the above coupling is described by one coupling constant. Notice that the fermion family classified as $\underline{1}$ gets no mass in the limit it is decoupled from the remaining n-1 families. It is attractive to suppose that $\underline{1}$ corresponds to (e, ν_e, u, d) and consider the limit where the masses and couplings of $\underline{1}$ are neglected compared to those of $\underline{(n-1)}$. Then for the n-1 families there is one unique Yukawa coupling! The only difference between each charge sector (2/3, -1/3 and -1) is the overall strength. Hence, the mass matrices are proportional, to wit:

$$M_d \propto M_\ell \propto M_u^* \tag{5}$$

Also, the M's are symmetric. Since M's are proportional, so are the eigenvalues and so[2])

$$m_c/m_t/m_{t'}, \ldots$$
$$= m_\mu/m_\tau/m_L \ldots \tag{6}$$
$$= m_s/m_b/m_{b'}, \ldots$$

This is satisfied if $m_s \sim 300$ MeV and predicts $m_t \sim 26$ GeV. Since $M_d \propto M_u^*$, the K-M matrix is symmetric (before phases are absorbed): $U_{KM} = U_{KM}^T$ and hence $|U_{cb}| = |U_{ts}|$, etc. Couplings of neutral Higgses change flavor as expected and processes such as $K_L \to e\mu$, $\Sigma \to p\mu e$ are expected at some level. So far we have not introduced either ν_R or $I = 1$ Higgses; hence leaving the neutrinos massless. We will return to ν-masses later.

The proportionality of masses found above predicts the fourth generation quark masses if the fourth charged lepton mass is known:

$$m_{t'} = m_t \frac{m_L}{m_\tau}$$
$$m_{b'} = m_b \frac{m_L}{m_\tau} \tag{7}$$

e.g., if m_L is near 30 GeV then $m_{b'} \sim 80$ to 100 GeV and $m_{t'} \sim 300$ to 400 GeV. These values have two interesting features: a) the near degeneracy of b' with W and Z which makes its detection difficult and b) the saturation of various bounds on fermion masses[9] in SU(2)⊗U(1) suggesting that this is the last sequential family. Why the choice of $m_L \sim 30$ GeV? We already know $m_L > 18$GeV[10] and perhaps L and t are nearly degenerate á la τ and c. Or else there is geometric scaling in lepton masses as once speculated[11], i.e., $m_L/m_\tau = m_\tau/m_\mu$. In addition, a quark flavor in this mass range is a very useful factory for W's, Z's, and Higgses.

THE S_4 MODEL

For just three families a somewhat different assignment for fermions in S_4 is very interesting. S_4 has irreducible representations of dimensions 3, 3', 2, 1', and 1. The Higgs doublets are assigned to $\underline{3}$. The fermions are assigned as follows:

$$\begin{pmatrix} \nu_i \\ \ell_i \end{pmatrix}_L \qquad \ell_{1R} \qquad \{\ell_{2R}, \ell_{3R}\}$$

$$\underline{3} \qquad \underline{1} \qquad \underline{2} \tag{8}$$

$$\begin{pmatrix} u_i \\ d_i \end{pmatrix}_L \qquad \begin{matrix} u_{1R} \\ d_{1R} \end{matrix} \qquad \begin{matrix} \{u_{2R}, u_{3R}\} \\ \{d_{2R}, d_{3R}\} \end{matrix}$$

$$\underline{3} \qquad \underline{1} \qquad \underline{2}$$

The Higgs potential invariant under $S_4 \otimes SU(2) \otimes U(1)$ is

$$\begin{aligned} V = {} & \mu_o^2(\bar{\phi}_o\phi_o+\bar{\phi}_1\phi_1+\bar{\phi}_2\phi_2) + \alpha(\bar{\phi}_o\phi_o+\bar{\phi}_1\phi_1+\bar{\phi}_2\phi_2)^2 \\ & + \beta\{\tfrac{1}{2}(\bar{\phi}_o\phi_1+\bar{\phi}_1\phi_o)^2 + \tfrac{1}{6}(\bar{\phi}_o\phi_o+\bar{\phi}_1\phi_1-2\bar{\phi}_2\phi_2)^2\} \\ & + \gamma\{\tfrac{1}{2}(\bar{\phi}_o\phi_2+\bar{\phi}_2\phi_o)^2 + \tfrac{1}{2}(\bar{\phi}_1\phi_2+\bar{\phi}_2\phi_1)^2 + \tfrac{1}{2}(\bar{\phi}_o\phi_o-\bar{\phi}_1\phi_1)^2\} \\ & + \delta\{\tfrac{1}{6}(\bar{\phi}_o\phi_1-\bar{\phi}_1\phi_o + \sqrt{2}(\bar{\phi}_o\phi_2-\bar{\phi}_2\phi_o))^2 + \tfrac{1}{2}(\bar{\phi}_1\phi_2-\bar{\phi}_2\phi_1)^2\} \\ & + \tfrac{1}{6}(\sqrt{2}(\bar{\phi}_o\phi_1-\bar{\phi}_1\phi_o) - (\bar{\phi}_o\phi_2-\bar{\phi}_2\phi_o))^2\} \end{aligned} \tag{9}$$

Denoting the vacuum expectation values of neutral members of the three Higgs doublets by

$$\begin{aligned} \langle\phi_2^o\rangle &= \xi_2 \\ \langle\phi_o^o\rangle &= \xi \cos\alpha e^{i\phi} \\ \langle\phi_1^o\rangle &= \xi \sin\alpha e^{i\psi} \end{aligned} \tag{10}$$

the potential is minimized when $\alpha = \frac{\pi}{4}$, $\phi + \psi = 0$,

$$\cos 2\phi = - \frac{\gamma+\delta}{\beta+\delta} \left(\frac{\xi_2}{\xi}\right)^2 \tag{11}$$

The minimum is stable if

$$\gamma + \delta > \beta + \delta > 0$$

$$|\xi_2| < |\xi|$$

The Higgs Yukawa coupling invariant under $S_4 \otimes SU(2) \otimes U(1)$ is

$$\mathcal{L}_Y = F \left\{ \overline{(\nu_1, \ell_1)}_L \phi_o + \overline{(\nu_2, \ell_2)}_L \phi_1 + \overline{(\nu_3, \ell_3)}_L \phi_2 \right\} \ell_{1R}$$

$$+ H \left[\frac{1}{\sqrt{2}} \left\{ \overline{(\nu_1, \ell_1)}_L \phi_1 + \overline{(\nu_2, \ell_2)}_L \phi_o \right\} \ell_{2R} \right.$$

$$\left. + \frac{1}{\sqrt{6}} \left\{ \overline{(\nu_1, \ell_1)}_L \phi_o + \overline{(\nu_2, \ell_2)}_L \phi_1 - 2 \overline{(\nu_3, \ell_3)} \phi_2 \right\} \ell_{3R} \right]$$

+ similar terms for quarks with

$$\begin{pmatrix} F \\ H \end{pmatrix} \longrightarrow \begin{pmatrix} f_+, & f_- \\ h_+, & h_- \end{pmatrix}$$

$$\ell_{iR} \longrightarrow (u_{iR}, d_{iR}) \tag{12}$$

$$\phi_i \longrightarrow (\tilde{\phi}_i, \phi_i)$$

We expect ε_i (where $\varepsilon_\ell = \sqrt{2}\, F/H$, $\varepsilon_\pm = \sqrt{2}\, f\pm / h_\pm$) to be small since ε_i vanish in the limit $m_e, m_n, m_d \longrightarrow o$. At tree level the mass matrices of quarks and leptons are:

$$\overline{(\ell_1, \ell_2, \ell_3)}_L \; M_\ell \begin{pmatrix} \ell_1 \\ \ell_2 \\ \ell_3 \end{pmatrix}_R + \overline{(u_1, u_2, u_3)}_L \; M_+ \begin{pmatrix} u_1 \\ u_2 \\ u_3 \end{pmatrix}_R$$

$$+ \overline{(d_1, d_2, d_3)}_L \; M_- \begin{pmatrix} d_1 \\ d_2 \\ d_3 \end{pmatrix}_R \tag{13}$$

where

$$M_\ell = \xi H/2 \begin{pmatrix} \varepsilon_\ell\, e^{i\phi}, & e^{-i\phi}, & \frac{1}{\sqrt{3}} e^{i\phi} \\ \varepsilon_\ell\, e^{-i\phi}, & e^{i\phi}, & \frac{1}{\sqrt{3}} e^{-i\phi} \\ \kappa\, \varepsilon_\ell, & 0, & \frac{-2}{\sqrt{3}} \kappa \end{pmatrix} \tag{14}$$

$M_{\mp}$ have the same form as M_{ℓ} with $h_{\mp}$ replacing H, $\varepsilon_{\mp}$ replacing ε_{ℓ} and $\phi \to \pm\phi$. Here $\kappa = \sqrt{2}\,\xi_2/\xi$. Diagonalizing M_i to $O(\varepsilon_i^2)$ and eliminating the parameters $\varepsilon_{\ell}, \varepsilon_+, \varepsilon_-, \kappa$ and ϕ one relation[2),12)] among the nine masses is found:

$$\begin{vmatrix} m_e^2 & m_\mu^2 & m_\tau^2 \\ m_u^2 & m_c^2 & m_t^2 \\ m_d^2 & m_s^2 & m_b^2 \end{vmatrix} = 0 \qquad (15)$$

Using the known values of other values of other masses, this gives $m_t \sim 26$ GeV. The value of m_t is rather insensitive to the actual quark masses used. Also in the approximation where m_e, m_u and m_d vanish one recovers Eq. (6)

$$m_t \simeq m_c\,(m_\tau/m_\mu)$$

To interpret the quark mass for a confined quark one may use the Georgi-Politzer[13)] definition viz. writing quark propagator as $i\,S^{-1}(p)\Big|_{p^2=-M^2} \sim \not{p} - m\,(M)$, quark mass is m_Q $(M = 2m_Q)$. Then if a mass relation such as Eq. (15) holds at some high energy scale (e.g. $\mu_o \gtrsim 10^3$ GeV where S_4 may be unbroken) the predicted quark mass can be calculated by the lowest order QCD correction:

$$m_i\,(\mu) = m_i\,(\mu_o)\left[\frac{\alpha_s(\mu)}{\alpha_s(\mu_o)}\right]^{\frac{4}{11-2/3f}} \qquad (16)$$

where $\mu = 2m_i$ and f = the number of flavors. As found by a number of authors,[14)] this makes $m_t \sim 20$ GeV. However, this procedure is not gauge invariant. Although this is usually regarded as a small effect, in principle it renders the result arbitrary. A gauge invariant propagator was suggested by Sugawara recently and a calculation of the resulting corrected value of m_t is now in progress.[15)]

Returning to diagonalization of M_i, the parameters ϕ and κ satisfy $1 \gg \sin^2 2\phi \gg \kappa^2$ and are given by

$$\sin^2 2\phi = \frac{16}{3}\left(\frac{m_c^2 m_d^2 - m_u^2 m_s^2}{m_t^2 m_d^2 - m_u^2 m_b^2}\right) .$$

$$\kappa^2 = \frac{2}{3}\left(\frac{m_c^2 m_d^2 - m_u^2 m_s^2}{m_t^2 m_s^2 - m_c^2 m_b^2}\right) \tag{17}$$

The diagonalizing matrices are defined as

$$U_\ell M_\ell V_\ell^+ = \begin{pmatrix} m_e & & \\ & m_\mu & \\ & & m_\tau \end{pmatrix} \tag{18}$$

and similarly for U_+, V_+; U_-, V_-. Then Kobayashi-Maskawa matrix U_{KM} is

$$U_{KM} = U_+ U_-^+$$

To order ϵ_i^2, $U_-^+ = U_+^T$ and hence

$$U_{KM} = U_{KM}^T$$

Explicitly U_+ is given by:

$$U_+ = \begin{bmatrix} \frac{e^{i\eta_u}}{N_1} \left(e^{-i\phi}, \quad e^{-i\phi}, \quad -i\sin 2\phi/\kappa\right) \\ \frac{e^{i\eta_c}}{N_2} \left(-e^{-i\phi}\left\{e^{-2i\phi} + (\lambda_2-1)e^{2i\phi}\right\},\ e^{-i\phi}\left\{e^{2i\phi} + (\lambda_2-1)e^{-2i\phi}\right\},\ 2\kappa(\lambda_2-2)\right) \\ \frac{e^{i\eta_t}}{N_3} \left(-e^{-i\phi}\left\{e^{-2i\phi} + (\lambda_3-1)e^{2i\phi}\right\},\ e^{-i\phi}\left\{e^{2i\phi} + (\lambda_3-1)e^{-2i\phi}\right\},\ 2\kappa(\lambda_3-2)\right) \end{bmatrix} \tag{19}$$

U_- has the same form with $\phi \to -\phi, \eta_u \to \eta_d$, etc. The η_i's are phases to bring U_{KM} to the KM form. N_i are given by

$$N_1^2 = 2 + \sin^2 2\phi/\kappa^2$$

$$N_i^2 = \frac{8}{3} \left[\left\{(2+\kappa^2)\lambda_i - (\sin^2 2\phi + 2\kappa^2)\right\}(1+2\kappa^2) - 3(\lambda_i - 1)(\sin^2 2\phi + 2\kappa^2)\right] \qquad (i = 2,3)$$

and λ_i are given

$$\lambda_{\substack{2\\3}} = \frac{2}{3} \left[(2+\kappa^2) \mp \sqrt{(2+\kappa^2)^2 - 3(2\kappa^2 + \sin^2 2\phi)}\right]$$

Now all elements of U_{KM} are completely fixed in terms of ϕ and κ and hence in terms of masses. We find:

$$\begin{aligned} |U_{us}| &= \sin\theta_1 \sin\theta_3 \sim \sin\theta_c \\ &= \sqrt{\frac{m_t^2 m_d^2 - m_u^2 m_b^2}{m_t^2 m_s^2 - m_c^2 m_b^2}} \\ &\cong m_d/m_s \left[1 - \frac{m_\mu^2}{m_\tau^2}\frac{m_b^2}{m_s^2}\right]^{-1/2} \end{aligned} \qquad (20)$$

For $m_d/m_s \sim 1/18$, $m_b \sim 5$ GeV and $m_s \sim 300$ MeV, this gives $\theta_c \sim 0.22$. The square root factor can be regarded as a corrective factor to the result m_d/m_s obtained[1] in S_3. We also find

$$\begin{aligned} \sin\theta_2 &= \sin\theta_3 \\ &= \frac{1}{\sqrt{3}}\left(\frac{m_c^2 m_d^2 - m_u^2 m_s^2}{m_t^2 m_d^2 - m_u^2 m_b^2}\right)^{1/2} \\ &\cong \frac{1}{\sqrt{3}}\frac{m_\mu}{m_\tau} \cong 0.034 \end{aligned} \qquad (21)$$

$$\sin\delta \sim 10^{-2}$$

Hence we predict

$$
\begin{aligned}
|U_{cd}| &= |U_{us}| \cong 0.22 \\
|U_{cb}| &= |U_{ts}| \cong 0.067 \\
|U_{ub}| &= |U_{td}| \cong 0.0075 \\
\alpha &= |U_{ub}| \big/ |U_{cb}| \cong 0.11
\end{aligned}
\tag{22}
$$

The phase δ is predicted to be too small to account for $K_L \to 2\pi$ from the KM matrix. The full predicted matrix is then

$$
U_{KM} = \begin{pmatrix} 0.975 & -0.22 & -0.0075 \\ 0.22 & 0.973 & 0.067 \\ 0.0075 & 0.067 & -0.998 \end{pmatrix} \tag{23}
$$

With this vaule of $|U_{cb}|$, lifetime of b is expected to be $\tau_B \sim (2 \text{ to } 3) \times 10^{-13}$ sec. Notice that since $\delta \sim o$ the amount of CP-nonconservation in the weak current is too small to account for the observed $K_L \to 2\pi$ rate. It is expected that the flavor-changing CP non-conserving couplings of neutral Higgses would account[16] for $K_L \to 2\pi$. The neutral Higgses have to be quite massive (of order TeV), even then rates for some processes such as $K_L \to \mu e$ have to be as large[16] as ~ O(1/10 present limits).

NEUTRINO MASSES AND MIXINGS

The pattern of neutrino masses and mixings depends on whether right-handed neutrinos exist, on their S_4 properties and on the pattern of Majorana mass terms.

First consider the possibility that ν_R^i exist. Let there be as many ν_R^i as $(\nu_i, \ell_i)_L$. (An interesting phenomenological possibility is that the number of ν_R is not equal to that of ν_L; but in the spirit of the sequential model we will not consider it further here.) Then how do the ν_R^i transform under S_4? The simplest possibility (A) is that all of ν_R^i are invariant under S_4. This has some logic to it in that ν_R^i have no gauge interactions at SU(2)⊗U(1) level. In this case the Yukawa coupling is

$$
G \left\{ \overline{(\nu_1, \ell_1)}_L \tilde{\phi}_o + \overline{(\nu_2, \ell_2)}_L \tilde{\phi}_1 + \overline{(\nu_3, \ell_3)}_L \tilde{\phi}_2 \right\} \nu_{\alpha R} \tag{24}
$$

and only one neutrino picks up a (Dirac) mass term. Then there is only one δm^2 relevant to ν-oscillations and the mixing matrix is given by precisely

$$\begin{pmatrix} \nu_e \\ \nu_\mu \\ \nu_\tau \end{pmatrix}_L = U \begin{pmatrix} \nu_1 \\ \nu_2 \\ \nu_3 \end{pmatrix}_L = \begin{pmatrix} 1/\sqrt{3} & -1/\sqrt{2} & 1/\sqrt{6} \\ 1/\sqrt{3} & -1/\sqrt{2} & 1/\sqrt{6} \\ 1/\sqrt{3} & 0 & -2/\sqrt{6} \end{pmatrix} \begin{pmatrix} \nu_1 \\ \nu_2 \\ \nu_3 \end{pmatrix}_L \tag{25}$$

where $m_1 = m_2 = 0$.

Another assignment is to let ν_i^R transform as ℓ_i^R under S_4, i.e., as $\underline{2} \oplus \underline{1}$. In this case (B) the (Dirac) masses of neutrinos satisfy determinantal mass formulas as for charged fermions viz.

$$\begin{vmatrix} m_{\nu_1}^2 & m_{\nu_2}^2 & m_{\nu_3}^3 \\ m_e^2 & m_\mu^2 & m_\tau^2 \\ m_d^2 & m_s^2 & m_b^2 \end{vmatrix} = 0 \tag{26}$$

This gives approximately

$$2.7\text{x}10^4 \; m_{\nu_1}^2 + 300 \; m_{\nu_2}^2 \simeq m_{\nu_3}^2 \tag{27}$$

In either case (A) or (B) we expect ν_R^i to get "large" $I = 0$, S_4 singlet Majorana masses. (This is an example of a large mass scale entering.) Assuming for simplicity only one large mass say M, the light LH Majorana neutrinos masses become m_i^2/M ($m_i = m_{\nu_i}$ (Dirac)) by the now famous "See-Saw" mechanism.[17)] In the case (B) these masses now satisfy the linearized version of Eq. (27):

$$2.7\text{x}10^4 \; m_{\nu_1} + 300 \; m_{\nu_2} \simeq m_{\nu_3} \tag{28}$$

The mixing matrix in various cases is: Case A: U given by Eq. (25) for either Dirac ν's or Majorana ν's (as long as only one heavy mass scale). Case B: $U \equiv U_{KM}$ for Dirac ν's but not in general.

A second possibility, in a sense simpler, is that ν_R^i do not exist. The only way for ν_L^i to get masses is by coupling to a $I = 1$ Higgs[18)] which may be regarded as being made up of two $I = 1/2$ Higgses. Then the effective coupling looks like

$$f/M \quad \overline{\psi_{L_\kappa}^{c}}\, \underline{\tau}\, \psi_{L_\ell} \quad \phi_i\, \underline{\tau}\, \phi_j; \tag{29}$$

Once again a new mass scale is called for. In general, there are 4 parameters since $3 \otimes 3 = \underline{3} \oplus \underline{3}' \oplus \underline{2} \oplus \underline{1}$. But, for example, if the effective I = 1 field transforms as S_4 singlet then all m_{ν_i} are equal. This is an interesting possibility because here is a case in which $m_{\nu_i} \neq 0$ but $\delta m^2_{ij} = 0$ and there is no mixing and no oscillations!

SUMMARY

We have shown how interpreting the observed sequential structure of fermions in terms of permutation symmetry leads to strong constraints on the fermion masses and mixings. The results for mass of the t-quark and for the K-M mixing angles will test the proposal in the near future. Further details on the flavor and CP-nonconserving processes due to Higgs exchange and on neutrino mass matrix will be given in a forthcoming publication.[16)]

ACKNOWLEDGMENT

I would like to thank Hirotaka Sugawara and the theory group at KEK for their warm hospitality and the Japan Society for Promotion of Science for their support during this stay. This work was also partially supported by the U.S. Department of Energy under Contract No. DE-AM03-76SF00235.

References and Footnotes

1. S. Pakvasa and H. Sugawara, Phys. Lett. 73B, 61 (1978).
2. S. Pakvasa and H. Sugawara, Phys. Lett. 82B, 105 (1979).
3. E. Derman, Phys. Rev. D19, 317 (1979); E. Derman and H. S. Tsao, Phys. Rev. D20, 1207 (1979).
4. D. Wyler, Phys. Rev. D19, 330, 3369 (1979); D. D. Wu, Phys. Lett. 85B, 463 (1979); G. Segré and H. Weldon, Phys. Rev. Lett. 42, 1191 (1979); H. Sato, Nucl. Phys. B148, 433 (1979); K. Uehara, Progr. Theor. Phys. 61, 1426 (1979); H. Hayashi, M. J. Hayashi, and A. Murayama, Ecole Polytechnique Report A397 (1980); M. Murayama and N. Ohtsubo, Kanazawa Report DPKU-8002 (1980).
5. D. Nanopoulos, These Proceedings; P. Ramond, These Proceedings; A. Zee, These Proceedings.
6. T. Maehara and T. Yanagida, Lett. Nuov. Cim. 19, 424 (1977), Progr. Theor. Phys. 60, 822 (1978); M. A. B. Bég and A. Sirlin, Phys. Rev. Lett. 38, 1113 (1977); C. L. Ong, Phys. Rev. D19, 2738 (1979); F. Wilczek and A. Zee, Phys. Rev. Lett. 42, 421 (1979); A. Davidson, M. Koca, and K. C. Wali, Phys. Rev. Lett. 43, 92 (1979), Phys. Rev. D20, 1195 (1979).
7. Y. Chicasige, R. Mohapatra, and R. Peccei, Phys. Lett. 98B, 265 (1981); R. Gelmini and G. Roncadelli, Phys. Lett. 98B, 265 (1981); H. Georgi, S. L. Glashow, and S. Nussinov, HUTP-81/A026; R. Peccei, These Proceedings; M. Dine, W. Fischler, and M. Srednicki, Princeton Report (1981) to be published; H. Georgi, S. L. Glashow, and M. Wise, Phys. Rev. Lett. 47, 402 (1981); D. Nanopoulos, These Proceedings; H. Georgi, These Proceedings.
8. R. Barbieri, R. Gatto, and F. Strocchi, Phys. Lett. 74B, 344 (1978); R. Gatti, G. Morchio, and F. Strocchi, Phys. Lett. 80B, 265 (1979); ibid. 83B, 348 (1979); G. Sartori, Phys. Lett. 82B, 255 (1979); R. Gatto, et al., Nucl. Phys. B163, 221 (1980); G. Segré and H. A. Weldon, Ann. Phys. 124, 37 (1980).
9. M. Veltman, Phys. Lett. 70B, 252 (1977); M. Chanowitz, M. Furman, and I. Hinchliffe, Phys. Lett. 78B, 285 (1978); P. Q. Hung, Phys. Rev. Lett. 42,

873 (1979); H. D. Politzer and S. Wolfram, Phys. Lett. 82B, 242, 421 (1979).

10. J. Bürger, Proceedings of the 1981 International Symposium on Leptons and Photons, Bonn, Aug. 24-31, 1981, to be published.

11. W. Heisenberg, et al., Z.für. Natur. 10A, 425 (1955); H. Mitter, Nuov. Cim. 32, 1789 (1964); K. Tennakone and S. Pakvasa, Phys. Rev. Lett. 13, 757 (1971); S. Blaha, Phys. Lett. 84B 116 (1979); J. D. Bjorken, SLAC-PUB-2195-1978; E. Lipmanov, Yad. Fiz. 32 1077 (1980).

12. Such a formula also obtains in SO(10) Grand Unified Model with a single 10 and a single 126 of Higgses: S. L. Glashow, Proceedings of Neutrino-79, international conference on neutrinos, weak interactions, and cosmology, Bergen, June 18-22, 1979; ed. A. Haatuft and C. Jarlskog, vol. 1, p. 518.

13. H. Georgi and H. D. Politzer, Phys. Rev. D14, 1829 (1976).

14. T. Yanagida, Phys. Rev. D20, 2986 (1979); G. Segré, H. Weldon, and J. Weyers, Phys. Lett. 83B, 851 (1979); D. Grosser, Phys. Lett. 83B, 855 (1979); M. Voloshin, V. Zakharov, and M. Krammer, Phys. Lett. 93B, 447 (1980); S. L. Glashow, Phys. Rev. Lett. 45 1914 (1980).

15. K. Kanaya, H. Sugawara, S. F. Tuan, and S. Pakvasa (in preparation).

16. S. Pakvasa, H. Sugawara, and Y. Yamanaka (in preparation).

17. T. Yanagida, Proceedings of the Workshop on Unified Theory and Baryon Number in the Universe, ed. by O. Sawada and A. Sugamoto, KEK (1979); M. Gell-Mann, P. Ramond, and R. Slansky, in Supergravity, ed. By P. Van Nieuvenhuizen, D. Z. Freedman, and N. Holland (1979).

18. This is a case in which B-L symmetry is broken explicitly. This may not be to everybody's taste but the important point is that it is not devoid of predictive power.

CP VIOLATION: SOFT OR HARD BREAKING?

Roberto D. Peccei
Max-Planck-Institut für Physik und Astrophysik
Fohringer Ring 6, Postfach 40 12 12, D-8000 Munich 40, Fed. Rep. Germany

in
Proceedings of the Fourth Kyoto Summer Institute on Grand Unified Theories and Related Topics, Kyoto, Japan, June 29 - July 3, 1981, ed. by M. Konuma and T. Maskawa (World Science Publishing Co., Singapore, 1981).

Contents

CP VIOLATION: SOFT OR HARD BREAKING?

R. D. Peccei

Max-Planck-Institut für Physik und Astrophysik, Munich, Fed. Rep. Germany

Abstract

I discuss the problem of strong CP violation and consider two alternative possibilities for its resolution: either the imposition of an extra chiral symmetry on the theory or the supposition that CP is conserved at the Lagrangian level, but broken by the vacuum. The former scheme leads to the appearance of axions, and I briefly review the present experimental status of the standard axion. Recent theoretical suggestions that might make axions either very heavy or extremely light are also discussed. The soft CP option is analyzed within the context of a particular left-right symmetric model, but its relevance for the case of composite models or technicolor models is also examined. Finally a brief discussion is also given of some of the cosmological difficulties that soft CP models may encounter.

1. The θ-problem

The θ -problem, or the problem of strong CP violation, arises as a consequence of the non-trivial topological properties associated with the QCD vacuum. 't Hooft[1)] showed that because the correct vacuum state of QCD is the, so-called, θ -vacuum, then effectively the QCD Lagrangian need not conserve CP. That is, one can add an additional CP violating piece to the Lagrangian of QCD of the form

$$\mathcal{L}_{\not{CP}} = \theta \frac{g^2}{32\pi^2} F_a^{\mu\nu} \tilde{F}_{a\mu\nu} \tag{1}$$

where $F_a^{\mu\nu}$ is the usual color field strength and

$$\tilde{F}_a^{\mu\nu} = \frac{1}{2} \epsilon^{\mu\nu\rho\sigma} F_{a\rho\sigma} \tag{2}$$

is its dual, with θ an arbitrary parameter which labels the QCD vacuum. Thus it appears that QCD is only CP-conserving if $\theta = 0$ (or π).

Actually the problem is even a little more complicated when one includes the weak interactions, since in this case the coefficient in front of the CP violating $F\tilde{F}$ term not only involves a parameter associated with the vacuum structure of QCD, but also involves a term arising directly from the weak interactions. Before discussing the full case, however, it is helpful to very briefly recall, at least qualitatively, how the CP violating term (1) enters. Then it will be much clearer why the inclusion of the weak interactions brings additional contributions to (1), besides those parametrized by θ.

The simplest way to see the origin of (1) is to consider the formal expression for the vacuum to vacuum amplitude in QCD. In terms of the usual path integral representation

$$_{+}\langle vac. | vac. \rangle_{-} \sim \int \delta A_\mu \, e^{iS[A]} \tag{3}$$

one is instructed to integrate over all gauge field configurations starting from configurations which are pure gauge fields (i.e., gauge equivalent to

$A_\mu = 0$) at $t = -\infty$ and ending up with pure gauge fields at $t = +\infty$. To see what is going on, it is very convenient, following Callan, Dashen and Gross[2], to adopt the $A^0 = 0$ gauge and work for simplicity with SU(2). In this gauge, pure gauge fields A^i are given by

$$A^i = \frac{i}{g} \Omega^{-1} \nabla^i \Omega \tag{4}$$

with

$$\Omega = \exp i \vec{\tau}\cdot\hat{r} F(r) \tag{5}$$

The quantity F(r) is in principle arbitrary. However, one imposes a further constraint on the pure gauge fields at spatial infinity, namely, that

$$\Omega \to 1 \qquad \text{as} \qquad r \to \infty \tag{6}$$

This is where the topology comes in, because this requirement, which is a compactification requirement, allows for mappings of the hypersphere S_3 onto the hypersphere in group space[3]. At any rate the boundary condition (6) implies that the functions F(r), and hence the pure gauge fields, can be classified by indices n which characterize how F(r) behaves at infinity

$$F_n(r) \underset{r\to\infty}{\longrightarrow} 2\pi n \tag{7}$$

Furthermore, one can obtain a formula for n in terms of the gauge fields themselves[4]:

$$n = \frac{i g^3}{24\pi^2} \int d^3x \, \mathrm{Tr}\, \epsilon_{ijk} A^i A^j A^k \tag{8}$$

where the trace is over SU(2) space.

In the $A^0 = 0$ gauge, one may think of the vacuum to vacuum amplitude as being a trasition from vacua at $t = -\infty$, labeled by the indices n_- of the pure gauge fields $A^i_{n_-}\Big|_{t=-\infty}$, to vacua at $t = +\infty$, labeled by the indices n_+ of the pure gauge fields $A^i_{n_+}\Big|_{t=+\infty}$. Pictorially one must integrate over all functional paths that join these configurations in the space of gauge fields, weighted of course by the usual action factor. Some schematic paths are shown below in Fig. 1.

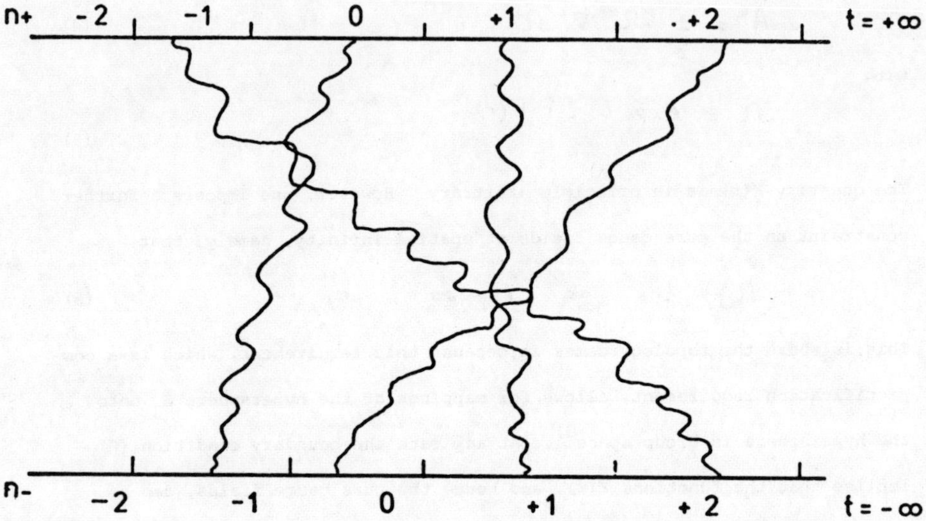

Fig. 1: Some paths contributing to the vacuum to vacuum amplitude

The change $\nu = n_+ - n_-$ of the paths above has in fact a simple interpretation and it can be related to a space-time integral over $F\tilde{F}$. To see this, one needs to use Bardeen's identity[5)] which expresses $F\tilde{F}$ as a total divergence. One has

$$\partial_\mu K^\mu = F_a^{\mu\nu} \tilde{F}_{a\mu\nu} \tag{9}$$

with

$$K^\mu = \epsilon^{\mu\alpha\beta\gamma} A_{a\alpha} \left\{ F_{a\beta\gamma} - \frac{1}{3} \epsilon_{abc} A_{b\beta} A_{c\gamma} \right\} \tag{10}$$

Clearly then

$$\frac{g^2}{32\pi^2}\int d^4x\, F_a^{\mu\nu}\tilde{F}_{a\mu\nu} = \frac{g^2}{32\pi^2}\int d\sigma_\mu K^\mu$$

$$\rightarrow \frac{g^2}{32\pi^2}\int d^3x\, K^0 \Big|_{t=-\infty}^{t=+\infty} \tag{11}$$

where the second line follows if we are in the $A^0 = 0$ gauge and if, as assumed, on the surfaces one has only pure gauge fields. Inspection of Eqs. (10) and (8) yields immediately the desired result

$$\nu = n_+ - n_- = \frac{g^2}{32\pi^2}\int d^4x\, F_a^{\mu\nu}\tilde{F}_{a\mu\nu} \tag{12}$$

Now, clearly, the correct vacuum of QCD cannot be just one of the $|n\rangle$ vacua, since by a gauge transformation $\mathcal{U}$ one can change the index n (That is, one can go from a configuration A^i_n labeled by F_n to one labeled by F_{n+1}):

$$\mathcal{U}|n\rangle = |n+1\rangle \tag{13}$$

The QCD vacuum must be a linear superposition of these states (θ-vacuum)[2].

$$|\theta\rangle = \sum_n e^{-in\theta}|n\rangle \tag{14}$$

Obviously $|\theta\rangle$ is gauge invariant since

$$\mathcal{U}|\theta\rangle = e^{i\theta}|\theta\rangle \tag{15}$$

The vacuum to vacuum amplitude then is

$$_+\langle \text{vac.}|\text{vac.}\rangle_- \equiv {}_+\langle\theta|\theta\rangle_- = \sum_m\sum_n e^{im\theta}e^{-in\theta}\,{}_+\langle m|n\rangle_-$$

$$= \sum_\nu e^{i\nu\theta}\left\{\sum_n {}_+\langle n+\nu|n\rangle_-\right\} \tag{16}$$

The quantity in the curly bracket is just given by a path integral over configurations in which the difference $n_+ - n_- = \nu$ is fixed. Having computed fixed ν path integrals one needs then to add all the contributions of each sector, weighted by $e^{i\nu\theta}$, to get the full vacuum amplitude. Recalling (3) and the result (12), it is clear that the phases $e^{i\nu\theta}$ can be viewed as arising from an extra term in the QCD Lagrangian, which is precisely the CP violating contribution given in Eq. (1). Thus the existence of θ-vacua forces CP violation in QCD unless the vacuum angle θ vanishes ($\theta = \pi$ is also CP conserving).

The quantity $F\tilde{F}$ plays another role in QCD and this is connected with the anomalous divergence of the axial current, discovered originally by Adler[6] and by Bell and Jackiw[7]. The current

$$J_5^\mu = \sum_{i=1}^{N_f} \bar{q}_i \gamma^\mu \gamma_5 q_i \quad . \tag{17}$$

associated with chiral $U_A(1)$ transformations of the quark fields q_i, even in the zero quark mass limit, is not conserved, but has an anomaly:

$$\partial_\mu J_5^\mu = 2N_f \left(\frac{g^2}{32\pi^2}\right) F_a^{\mu\nu} \tilde{F}_{a\mu\nu} + \text{mass terms} \tag{18}$$

Because of this anomaly one cannot really use the charge

$$Q_5 = \int d^3x \, J_5^0 \tag{19}$$

as a generator of chiral transformations. However, using Bardeen's identity[5], one can always find a new current

$$\tilde{J}_5^\mu = J_5^\mu - 2N_f \left(\frac{g^2}{32\pi^2}\right) K^\mu \tag{20}$$

whose divergence vanishes in the zero quark mass limit

$$\partial_\mu \tilde{J}_5^\mu = \text{mass terms} \tag{21}$$

From $\tilde{J}_5^\mu$ one can construct a perfectly suitable generator for chiral transformations:

$$\tilde{Q}_5 \doteq \int d^3x \, \tilde{J}_5^0 \tag{22}$$

since, when $m_i \to 0$, one has $\dot{\tilde{Q}}_5 = 0$.

It is important to note that $\tilde{Q}_5$, although a perfectly good generator for $U_A(1)$ transformations, is not itself gauge invariant (This follows since K^0 depends explicitly on the gauge fields.). In fact, if one considers the effect of the operator $\mathcal{U}$ which changes an $|n\rangle$ vacua to an $|n+1\rangle$ vacua, it is easy to check that[8)]

$$\mathcal{U}^{-1} \tilde{Q}_5 \, \mathcal{U} = \tilde{Q}_5 - 2N_f \, , \tag{23}$$

since $\int d^3x \, K^0$ is related to n (cf Eqs. (8), (10) and (11)). This equation is very important since it tells us that a chiral transformation changes a $|\theta\rangle$ vacua into a different vacuum state. Consider in fact

$$\begin{aligned} \mathcal{U}\left[e^{-i\tilde{Q}_5\alpha} |\theta\rangle \right] &= \mathcal{U}\left[e^{-2i\alpha N_f} \, \mathcal{U}^{-1} e^{-i\tilde{Q}_5\alpha} \, \mathcal{U} |\theta\rangle \right] \\ &= e^{i(\theta - 2N_f\alpha)} \left[e^{-i\tilde{Q}_5\alpha} |\theta\rangle \right] \end{aligned} \tag{24}$$

where we made use of Eqs. (23) and (15). Thus it follows that rotating chirally by α the vacuum state $|\theta\rangle$ we obtain the state $|\theta - 2N_f\alpha\rangle$:

$$|\theta - 2N_f\alpha\rangle = e^{-i\tilde{Q}_5\alpha} |\theta\rangle \tag{25}$$

The above remark is particularly useful when we consider QCD in the presence of the weak interactions, which we shall assume are given by the standard Glashow-Weinberg-Salam model[9)]. In general the weak interactions produce mass terms for the quarks which are non-diagonal and not γ_5 free. If q_L and q_R are the usual left and right projections of the quark fields, then one has, in general,

$$\mathcal{L}_{mass} = - \left\{ \bar{q}_{iL} M_{ij} q_{jR} + \bar{q}_{iR} (M^{\dagger})_{ij} q_{jL} \right\} \tag{26}$$

with M arbitrary but charge diagonal. What enters in the usual QCD Lagrangian is not M but its diagonal eigenvalues, which one can always obtain by a bi-unitary transformation:

$$(U_L^{\dagger} M U_R)_{ij} = \delta_{ij} m_i \tag{27}$$

Now in performing this diagonalization to bring the total Lagrangian to standard form (i.e., with real, diagonal and γ_5 free masses) one changes the basis of the quark fields. This basis change involves a chiral U(1) transformation, unless Arg det M vanishes. Thus, because of the presence of the weak interactions, it is easy to see that the total $F\tilde{F}$ term in the theory, when QCD is written in standard form, is given by

$$\mathcal{L}_{\not{CP}} = \bar{\theta} \frac{g^2}{32\pi^2} F_a^{\mu\nu} \tilde{F}_{a\mu\nu} \tag{28}$$

with

$$\bar{\theta} = \theta + \text{Arg det } M \tag{29}$$

The θ-problem is why $\bar{\theta}$, which is this funny combination of pure QCD information - the vacuum angle θ - and of weak interaction information - Arg det M - is so small. The point is that a non-vanishing $\bar{\theta}$ contributes to the dipole moment of the neutron. In fact, Baluni[10)], Crewther, di Vecchia, Veneziano and Witten[11)] and recently Otha and Kawarabayashi[12)] have estimated bounds on $\bar{\theta}$ following from the experimental bound on the neutron electric dipole moment. They find that

$$\bar{\theta} \lesssim 10^{-8} - 10^{-9} \tag{30}$$

so as not to violate experiment.

To my knowledge there are only two plausible solutions to the θ-problem, although in either case a truly satisfactory model does not exist. One solution - the chiral solution - makes $\bar{\theta} = 0$ naturally, but at the price of imposing extra symmetries on the weak interactions which lead to the appearance of pseudo-Goldstone bosons - axions. The other solution hopes to calculate $\bar{\theta}$ from the theory directly and find that it naturally is a small number. Calculability implies that CP must be softly broken in the weak interactions and this again leads to constraints which are not so easily satisfied. I shall discuss both of these options in the next two sections. Before closing this section, however, I should point out that perhaps there may be totally different solutions to the θ-problem, not immediately related to particle physics. For instance, recently, Linde[13] has suggested that perhaps $\bar{\theta}$ is set to zero by some boundary condition effect occuring in the very early universe. I do not believe that this actually works, but it is an option one may have to keep in mind. In a more pessimistic vein it may be that $\bar{\theta}$ of $O(10^{-9})$ may be unexplainable, just like $\frac{m_e^2}{m_t^2} \lesssim 10^{-9}$ is at present unexplainable.

2. The Chiral Solution

We saw earlier that under a chiral $U_A(1)$ transformation, $e^{-i\tilde{Q}_5\alpha}$, one could change the θ angle:

$$\theta \to \theta - 2N_f\alpha \tag{31}$$

Obviously if the total Lagrangian of the theory is such as to admit such a transformation as a symmetry transformation, then $\bar{\theta}$ can be rotated to zero. That is, in such a theory all values of $\bar{\theta}$ are equivalent and there is no CP violation in the strong sector. This observation, suggested by Helen Quinn and myself[14], is the basis for a solution to the θ-problem. The immediately apparent problem with this solution is that if quarks have masses it does not appear that one can ask for a chiral invariance of the Lagrangian. However, recall that in the standard model masses arise from vacuum expectation values of Higgs fields:

$$M = \Gamma\langle\phi\rangle \tag{32}$$

Because of this, chiral symmetry can be imposed on the theory by demanding that the chiral rotations on the fermions be appropriately absorbed by phase rotations of the Higgs fields. For example, consider the U(1) case where

$$\mathcal{L}_{Yukawa} = -\left\{ \Gamma \bar{\psi}_R \phi \psi_L + \Gamma^* \bar{\psi}_L \phi^\dagger \psi_R \right\} \tag{33}$$

If $\langle\phi\rangle \neq 0$, then the fermion field ψ has a mass. Yet $\mathcal{L}$ is invariant under chiral transformations:

$$\psi_L \to e^{-i\frac{\alpha}{2}} \psi_L \tag{34a}$$

$$\psi_R \to e^{i\frac{\alpha}{2}} \psi_R \tag{34b}$$

provided also

$$\phi \to e^{i\alpha} \phi \tag{34c}$$

Of course although $\mathcal{L}$ is chiral invariant, the fact that $\langle\phi\rangle \neq 0$ breaks chirality spontaneously.

In the standard model for the weak interactions[9)] one can implement a $U_A(1)$ symmetry very simply[14)]. All one needs to do is to introduce two distinct Higgs fields, ϕ_1 and ϕ_2, instead of the usual doublet ϕ and it charge conjugate $\tilde{\phi} = \tau_2 \phi^*$. The field ϕ_1 has charges so that it can only give mass to the u-like quarks, while ϕ_2 is responsible for the d-like quark masses. In this way the Yukawa Lagrangian is invariant under chiral $U_A(1)$ transformations of the quark fields, provided that ϕ_1 and ϕ_2 respond appropriately. For instance,

$$u \to e^{i\frac{\alpha}{2}\gamma_5} u \; ; \quad d \to e^{i\frac{\alpha}{2}\gamma_5} d \tag{35a}$$

requires

$$\phi_i \to e^{i\alpha} \phi_i \tag{35b}$$

(Note that it is because of (35b) that ϕ and $\tilde{\phi}$ will not do, since $\tilde{\phi}$ transforms oppositely to ϕ under phase transformations.) Of course to guarantee the total $U_A(1)$ invariance of the model, the Higgs potential must also respect this symmetry. This means, for instance, that perfectly good invariant terms, under SU(2) x U(1), like $\lambda(\phi_1\phi_2)^2$ are forbidden to exist in $V(\phi_1,\phi_2)$.

Such a model of weak interaction has effectively $\bar{\theta}=0$, and hence no strong CP problem. However, there is a price to pay. Since we have imposed an extra chiral $U_A(1)$ symmetry at the Lagrangian level and since $\langle\phi_i\rangle\neq 0$ breaks the symmetry spontaneously, we expect that a Goldstone boson - the axion - should appear in the theory. The existence of such an excitation was first pointed out by Weinberg[15)] and Wilczek[16)]. In fact the axion is not a Goldstone boson because the chiral current J_μ^5 has an anomaly, and a careful analysis shows that the axion picks up a small mass.

To see how this comes about, it is useful to study the current which is coupled to the axion. Because the potential $V(\phi_1,\phi_2)$ is SU(2) x U(1) x $U_A(1)$ symmetric, when we assume that

$$\langle\phi_1\rangle = \frac{1}{\sqrt{2}}\begin{pmatrix} f_1 \\ 0 \end{pmatrix} \tag{36a}$$

$$\langle\phi_2\rangle = \frac{1}{\sqrt{2}}\begin{pmatrix} 0 \\ f_2 \end{pmatrix} \tag{36b}$$

there will appear now two Goldstone bosons in the zero charge sector. If we write the zero charge fields as

$$\phi_i^0 = \frac{1}{\sqrt{2}}\left\{ f_i + \rho_i + i\sigma_i \right\} \tag{37}$$

it is easy to check that the combination

$$\xi_Z = \frac{1}{\sqrt{f_1^2+f_2^2}}\left\{ f_1\sigma_1 - f_2\sigma_2 \right\} \tag{38}$$

is the one which is eventually "eaten" by the Z^0. The axion is the orthogonal field to ξ_Z and thus one has

$$a = \frac{1}{\sqrt{f_1^2+f_2^2}}\left\{ f_2\sigma_1 + f_1\sigma_2 \right\} \tag{39}$$

The chiral current which only contains the axion but not the ξ_Z field is a combination of the currents generated by the $U_A(1)$ and the SU(2) x U(1) transformations, if $f_1 \neq f_2$. It can be taken, for instance, as[17)]

$$J_5^\mu = \left[\frac{f_1^2 + f_2^2}{2 f_1 f_2}\right] J^\mu_{U_A(1)} + \left[\frac{f_2^2 - f_1^2}{f_1 f_2}\right] J^\mu_{U(1)} , \tag{40}$$

which yields

$$J_5^\mu = - f \partial_\mu a + x \sum_f \bar{u}_f \gamma_\mu \frac{\gamma_5}{2} u_f + + \frac{1}{x} \sum_f \bar{d}_f \gamma_\mu \frac{\gamma_5}{2} d_f + \frac{1}{x} \sum_f \bar{\ell}_f \gamma_\mu \frac{\gamma_5}{2} \ell_f + \dots \tag{41}$$

In the above x is a free parameter:

$$x = \frac{f_2}{f_1} \tag{42}$$

associated with the ratio of the Higgs vacuum expectation values. However, the other parameter f can be related to the Fermi constant :

$$f = \sqrt{f_1^2 + f_2^2} = (\sqrt{2}\, G_F)^{-\frac{1}{2}} \simeq 250 \text{ GeV} \tag{43}$$

In the absence of the strong interactions the current J_5^μ is, of course, conserved. However, when the strong interactions are included this current has an anomalous divergence[5,6,7)]

$$\partial_\mu J_5^\mu = (x + \frac{1}{x}) N \left\{ \frac{g^2}{32\pi^2} F_a^{\mu\nu} \tilde{F}_{a\mu\nu} \right\} \tag{44}$$

where $N = 1/2N_f$ is the number of quark doublets in the theory. Because of the anomaly (44), it is no longer clear that the axion remains at zero mass. Indeed, once one countenances the inclusion of strong interaction effects it is possible to generate a mass for the axion through mixing with quark-antiquark bound states, as shown schematically in Fig. 2

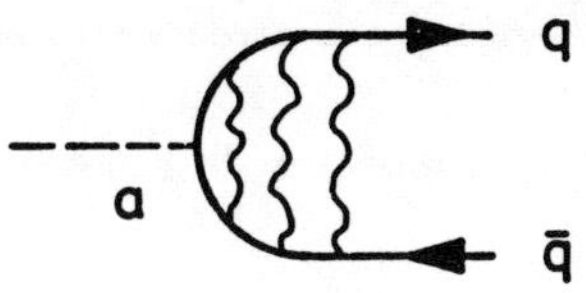

Fig. 2: Axion mixing with $q\bar{q}$ states

To compute the axion mass one needs to construct an anomaly free current, coupled to the axion, whose divergence is "soft", so that the usual current algebra methods can be applied. This construction has been carried out by Bardeen and Tye[18], who studied the current:

$$\tilde{J}^{\mu}_{5} = J^{\mu}_{5} - \frac{1}{2}\left(x+\frac{1}{x}\right) N \left\{ \frac{m_u}{m_u+m_d}\, \bar{u}\gamma^{\mu}\gamma_5 u + \frac{m_d}{m_u+m_d}\, \bar{d}\gamma^{\mu}\gamma_5 d \right\} \tag{45}$$

It is easy to check that the divergence of $\tilde{J}^{\mu}_{5}$ is anomaly free and that it vanishes in the SU(2) chiral limit ($m_u \to 0, m_d \to 0$):

$$\partial_{\mu} \tilde{J}^{\mu}_{5} = -\left(x+\frac{1}{x}\right) N \frac{m_u m_d}{(m_u+m_d)} \left\{ \bar{u} i\gamma_5 u + \bar{d} i\gamma_5 d \right\} \tag{46}$$

Armed with $\tilde{J}^{\mu}_{5}$ one can then obtain the mass of the axion in an analogous way to that by which one computes the pion mass in current algebra in terms of $\langle \bar{u}u \rangle$ and $\langle \bar{d}d \rangle$. In this latter case the SU(2) chiral current, which couples to the pion, is

$$J^{\mu}_{53} = \frac{1}{2} \left\{ \bar{u}\gamma^{\mu}\gamma_5 u - \bar{d}\gamma^{\mu}\gamma_5 d \right\} \tag{47}$$

and its divergence is given by

$$\partial_{\mu} J^{\mu}_{53} = m_u\, \bar{u} i\gamma_5 u - m_d\, \bar{d} i\gamma_5 d \tag{48}$$

A straightforward computation[19] yields the familiar formula

$$m_\pi^2 f_\pi^2 = -\left\{ m_u \langle \bar{u} u \rangle + m_d \langle \bar{d} d \rangle \right\} \tag{49}$$

where $f_\pi \simeq 93$ Mev is the usual pion decay constant, which typifies how J^μ_{53} couples to the π^0:

$$J^\mu_{53} \sim -f_\pi \partial^\mu \pi^0 \tag{50}$$

Using Eqs. (45) and (46) and an analogous computation, Bardeen and Tye[18] obtained

$$m_a^2 f^2 = -\left(x+\frac{1}{x}\right)^2 N^2 \frac{m_d m_u}{(m_d+m_u)^2} \left\{ m_u \langle \bar{u} u \rangle + m_d \langle \bar{d} d \rangle \right\} \tag{51}$$

Hence it follows that the axion mass is given by

$$m_a = m_\pi \frac{f_\pi}{f} \left(x+\frac{1}{x}\right) N \sqrt{\frac{m_d m_u}{(m_d+m_u)^2}} \tag{52}$$

This result was also derived by using effective Lagrangian techniques by Weinberg[15] and by Kandaswamy, Salomonson and Schechter[20].

Using for $m_u/m_d \simeq 0.6$, which follows from the usual current algebra analysis[19], and 3 generations of quarks and leptons yields for the axion mass, approximately,

$$m_a \simeq 75\left(x+\frac{1}{x}\right) \text{ KeV} \tag{53}$$

If x if of O(1), then the axion will be stable under decay into e^+e^- and its only decay possibility will be into two photons. The decay rate for this process can be directly calculated by using the electromagnetic anomaly[18] and one finds

$$\tau(a\to 2\gamma) \simeq 0.8 \left(\frac{100 \text{ Kev}}{m_a}\right)^5 \text{ sec} \tag{54}$$

The factor of m_a^5 arises from a factor of m_a^3 coming from phase space and a counting factor of N^2, which via Eq. (52) is then turned into m_a^2.

Finally, by again using current algebra techniques, one can compute the coupling of axions to nucleons. (The coupling to leptons follows from the original Lagrangian.) One finds[21)]

$$\mathcal{L}^{eff}_{NNa} = i \bar{N} \gamma_5 [g_0 + g_1 \tau_3] N a \tag{55}$$

with

$$g_0 = -\frac{M_N}{f} F_A^0 \left\{ \left(\frac{N-1}{2}\right)\left(x + \frac{1}{x}\right)\right\} \tag{56a}$$

$$g_1 = \frac{M_N}{f} F_A^1 \left\{ \frac{x}{2}\left[1 - N\frac{(m_d - m_u)}{(m_d + m_u)}\right] - \frac{1}{2x}\left[1 + N\frac{(m_d - m_u)}{(m_d + m_u)}\right]\right\} \tag{56b}$$

Here $F_A^{0,1}$ are, respectively, the isoscalar and isovector axial form factors, with $F_A^1 \simeq -1.23$ and, in a quark model estimate, $F_A^0/F_A^1 = 3/5$. Since the usual Goldberger-Treiman relation informs us that

$$g_{\pi NN} = \frac{M_N F_A^1}{f_\pi} \tag{57}$$

the parameter of merit here for the axion coupling is

$$g^2_{aNN} = g^2_{\pi NN}\left(\frac{f_\pi}{f}\right)^2 \simeq 1.45\times 10^{-7}\, g^2_{\pi NN} \tag{58}$$

However, note that according to Eqs. (56), while the isoscalar coupling of the axion to nucleons never vanishes, the isovector coupling can be zero. This point will be of some relevance soon.

The results (53), for the axion mass, (54), for its 2γ lifetime and (56), for its coupling to nucleons, along with its coupling to charged leptons

$$g_{a\ell\ell} = \frac{m_\ell}{x f} \tag{59}$$

are properties of what one may call the standard axion. The only free parameter in all these equations is the parameter x and one can imagine analyzing experiments which search for axions as a function of this parameter. In 1978, at the Tokyo Conference, I reviewed the status of the standard axion[22]. I concluded there that the experimental evidence against its existence seemed rather strong, but that really no directed experiments had been performed to search for axions (except for high energy beam dump experiments) and that perhaps one ought to wait for the results of these experiments. I do not want here to discuss again the evidence against axions presented in Tokyo, but I will make just two remarks: 1) The null results of the SLAC beam dump analysis in Ref. 21 implies that x cannot be small ($x \gtrsim 1$). This conclusion was confirmed later on by the null result of the directed axion search experiment of Bechis et al.[23]; 2) High energy beam dump experiments can rule out axions by their null results only if one assumes that the production and detection cross section for axions can be scaled using the relation (58). Although this is a reasonable assumption, it is by no means imperative that axion production at high energy be as large as this[24].

Recently, interest in axions has been revived by Faissner and his collaborators at Aachen who presented evidence for the two photon decay of a light penetrating particle[25], which could be the axion. Their experiment, which consists of looking for penetrating radiation arising from a dumped proton beam at SIN, is a low energy experiment. Thus it may be reliably analyzed in terms of the effective couplings we just discussed. The excess of (14 ± 5) 2γ events seen above background[25], if interpreted as arising from the production and decay of a standard axion, implies that the parameter

$$x = 3 \pm 0.3 \tag{60}$$

which yields $m_a = (250 \pm 25)$ KeV.

One should note that for this value of x, the axion is essentially isoscalarly coupled to nucleons, since

$$\frac{g_0}{g_1} \simeq 12 \tag{61}$$

As remarked originally in Ref. 21, a predominantly isoscalar axion would probably not be in contradiction with the results for the search for the process $\bar{\nu}_e d \rightarrow np\bar{\nu}_e$ done at the Savannah River reactor by Gurr et al[26]. This experiment provided one of the strongest evidences against axions in 1978, since, if the axion were isovector-coupled, one expected a background[21,22] of approximately 4×10^3 events/day, while no events were found. With a predominantly isoscalar axion this number would be dramatically reduced, both because the deuteron would not be disintegrated by axions and because the rate of axions produced in reactors would be below the canonical estimate of Ref. 21[27].

Unfortunately, the positive evidence for the standard axion's existence has to be tempered by equally positive evidence for its non-existence! Zehnder[28], recently, has looked for the 2γ decay of axions produced by the 662 KeV nuclear transition from the exicted $h_{11/2}^{+}$ state of $^{137}B_a$ to its $d_{3/2}^{+}$ ground state. One can estimate rather reliably the relative probability that nuclear deexcitation occurs by axion rather than γ-emission, since again all the relevant couplings are given and most nuclear uncertainties vanish in the ratio[21,29]. For the particular transition in question, Barroso and Mukhopadhyay[30] estimate, using x = 3,

$$\frac{\Gamma_a}{\Gamma_\gamma} \simeq 1.3\times 10^{-4} \tag{62}$$

Because the excited state of $^{137}B_a$ can be copiously populated by a 1 kCi $^{137}C_s$ source ($\sim 3 \times 10^{13}$/sec), one expects of the order of 3×10^9 axions produced per second! Taking into account various experimental conditions, Zehnder[28] expected to see something like 10^{-2} 2γ coincidences per second. (Again for x = 3.) In fact, the observed experimental rate is $< 3 \times 10^{-5}$. Thus this experiment is in contradiction to the observations of Faissner et al[25]. We note, however, that if x is sufficiently near one, then one expects that the 2γ rate becomes sufficiently small so that no effects should have been seen by Zehnder[28,30]. But $x \simeq 1$ would enhance the isovector content of the axion and again create problems with the Gurr[26] deuteron experiment! One should remark also that the new experimental limit from KEK[31]

$$\frac{\Gamma(K^+\to\pi^+ a)}{\Gamma(K^+\to \text{all})} \leq 4.8\times 10^{-8} \tag{63}$$

does not bode very well for the axion - especially if it is mostly isoscalar - since various recent calculations[32,33] have obtained much larger values for this ratio. Clearly we are at an experimental impasse and it will require some time before the situation regarding the standard axion clarifies. All that I can do here it to quote an old Italian proverb: "Se sono rose fioriranno"[34].

While the standard axion is in the midst of experimental travail, the theorists have not been idle. Two new ideas have emerged recently which still retain the chiral solution proposed by Quinn and myself[14] but render the axion even less accessible experimentally. Recall the formula

$$m_a \sim \frac{m_\pi f_\pi}{f} \tag{64}$$

Tye[35] has suggested that perhaps the axion is very heavy (in the GeV range) because its mass does not arise from mixing with ordinary pions but from mixing with some superheavy pions of another theory. Thus $m_\pi f_\pi \to M_\pi F_\pi$ and the axion is rendered (temporarily) invisible. On the other hand, Dine, Fischler and Srednicki (DFS)[36], have adopted an old idea of Kim[37] and made the axion extremely light by essentially making $f \gg (\sqrt{2}\, G_F)^{-1/2}$. Since, as it turns out all axion's couplings scale with f^{-1} in the DFS model, the axion is also essentially invisible.

Tye's[35] scheme is rather extravagant. The weak interactions are as in the Peccei-Quinn model, but he assumes that besides QCD there is another confining theory QCD' with a different scale $\Lambda' \gg \Lambda_{QCD}$. Further, at some stage he must also assume that QCD and QCD' unify. If QCD' has quark doublets $\begin{pmatrix} U \\ D \end{pmatrix}$, which transform under the weak interactions as $\begin{pmatrix} u \\ d \end{pmatrix}$ do, then the axial current J^μ_5 will contain terms involving these doublets. The divergence of J^μ_5 now will not only contain terms proportional to the usual chiral anomaly $F\tilde{F}$ but also terms proportional to $F'\tilde{F}'$. To make a soft current then one must now subtract from J^μ_5 also terms involving the new doublets of QCD' quarks (cf Eq. (45)). Whence, for Tye's model the relevant divergence equation, instead of Eq. (46) is

$$\partial_\mu \tilde{J}^\mu_5 = -\left(x+\frac{1}{x}\right) N \frac{m_u m_d}{(m_u+m_d)} \left\{ \bar{u} i\gamma_5 u + \bar{d} i\gamma_5 d \right\} - \left(x+\frac{1}{x}\right) N' \frac{M_U M_D}{(M_U+M_D)} \left\{ \bar{U} i\gamma_5 U + \bar{D} i\gamma_5 D \right\} \tag{65}$$

If $\Lambda' \gg \Lambda_{QCD}$ in doing current algebra only the second term is important, since effectively, for QCD' one expects π'-mesons with much larger masses and decay couplings. Thus in Tye's model one gets axions with mass

$$m_a^{Tye} = \frac{M_\pi F_\pi}{m_\pi f_\pi} \left(\frac{N'}{N}\right) m_a \tag{66}$$

which can be taken to be above present experimental possibilities by tuning appropriately M_π and F_π (i.e., Λ').

I should note that Tye's introduction of a QCD' does not solve the strong CP problem, since although there is a chiral symmetry, there are now two θ-terms and one can show that the chiral symmetry rotates away only one combination of these angles. What remains are CP violating $F\tilde{F}$ terms proportional to $\theta - \theta'$. Only if one assumes that QCD and QCD' are unified in a still larger group, is there no strong CP violation, since then $\theta = \theta'$.[38)]

To my mind Tye's model is too complicated to be realistic. However, the DFS scheme, of a very light axion, is more appealing. The DFS idea was essentially exploited earlier by Kim[37)] in a scheme which although less attractive is well worth recalling. What Kim assumed was that one had the usual weak interactions (without a $U_A(1)$ symmetry) but that in addition one had a heavy quark Q coupled in a chirally invariant way to an SU(2) x U(1) singlet Higgs field σ, which had a large vacuum expectation value.

$$\mathcal{L}^{\rm Kim}_{Q-\sigma} = h\{\bar{Q}_L \sigma Q_R + \bar{Q}_R \sigma^+ Q_L\} \qquad (67)$$

Chiral rotations of the Q-field allowed one to get rid of the θ-problem and the axion was essentially $a = \mathrm{Im}\,\sigma$. The Kim axion had no direct coupling to the usual quarks, but could eventually couple in higher order with a coupling proportional to $1/\langle\sigma\rangle$. Similarly, the axion's mass also scaled with $1/\langle\sigma\rangle$ instead of 1/f. Clearly by making $\langle\sigma\rangle$ sufficiently large, and hence making the quark Q very heavy, the axion got effectively invisible. In fact, the only constraint on $\langle\sigma\rangle$ (or equivalently $m_a^{\rm Kim}$) came from stellar energy loss considerations. The process $\gamma e \to ae$ can remove energy from stars with a rate[39,40)] which, for small mass axions, is

$$Q \sim m_a^2 \qquad (68)$$

This rate is not in contradiction with observation (i.e., hot stars exist!) if m_a is sufficiently small ($\langle\sigma\rangle$ sufficiently large). In fact, Dicus et al[40)] found that stellar energy loss considerations implied

$$m_a^{Kim} \lesssim 10^{-2}\ eV \tag{69}$$

The unattractive part of Kim's scheme was the need for a heavy quark Q. In fact, what Dine, Fischler and Srednicki[36)] show is that this quark is not necessary. The idea is to consider again as the relevant weak Lagrangian the one considered by Peccei and Quinn[14)], with two Higgs doublets and an extra $U_A(1)$ symmetry, but in addition introduce an extra SU(2) x U(1) singlet field ϕ, which is active under chiral rotations. That is

$$\phi_i \rightarrow e^{i\alpha} \phi_i \tag{70a}$$

$$\phi \rightarrow e^{i\frac{\alpha}{2}} \phi \tag{70b}$$

Because the field ϕ responds to chiral transformations, the axion field will contain it. Further if $f_\phi = \sqrt{2}\langle\phi\rangle \gg f$, then to a good approximation the axion field is just Im ϕ. Indeed, writing

$$\phi = \frac{1}{\sqrt{2}}\{f_\phi + \rho + i\sigma\} \tag{71}$$

and using for the neutral components of ϕ_i the representation of Eq. (37), one immediately learns, using Eq. (40) for the chiral current, that the axion is

$$a = \left[f\sqrt{f^2 f_\phi^2 + 16 f_1^2 f_2^2}\right]^{-1}\left\{4f_1f_2(f_2\sigma_1 + f_1\sigma_2) + f^2 f_\phi \sigma\right\} \tag{72}$$

$$\simeq \sigma + \frac{4f_1f_2}{f_\phi f^2}\left\{f_2\sigma_1 + f_1\sigma_2\right\}$$

and

$$f_A \simeq \frac{f_\phi}{2}\left(x + \frac{1}{x}\right) \tag{73}$$

where

$$J^\mu_5 = -f_A\, \partial^\mu a + \cdots \tag{74}$$

The DFS axion has a mass given by Eq. (52) but with $f \to f_A$. Whence

$$m_a^{DFS} = \frac{2 m_\pi f_\pi N}{f_\phi} \sqrt{\frac{m_d m_u}{(m_d + m_u)^2}}$$

$$\simeq 75 \left(\frac{500\ \text{GeV}}{f_\phi} \right) \text{KeV} \tag{75}$$

The coupling of the DFS axion to matter occurs only though the small admixture of ϕ_1 and ϕ_2 in the axion field, since ϕ itself has no couplings to fermions. Clearly then

$$g^{DFS}_{ffa} \sim \frac{m_f}{f_\phi} \tag{76}$$

which is down by a factor $\frac{f}{f_\phi} \ll 1$ from the coupling of the standard axion. Thus the DFS axion is for all practical purposes invisible! Its only constraint comes from the stellar energy loss limit[39,40]. Dine, Fischler and Srednicki find that this implies

$$f_\phi \gtrsim 10^9\ \text{GeV} \tag{77}$$

Obviously, to have such a large vacuum expectation value in the theory is not particularly natural, unless one can relate it in some ways to the grand unification scale. This has been done by Wise, Georgi and Glashow[41] who observed that the field ϕ can be neatly fit in the <u>24</u> of SU(5)[42] so that $f_\phi = M_{GUT}$. I shall not further discuss this grand unified axion since Georgi[43], in his lectures here, will touch upon it. Rather, I will close this section by making three comments:

(1) The invisible axion is a nice theoretical solution but it is philosophically troublesome that it cannot be tested experimentally.

(2) Even if ϕ can fit in the $\underline{24}$ of SU(5), it should be remarked that having two $\underline{5}$'s of Higgs (i.e., those that contain ϕ_i) may not be enough for the baryon asymmetry of the universe. The point is that the $\underline{5}$'s associated with ϕ_1 and ϕ_2 act, for the baryon asymmetry problem, effectively as one $\underline{5}$ and one would need to extend the model further to obtain a realistic value for the asymmetry[44].

(3) Finally, one really ought to ask oneself what, besides the desire of solving the θ-problem, is the reason for having the extra $U_A(1)$ global chiral symmetry[43]?

3. Soft CP Violation

If no chiral symmetry of the type discussed in the previous section is present in nature, it may still be possible that the θ-problem receives a natural solution if CP is broken softly in the weak interactions. Remember that what one had to explain was why

$$\bar{\theta} = \theta + \text{Arg det } M \lesssim 10^{-8} - 10^{-9} \tag{78}$$

A satisfactory explanation could be provided by a theory in which:

1) $\theta = 0$, as a symmetry requirement.

2) Arg det M $\lesssim 10^{-8} - 10^{-9}$, from direct calculation.

The second point above requires that CP be broken by operators of dimension less than 4 in the weak interactions (soft CP-breaking), because only in this case can one guarantee that Arg det M is calculable (i.e., finite to all orders in perturbation theory):[45]

$$\text{Arg det } M < \infty \tag{79}$$

The original idea of solving the θ-problem via a soft CP model is due to Wilczek[16], and a first prototype model was constructed by Georgi[46].

Many other models of soft CP have appeared since[47)]. The usual strategy of these models was to impose discrete symmetries on the theory that guaranteed that at zero'th and first order Arg det M vanished. Then the second order contribution could well be expected to be $\lesssim 10^{-8} - 10^{-9}$. Unfortunately, at least for my taste, the imposition of discrete symmetries to forestall the appearance of a large Arg det M does not make these models very appealing. Recently, however, with Mohapatra and Masiero[48)] I constructed a soft CP model where the smallness of Arg det M follows not from a discrete symmetry but more dynamically. I shall briefly describe this model here, since I believe that the dynamical framework for obtaining small Arg det M in the model may in fact be more general than the model itself - which has far too many free parameters to be really predictive.

Before describing the model there are a couple of general comments that ought to be made concerning soft CP breaking. There are really two possible ways to break CP softly:

1) CP is broken explicitly in the Lagrangian by operators of dimension less than 4. For example, by complex mass mixing in the Higgs sector

$$\mathcal{L}_{\not{CP}} = \mu H_1 H_2 + \mu^* H_1^+ H_2^+ \tag{80}$$

2) CP is broken spontaneously by phases appearing in the vacuum expectation of Higgs fields:

$$\langle \phi \rangle = v\, e^{i\delta} \tag{81}$$

Only the second type of CP violation, which was originally suggested by T.D. Lee[49)], makes sense to me for the case of the θ-problem. Recall that to solve this problem we not only want Arg det M to be small, but must require that also $\theta = 0$. This can be imposed as a CP requirement on the QCD-part of the total Lagrangian, but this by itself is not particularly reasonable if there are pieces of $\mathcal{L}_{TOT}$ which break CP, as in the case 1) above. However, if CP is only broken by the vacuum, $\theta = 0$ can be thought of as a CP requirement on $\mathcal{L}_{TOT}$ and not just $\mathcal{L}_{QCD}$, which is much more appealing.

It may be worth mentioning in this context that if Higgs fields do not exist but are replaced by effective di-fermion condensates, either through some technicolor scheme or because all quarks and leptons are themselves composite, then most likely CP violation must occur spontaneously. The point is that, in these more complex dynamical schemes, the underlying fundamental theories are very likely to be based on gauge interactions and

$$\mathcal{L}_{int} = g\, \bar{\psi}\, \gamma^{\mu}\, T \cdot A_{\mu}\, \psi \tag{82}$$

always involves real coupling constants g. Thus the only phases that could appear in these theories must occur because the vacuum itself aligns in a complex way. Thus soft CP theories, due to complex vacuum expectation values, may therefore be of more general interest than just for trying to solve the θ-problem.

The soft CP model devised by Masiero, Mohapatra and me[48] uses as the weak group the left-right extension of the standard model[50]: $G_W = SU(2)_L \times SU(2)_R \times U(1)_{B-L}$, which if $SU(2)_R$ is broken down at sufficiently large scales ($M_{W_R} \gg M_{W_L}$) yields the usual weak interaction phenomenology. What we[48] were able to show, under some specific assumptions to be detailed below, is that if the breakdown of $SU(2)_R$ (as well as of B-L) also broke CP spontaneously, then the existence of the left-right symmetry breaking hierarchy naturally leads to a small Arg det M. Specifically we obtained that in zero'th order:

$$(\mathrm{Arg}\det M)_0 \sim O\left[\left(\frac{M_{W_L}}{M_{W_R}}\right)^2\right] \tag{83}$$

with this relation preserved in higher order,

$$(\mathrm{Arg}\det M)_1 < (\mathrm{Arg}\det M)_0 \tag{84}$$

Thus the existence of a hierarchy, along with spontaneous CP breakdown, provides a raison d'etre for a small $\bar{\theta}$.

The only relevant details of the model, for the present discussion, concerns the Higgs sector. Three types of Higgs fields enter, which transform under $SU(2)_L \times SU(2)_R$ as

$$\Delta_L \sim (1, 0) \tag{85a}$$

$$\Delta_R \sim (0, 1) \tag{85b}$$

$$\phi \sim (\tfrac{1}{2}, \tfrac{1}{2}) \tag{85c}$$

The Δ-fields have the quantum number of pairs of fermions, which in the model transform as (1/2,0) or (0,1/2), and thus one has both $\Delta_{qq} \sim qq$ and $\Delta_{ll} \sim ll$ (i.e., fields that transform as pairs of quarks or leptons). The ϕ-field transforms as a fermion-antifermion pair and thus its vacuum expectation value contributes to the quark mass matrix. It is easy to check that charge conservation permits only the following nonzero vacuum expectation values. (Here $\Delta = \tau_i \Delta_i$.)

$$\langle (\Delta_{ee})_L \rangle = \begin{pmatrix} 0 & 0 \\ V_L & 0 \end{pmatrix} \tag{86a}$$

$$\langle (\Delta_{ee})_R \rangle = \begin{pmatrix} 0 & 0 \\ V_R & 0 \end{pmatrix} \tag{86b}$$

$$\langle \phi \rangle = \begin{pmatrix} \kappa & 0 \\ 0 & \kappa' \end{pmatrix} \tag{86c}$$

The hierarchical breakdown of $G_W \rightarrow SU(2)_L \times U(1)$ is accomplished by having[51)]

$$V_R \gg \kappa, \kappa' \gg V_L \qquad \text{with} \qquad V_L V_R \simeq \kappa^2, \kappa'^2 \tag{87}$$

which is a possible minimum of the potential. Since $M_{W_R} \sim V_R$ and $M_{W_L} \sim \kappa, \kappa'$ it follows that

$$M_{W_R} \gg M_{W_L} \tag{88}$$

Masiero, Mohapatra and I[48)] investigated whether it is possible that the above vacuum expectation values be complex, thereby giving rise to spontaneous breakdown of CP. We found that only if one had two or more Δ_{11} fields (left and right) could the vacuum expectation values be complex and in that case then the relevant phases had to obey the following rules:

$$\text{phase}\langle\phi\rangle \sim O\left(\frac{v_L}{v_R}\right) \sim O\left(\frac{\kappa^2}{v_R^2}\right) \sim O\left(\frac{M^2_{W_L}}{M^2_{W_R}}\right) \tag{89a}$$

$$\text{phase}\langle\Delta_R\rangle \sim O\left(\frac{v_L^2}{v_R^2}\right) \sim O\left(\frac{\kappa^4}{v_R^4}\right) \sim O\left(\frac{M^4_{W_L}}{M^4_{W_R}}\right) \tag{89b}$$

$$\text{phase}\langle\Delta_L\rangle \sim O(1) \tag{89c}$$

where the approximate results in Eqs. (89a) and (89b) follow from Eqs. (87) and (88). Thus the hierarchy of vacuum expectation values causes a hierarchy of phases, with the largest phases belonging to those fields which have the smallest vacuum expectation values. We believe that this result is general and does not depend on the particular model in question, being essentially a reflection of the Appelquist-Carrazone decoupling theorem[52].

Because the quark mass matrix is essentially proportional to $\langle\phi\rangle$, Eq. (89a) implies that in the model, to lowest order

$$\text{Arg det } M \sim \text{phase}\langle\phi\rangle \sim O\left(\frac{M^2_{W_L}}{M^2_{W_R}}\right) \tag{90}$$

which, if M_{W_R} is sufficiently large, can be less than $10^{-8} - 10^{-9}$. It need be checked, however, that in higher order Arg det M does not pick up a larger contribution, since there are CP violating phases of O(1) in the model associated with $\langle\Delta_L\rangle$. The relevant graphs to consider are of the type shown below

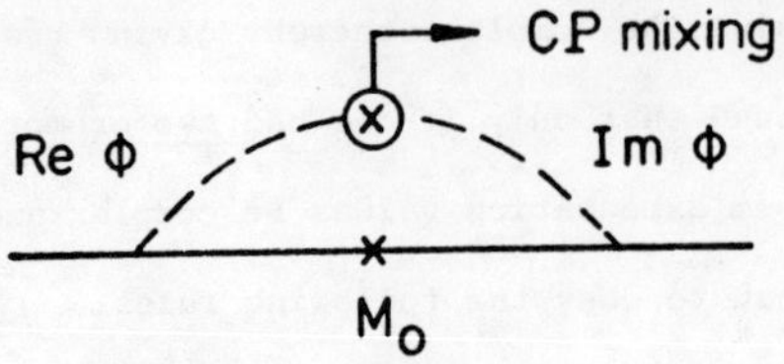

Fig. 3: Typical graph which contributes to Arg det M in first order

It can be shown[48)] that these graphs give

$$(\mathrm{Arg}\,\det M)_1 < (\mathrm{Arg}\,\det M)_0 \tag{91}$$

This follows essentially because, even though the CP mixing can be large,

$$\text{C}\not{\text{P}}\text{ mixing} \sim v_R V_L \sin\delta_L \sim K^2 \tag{92}$$

the integral itself is of $O(1/V_R^2)$, since always at least one of the Higgs exchanged is quite massive ($\mu \simeq V_R$). Since the lowest order estimate for Arg det M is preserved it follows that the θ -problem is solved, if the left-right hierarchy is sufficiently large.

The particular model studied by us in Ref. 48) is rather complex and not terribly predictive because of the large numbers of unknown parameters[53)]. Ordinary CP violation in K -decays is essentially of a superweak variety due to $\Delta S = 2$, Δ_{qq} exchanges and one expects that $\epsilon'/\epsilon \lesssim 10^{-5}$. The dipole moment of the neutron is determined by θ and in view of Eq. (90) one could interpret a non-zero d_n as setting a scale for W_R. However, the large number of Higgs fields present in the model cause some difficulties with flavor changing processes in the kaon sector and one cannot really say that the model is fully realistic[48,53)]. Nevertheless, I think the model illustrates clearly the point that if there are hierarchies present, perhaps it is not so surprising that small phases <u>may</u> appear in the theory, particularly if these phases are connected with the same dynamics which give rise to the hierarchies (spontaneous symmetry breaking).

Before closing I want to very briefly discuss some cosmological aspects of soft CP models which are troubling. If the breakdown of CP is due to vacuum misalignment, then at least two difficulties are apparent:

1) Baryon asymmetry may not occur at all.

2) Particle-antiparticle domains may form which, even if a baryon asymmetry could be established, would tend to wash out any produced asymmetry.

The first point is rather easy to understand. As is by now clear[54], there are three conditions necessary for the development of a cosmological baryon asymmetry:

a) Baryon number must be violated in the theory.

b) The baryon violating processes ought to have been out of thermal equilibrium in the epoch of the universe where the asymmetry ensued.

c) C and CP violation must accompany baryon number violation.

Grand unified theories (GUTS) in general have the possibility of having baryon violating interactions due to the exchange of very massive grand unified bosons X ($M_X \simeq 10^{15}$ GeV). Careful analysis[54] shows that, if GUTS esist, it is possible to have the baryon violating processes go out or equilibrium in the early universe at temperatures of order $T \sim M_X$. Thus the exciting possibility of actually being able to develop a matter-antimatter asymmetric universe from a symmetric starting point is open, provided that condition c) is satisfied. But this may not be so if CP is broken softly by the vacuum, since at high temperatures one expects that symmetries which are broken by the vacuum be restored[55].

If one has just a single Higgs field ϕ which at T = 0 has $\langle\phi\rangle = f \neq 0$, then one can show[55] that as the temperature grows $\langle\phi\rangle \to 0$, with the transition temperature being of order $T_C \sim f$. This is illustrated below

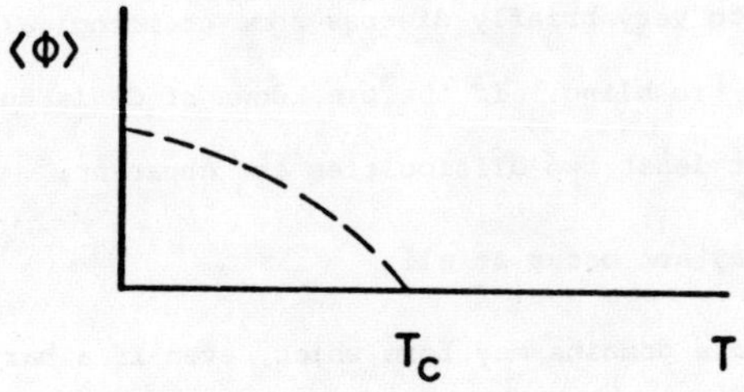

Fig. 4: Temperature dependence of $\langle\phi\rangle$

If CP violation were to be associated with non-zero vacuum expectation values which break the standard low energy weak interaction theory[9], then one would naturally expect that at $T \sim 10^{15}$ GeV (which is much above $\langle\phi\rangle \simeq 250$ GeV of the usual SU(2) x U(1) theory) no CP violation remains. In this scenario, then, no baryon asymmetry could ever be generated.

However, the above point of view may be too pessimistic. Although it is true that with a single Higgs field $\langle\phi\rangle \to 0$ at $T > T_C \sim \langle\phi\rangle_{T=0}$, this situation does not necessarily hold in a many Higgs theory. Mohapatra and Senjanovic[56] have shown that in a theory with many Higgs fields, with a potential whose parameters are appropriate, one can have

$$\begin{aligned} \langle\phi_i\rangle &= 0 \qquad i = 1,\ldots,n \\ \langle\phi_i\rangle &\neq 0 \qquad i = n+1,\ldots,m \end{aligned} \qquad T \gg T_c \tag{93}$$

where T_C is the nominal transition temperature of $O(\langle\phi_i\rangle_{T=0})$. Thus it may be possible to avoid symmetry restoration at high temperature, provided that the Higgs sector is rich enough. However, one must point out, in fairness, that in general to obtain (93) one requires a hierarchy of vacuum expectation values and one ends up, in the case of SU(2) x U(1), with various Higgs mesons of mass much greater than M_{W_L}.

In fact, if CP is broken spontaneously by the vacuum, it may be more reasonable that there should exist a hierarchy of symmetry breaking interactions, as for example in the left-right model just discussed. It is possible that in these classes of models one can get baryon asymmetry from temperatures T which are not those of grand unification, but much smaller ones. If this is the case, it may well be that T is low enough so that it is below some of the relevant vacuum expectation value scales. Then the question of symmetry restoration never enters. Such a model of low temperature baryon asymmetry generation based on $SU(2)_L \times SU(2)_R \times U(1)_{B-L}$ has recently been discussed by Masiero and Mohapatra[57)] and indeed they have $T_{asym} < \langle \Delta_R \rangle$[58)]. Their model is rather complex in that it requires a fine scale hierarchy between the various particle masses, whose decays eventually produce the required asymmetry[59)]. Thus perhaps the details of the model may not be totally believable. However, the underlying principles which could give rise to a low temperature baryon asymmetry (and thus for soft CP theories by-pass the symmetry restoration issue) are reasonably clear[57)]. One must have that:

1) $\Delta B \neq 0$ processes which give the dominant contribution to the asymmetry should have reaction rates which are given by effectively non-renormalizable interactions $\Gamma \sim T^n$ $n \gtrsim 5$.

2) Particles whose decay violate baryon number via renormalizable interactions ($\Gamma \sim T$) must be heavy enough so that they have essentially disappeared by the time the processes in 1) fall out of equilibrium.

These points are necessary since particles with mass much less than 10^{15} GeV will always be in equilibrium ($\Gamma \gg H \sim T^2/m_p$, where H is the expansion rate of the universe and $m_p \sim 10^{19}$ GeV is the Planck's mass) <u>if</u> their decay rate is renormalizable. If these particles decay in a baryon violating way, they will tend to delete any previously obtained asymmetry. On the other hand if $\Gamma \sim T^n$, it is possible, for low enough T, that $\Gamma \ll H$, so that these processes are out of equilibrium. Clearly, at that time there better be no more processes which serve to delete away any asymmetry produced.

The second cosmological difficulty that soft CP models may encounter is related to the appearance of domains[60,61]. Suppose that CP is violated softly by vacuum breaking and that the theory is such that CP is still not restored at temperatures needed to produce the baryon asymmetry but gets restored soon after at a temperature T*. At these high temperatures the casually connected regions in the universe are small, of size

$$\ell \sim t \sim \frac{m_P}{T^{*2}} \tag{94}$$

Presumably in each of these domains the sign of the violation of CP, connected with the sign of the phase of the vacuum expectation value which first appears at T*, is uncorrelated. At a later stage in the evolution of the universe these domains are casually connected. Because the baryon asymmetry is proportional to the CP sign, the superposition of results from various domains should effectively dilute the asymmetry obtained from any individual domain[62].

There are some possible solutions to this domain problem. The simplest, but perhaps the least realistic, is to suppose that the soft CP symmetry never gets restored until $T^* \gtrsim m_p$. Of course in this case all bets are off, since obviously we know nothing of the universe at these temperatures. An observation on the same vein, which perhaps is more sensible, is to point out that one would equally well expect that, in the early universe, outside of "domains" of order 2t (i.e., over distances of the order of the particle horizon[63]) the universe at time t should be inhomogenous, since these domains are again casually separated. Yet no trace of this inhomogeneity appears to survive. For all practical purpose one can treat the early universe as homogenous. This is the famous horizon problem[64]. Perhaps then the CP domain problem also is a nonproblem. Although domains exist, for some as yet unclear reason, one can generate a casual coherence over larger regions.

In this respect I should mention a very interesting proposal for solving the horizon problem due to Guth[65]. He suggests that perhaps the early universe goes through a first order phase transition with a very strong supercooling and associated release of a large quantity of latent heat. During the supercooling period the universe is expanding exponentially and this effectively erases the horizon problem. K. Sato[66] recently has applied this idea to the CP domain problem. He supposes that during the supercooling the allowed CP domains stretch exponentially. He then argues that these domains are big enough so that they can grow into superclusters of galaxies. If this scenario is correct, then the existence of CP domains may not vitiate the usual calculations of the baryon asymmetry of the universe.

4. Conclusions

I end this lecture by recalling again the two possible solutions to the θ-problem discussed here. Either there is a chiral symmetry, and if the standard axion does not exist perhaps the DFS mechanism is effective; or the weak interactions break CP spontaneously and a gauge hierarchy guarantees that $\bar{\theta}$ is small enough. In either case a proliferation of Higgs fields is needed and perhaps one should be a little wary. Also, even more preoccupying is the fact that both solutions are "constructed" to solve a problem and do not arise from some deeper principle. Thus one may really query why there should be a chiral symmetry at all to be imposed on the theory. Or, in the alternative scenario, why if the Lagrangian conserves CP should the vacuum want to break this symmetry spontaneously. Clearly these are very hard questions to answer. Perhaps we should be content for the moment with the progress made in 1981 and hope to eventually come to grips with these more profound questions.

Acknowledgements

I would like to thank Profs. M. Konuma and Z. Maki for the splendid hospitality afforded me in Kyoto. I also would like to acknowledge my collaborators, Rabi Mohapatra and Antonio Masiero for some very helpful discussions on the material presented here.

References

1) G. 't Hooft, Phys. Rev. Lett. 37 (1976) 8; Phys. Rev. D14 (1976) 3432.

2) C.G. Callan, R.F. Dashen and D.J. Gross, Phys. Lett. 63B (1976) 334.

3) For a very nice discussion of this point see R.J. Crewther, Proceedings of the XVII Internationale Universitätswochen für Kernphysik, Schladming (Austria), 1978.

4) This formula follows rather straightforwardly once one recognizes that ϵ_{ijk} Tr $A^iA^jA^k$ is essentially the Jacobian of the transformation from S_3 to the hypersphere in group space. Thus n measures the number of windings of S_3 onto the group space. For more detail, see Ref. 3).

5) W.A. Bardeen, Nucl. Phys. B75 (1974) 246.

6) S.L. Adler, Phys. Rev. 177 (1969) 2426.

7) J.S. Bell and R. Jackiw, Nuov. Cim. 60A (1969) 47.

8) R. Jackiw and C. Rebbi, Phys. Rev. Lett. 37 (1976) 172.

9) S. Weinberg, Phys. Rev. Lett. 19 (1967) 1264; A. Salam, Proceedings 8th Nobel Symposium, Stockholm, 1968 (ed. N. Svartholm, Almqvist and Wiksells, Stockholm, 1968); S.L. Glashow, J. Iliopoulos and L. Maiani, Phys. Rev. D2 (1970) 1285.

10) V. Baluni, Phys. Rev. D19 (1979) 2227.

11) R. Crewther, P. di Vecchia, G. Veneziano and E. Witten, Phys. Lett. 89B (1979) 123.

12) N. Ohta and K. Kawarabayashi, Tokyo Preprint 1981.

13) A. Linde, Phys. Lett. 93B (1980) 327.

14) R.D. Peccei and H.R. Quinn, Phys. Rev. Lett. 38 (1977) 1440, Phys. Rev. D16 (1977) 1791.

15) S. Weinberg, Phys. Rev. Lett. 40 (1978) 223.

16) F. Wilczek, Phys. Rev. Lett. 40 (1978) 279.

17) Possible vector contributions in J^{μ}_{5} can be eliminated by adding to the definition (40) appropriate amounts of conserved electromagnetic, baryonic and leptonic currents.

18) W.A. Bardeen and S.H. Tye, Phys. Lett. 74B (1978) 229.

19) For a review of some of these techniques, see, for example, R.D. Peccei in Particle Physics 1980, I. Andric, I. Dadic and N. Zovko editors.

20) J. Kandaswamy, P. Salomonson and J. Schechter, Phys. Rev. D17 (1978) 3051.

21) T.W. Donnelly, S.J. Freedman, R.S. Lytel, R.D. Peccei and M. Schwartz, Phys. Rev. D18 (1978) 1607.

22) R.D. Peccei, Proceedings of the XIX International Conference on High Energy Physics, Tokyo 1978.

23) D.J. Bechis et al, Phys. Rev. Lett. 42 (1979) 1511.

24) W.A. Bardeen, S.H. Tye and J. Vermaseren, Phys. Lett. 76B (1978) 580.

25) H. Faissner et al, Aachen Prepring PITHA 81/13, to be published in Phys. Lett. B.

26) H.S. Gurr, F. Reiner and H.W. Sobel, Phys. Rev. Lett. 33 (1974) 179.

27) H. Faissner, (private communication) and his group appear to have also observed an excess of 2γ radiation in an experiment at the Julich reactor (see H. Faissner, Proceedings of Neutrino '81, Hawaii, July 1981); J.L. Vuilleumier et al, Phys. Lett. 101B (1981) 341, on the other hand see no such excess above background in the Grenoble reactor experiment and, using the flux estimate of Ref. 21, obtain a limit for the axion mass: $m_a \lesssim 280$ KeV, consistent with Ref. 25.

28) A. Zehnder, SIN Preprint, submitted to Phys. Lett. B and Proceedings of Neutrino '81, Hawaii, July 1981.

29) S.B. Treiman and F. Wilczek, Phys. Lett. 74B (1978) 381.

30) A. Barroso and N. Mukhopadhyay, SIN Preprint PR-81-03.

31) Y. Nagashima, Proceedings of Neutrino '81, Hawaii, July 1981.

32) J.M. Frere, M.B. Gavela and J. Vermaseren, Phys. Lett. 103B (1981) 29.

33) I. Antoniadis and T.N. Truong, Ecole Polytechnique Preprint (1981).

34) Roughly translated it means: if they are roses, they will bloom.

35) S.H. Tye, Cornell Preprint CLNS/81-489.

36) M. Dine, W. Fischler and M. Srednicki, IAS Preprint 1981.

37) J.E. Kim, Phys. Rev. Lett. 43 (1979) 103.

38) This unification may be troublesome since then fermions in QCD and QCD' would appear to have to have the same masses.

39) K. Sato and H. Sato, Prog. Theo. Phys. 54 (1975) 912.

40) D.A. Dicus, E.W. Kolb, V.L. Teplitz and R.V. Wagoner, Phys. Rev. D22 (1980) 839.

41) M. Wise, H. Georgi and S.L. Glashow, Phys. Rev. Lett. 47 (1981) 402.

42) A similar observation is contained also in H.P. Nilles and S. Rabi, SLAC-PUB 2743, 1981.

43) See H. Georgi, these proceedings.

44) D. Nanopoulos, private communication.

45) J. Ellis and M.K. Gaillard, Nucl. Phys. B150 (1979) 41, have argued that certain theories in which CP is broken by d = 4 terms (like the minimal Kobayashi-Maskawa model) may "effectively" be considered as giving a finite Arg det M, because the first infinite contributions to this quantity occur at very high order (14th order in the minimal K-M model!). Presumably the right "theory of everything" could then take care of these "small" infinities. Although I have some sympathy with this type of approach, I shall not consider it further here.

46) H. Georgi, Hadronic Journal 1 (1978) 155.

47) M.A.B. Beg and H.S. Tsao, Phys. Rev. Lett. 41 (1978) 278;

R.N. Mohapatra and G. Senjanovic, Phys. Lett. 79B (1978) 283;
G. Segrê and H.A. Weldon, Phys. Rev. Lett. 42 (1979) 1191;
S.M. Barr and P. Langacker, Phys. Rev. Lett. 42 (1979) 1654.

48) A. Masiero, R.N. Mohapatra and R.D. Peccei, MPI-PAE/PTh 25/81, to be published in Nucl. Phys. B.

49) T.D. Lee, Phys. Rev. D8 (1973) 1226, Phys. Rept. C9 (1974) 148; see also P. Sikivie, Phys. Lett. 65B (1976) 141.

50) J.C. Pati and A. Salam, Phys. Rev. D10 (1974) 275;
R.N. Mohapatra and J.C. Pati, Phys. Rev. D11 (1975) 566, 2558;
G. Senjanovic and R.N. Mohapatra, Phys. Rev. D12 (1975) 1502.

51) R.N. Mohapatra and G. Senjanovic, Phys. Rev. D23 (1981) 165.

52) T.A. Appelquist and J. Carrazone, Phys. Rev. D11 (1975) 2856.

53) The predictions of the model are contrasted with those of the standard Kobayashi-Maskawa model in R.D. Peccei, Proceedings of the IV Warsaw Symposium on Elementary Particle Physics, Kazimierz, Poland, 1981.

54) See, for example, the lectures of M. Yoshimura in these proceedings.

55) For a review, see for example, A. Linde, Rep. Prog. Phys. 42 (1979) 389.

56) R.N. Mohapatra and G. Senjanovic, Phys. Rev. Lett. 42 (1979) 1651, Phys. Rev. D20 (1979) 3390, Phys. Lett. 89B (1979) 57, Phys. Rev. D21 (1980) 3470.

57) A. Masiero and R.N. Mohapatra, Phys. Lett. 103B (1981) 343.

58) Note that the phase of $\langle \Delta_R \rangle$ in the Masiero-Mohapatra model is not of $O(V_L^2/V_R^2)$ as in Eq. (89b), since here one is working at $T > V_L, \kappa, \kappa'$ Indeed, it turns out that then the phase of $\langle \Delta_R \rangle$ is of $O(1)$.

59) Indeed, T. Yanagida (private communication) has criticized some of the detailed numerics in the Masiero-Mohapatra scheme.

60) I.Yu. Kobsarev, L.B. Okun and Ya.B. Zeldovich, Phys. Lett. 50B (1974) 340.

61) R.W. Brown and F. Stecker, Phys. Rev. Lett. 43 (1979) 315.

62) J. Ellis, M.K. Gaillard and D.V. Nanopoulos, Proceedings of the First Europhysics Study Conference on Unified Theories of the Fundamental Interactions, Erice, 1980.

63) See, for example, S. Weinberg, Gravitation and Cosmology, (Wiley, 1972) p. 489.

64) W. Rindler, Mo. Not. R. Astron. Soc. 116 (1956) 663; see also S. Weinberg, Ref. 63), p. 525.

65) A. Guth, Phys. Rev. D23 (1981) 347.

66) K. Sato, Phys. Lett. 99B (1981) 66.

PROTON DECAY RATES

Yukio Tomozawa

The Harrison M. Randall Laboratory of Physics,

University of Michigan, Ann Arbor, MI 48109, U.S.A.

in

Proceedings of the Fourth Kyoto Summer Institute on Grand Unified Theories and Related Topics, Kyoto, Japan, June 29 - July 3, 1981, ed. by M. Konuma and T. Maskawa (World Science Publishing Co., Singapore, 1981).

Contents

PROTON DECAY RATES

Yukio Tomozawa

The University of Michigan

Abstract

The two-body decay rates of the nucleon in the SU(5) grand unified gauge theory are computed using field theoretical methods. The result obtained indicates that the rates for the vector meson modes are much smaller than those of the pionic modes.

I. Introduction

Proton instability is one of the most important predictions of grand unified gauge theories[1)-3)] and experimental efforts to detect its decay processes are in progress[4)]. Estimate of the decay rate and branching ratios is an important task for the design and analysis of experiments and much work has already been done in the context of the SU(5) model[2)]. However, in most calculations SU(6) wave functions or the MIT Bag model wave functions are used in the computation of the branching ratios for pseudoscalar and vector meson modes of nucleon decay. The effect of symmetry breaking has never been questioned in these computations. In this lecture, I will present a calculation based on field theoretical methods without using the SU(6) wave functions for mesons. It will be shown that the effect of symmetry breaking is significant and in particular the vector meson modes may be severely suppressed.

II. General Discussion

The Lagrangian relevant to nucleon decay in grand unified gauge theories (which include the SU(5) model) can be written as

$$\mathcal{L} = \frac{4G\lambda}{\sqrt{2}} \varepsilon_{ijk} \{\overline{u^c_{kL}} \gamma_\mu u_{jL} [r \overline{e^+_L} \gamma_\mu d_{iL} + \overline{e^+_R} \gamma_\mu d_{iR}$$

$$+ \overline{\mu^+_L} \gamma_\mu s_{iL} + \overline{\mu^+_R} \gamma_\mu s_{iR}] \quad (2.1)$$

$$+ \overline{u^c_{kL}} \gamma_\mu d_{jL} (\overline{\nu^c_{eR}} \gamma_\mu d_{iR} + \overline{\nu^c_{\mu R}} \gamma_\mu s_{iR})\}$$

where

$$G = g^2_{GUT}/(4\sqrt{2}\, m^2_X) \quad (2.2)$$

and

$$r = 2 \quad \text{for } SU(5), \quad (2.3)$$

g_{GUT}, m_X and λ being the grand unified coupling constant the unification mass and the enhancement factor due to the renormalization group analysis, respectively. Under the assumption of three generations for fermions, we have the values of the parameters[7].

$$g^2_{GUT}/4\pi = 0.024$$

$$m_X = (4\text{-}8) \times 10^{14} \text{ GeV} \quad (2.4)$$

$$\lambda = 3.5$$

The SU(6) or quark model predictions for the two body branching ratios are summarized as[6]

$p \to e^+\pi^\circ$	30~40%	
$\overline{\nu_e}\, \pi^+$	15%	
$e^+\omega$	20~30%	(2.5)
$e^+\rho^\circ$	~10%	
μ^+K°	5~10%	

$$\begin{array}{lll} n \to & e^+\pi^- & \sim 70\% \\ & \overline{\nu_e}\ \pi^\circ & 7\% \\ & e^+\rho^- & 10\sim 20\% \\ & \nu_\mu K^\circ & 1\% \end{array} \qquad (2.6)$$

The other decay modes have much smaller branching ratios. (Note that Ref. 8 claims that the three body branching ratios are comparable to those of the two body decay, e.g. $\Gamma(p\to e^+\pi^+\pi^-) \approx (p\to e^+\pi^\circ)$. We will not discuss this problem in this lecture). The two body decay rate for $m_X = 6\times10^{14}$ GeV is summarized as

$$\tau^{(2)}_{p\,(n)} = \xi_{p\,(n)} \times 10^{31} \text{ years,} \qquad (2.7)$$

where[6,7)]

$$\begin{array}{lll} \xi_p = & 0.27 & \text{Goldman \& Ross} \\ & 0.41 & \text{Din et. al} \\ & 0.56 & \text{Buras et. al} \\ & 1.0 & \text{Gavela et. al} \\ & 1.6\text{-}4.2 & \text{Arisue} \\ & 4.6 & \text{Donoghue} \end{array} \qquad (2.8)$$

and

$$\xi_n \approx (0.8 \sim 1.5)\ \xi_p \qquad (2.9)$$

(see also Table 5)

The above results (5)-(9) are subject to the following criticisms:

(1) SU(6) symmetry breaking is considered only through the phase space factors. What is the effect of symmetry breaking in the matrix element?

(2) In particular, how are the branching ratios of the pionic and the vector meson modes changed due to the fact that the pion and the vector mesons have vastly different masses?

(3) How reliable is the calculation of the absolute value?

Obviously these are difficult questions to answer.

I will give possible answers to these questions indirectly by presenting a calculation of the decay rate and the branching ratios which does not use the SU(6) symmetry. The methods which I will use are entirely field theoretical ones and have proved to be successful in low energy hadron physics: PCAC and current algebra for the pionic and kaonic modes, and vector dominance for the vector meson modes. The PCAC relation

$$\phi^{\alpha} = (F_{\alpha}\mu_{\alpha}^{2})^{-1} \frac{\partial A_{\mu}^{\alpha}}{\partial x_{\mu}}, \qquad \alpha = 1,2,3, \text{ for pion; } 4\text{-}7 \text{ for kaon; } 8 \text{ for } \eta \tag{2.10}$$

together with current algebra and the soft pion limit lead to the well known pion nucleon scattering length formula[9,10)]

$$a_{1/2} = -2a_{3/2} = \frac{1}{4\pi} \frac{\mu_{\pi}}{F_{\pi}^{2}} \frac{1}{1+\mu_{\pi}/M_{N}} \tag{2.11}$$

the index I=1/2, 3/2 being the isospin of the πN system. In Eqs. (10) and (11), $\sqrt{2}\, F_{\pi}$ is the pion decay constant of $\pi^{\pm}\to\mu^{\pm}\nu$ ($\sqrt{2}\, F_{\pi} = 0.945\ \mu_{\pi}$). Eq. (11) gives $a_{1/2(3/2)} = 0.156\mu_{\pi}^{-1}$ ($-0.078\ \mu_{\pi}^{-1}$) which should be compared with experimental data[11)] $0.171\pm0.005\ \mu_{\pi}^{-1}$ ($-0.088\pm0.004\ \mu_{\pi}^{-1}$). The agreement is of order 10%. For the KN scattering length a_I, the prediction is[9)]

$$\begin{aligned} a_{0}(KN) &= 0 \\ a_{1}(KN) &= -\frac{1}{4\pi} \frac{\mu_{K}}{F_{K}^{2}} \frac{1}{1+\mu_{K}/M_{N}} = -0.264\ \mu_{\pi}^{-1} \end{aligned} \tag{2.12}$$

where

$$F_K = 1.25\ F_\pi \tag{2.13}$$

These values are to be compared with experimental data

$$a_1(KN) = a_{k^+p} = -0.220 \pm 0.004\ \mu_\pi^{-1}$$

and (2.14)

$$a_0(KN) = -0.007\ (-0.025)$$

which corresponds to an accuracy of 15%. The situation of the K^-N scattering is more complicated due to the existence of the reactions such as $K^-N \to \pi\Lambda,\ \pi\Sigma$ below the threshold. A surprising success of the soft kaon limit for the scattering length may be seen as the coefficients of the expansion in μ_k^2/M_N^2 ($\approx$ 0.25) being reasonably small.

Before closing this section, let us write down the general formula for two body decay rate for a proton. For the process

$$\begin{matrix} p & \to & \ell & + \text{ ps meson} \\ (M) & & (m) & (\mu) \\ p & & k & q \end{matrix}, \tag{2.15}$$

the amplitude and the decay rate are given by

$$f = g\frac{1}{\sqrt{2q_o}}\sqrt{\frac{mM}{k_op_o}}\ \bar{u}^{(\ell)}(k)\ (A+B\gamma_5)\ u(p) \tag{2.16}$$

$$\Gamma(p\to\ell+\text{ps meson}) = \frac{g^2M}{16\pi}\left(|A|^2+|B|^2\right)\left(1-\frac{\mu^2}{M^2}\right)^2, \tag{2.17}$$

where lepton masses are neglected in the last equation.

For a vector meson process,

$$\begin{matrix} p & \to & \ell^+ & \text{vector meson} \\ (M) & & (m) & (\mu) \\ p & & k & q,\epsilon_\lambda \end{matrix} \tag{2.18}$$

we have

$$f = g \frac{1}{\sqrt{2q_o}} \sqrt{\frac{mM}{k_o p_o}} \varepsilon_\mu \bar{u}^{(\ell)}(k) \left((C+D\gamma_5) \frac{ip_\mu}{M} \right. \tag{2.19}$$

$$\left. + (E+F\gamma_5)\gamma_\mu \right) u(p)$$

and

$$\Gamma(p\to\ell + \text{vector meson})$$

$$= \frac{g^2 M}{16\pi} \left[\frac{1}{4} (|C|^2+|D|^2) \left(1 - \frac{\mu^2}{M^2}\right)^2 \right. \tag{2.20}$$

$$\left. +(|E|^2+|D|^2)\left(1 + \frac{2\mu^2}{M^2}\right) \right] \frac{\left(1 - \frac{\mu^2}{M^2}\right)^2}{\frac{\mu^2}{M^2}}$$

III. The Pionic Decay

A typical pionic mode is $p\to e^+ + \pi^\circ$. We use PCAC and current algebra to compute the decay amplitude. The basic idea of the calculation is given in Ref. 12. The point which is discussed there is to use the rest frame of the pion and then soft pion techniques. Such a method was used in nonleptonic hyperon decays to obtain sum rules for s-wave as well as p-wave amplitudes[13]. For a pion emission process $A\to B\pi$, the energy of the incident particle A is given by

$$\sqrt{p_A^2 + m_A^2} = \frac{m_A^2 - m_B^2 + \mu_\pi^2}{2\mu_\pi} \tag{3.1}$$

in the reference frame where the pion is at rest. In the soft pion limit $\mu_\pi \to 0$, the momentum p_A and p_B become infinite. This may suggest that the most natural reference frame for the soft pion limit described above is the infinite momentum reference frame. It was shown in Refs. 12 and 13. however, that the results of computations in the infinite momentum reference frame and the pion rest frame differ very little.

a) $p \to e^+\pi^\circ$

The standard technique of the reduction formula gives

$$(2q_o)^{1/2} \langle e(\vec{p})\ \pi^\circ(\vec{0})|\mathcal{L}|p(\vec{p})\rangle$$

$$= -i\,\frac{\mu_\pi^2 - q_o^2}{F_\pi \mu_\pi^2}\Big\{\langle e|[Q_5^3,\mathcal{L}]|p\rangle - q_o \sum_\ell \Big[\frac{\langle e|A_0^3(0)|\ell\rangle\langle \ell|\mathcal{L}|p\rangle}{q_o + p_o^e - p_o^\ell + i\varepsilon} - \frac{\langle e|\mathcal{L}|\ell\rangle\langle \ell|A_0^3(0)|p\rangle}{q_o + p_o^\ell - p_o^p - i\varepsilon}\Big]_{\vec{p}^\ell = \vec{p}}\Big\} \tag{3.2}$$

The limit $q_o \to 0$ selects the intermediate state ℓ in the summation of Eq. (2) to be equal to either the initial or the final state. The resulting formula is

$$f = (2q_o)^{1/2} \langle e\pi^\circ|\mathcal{L}|p\rangle = \frac{-i}{F_\pi}\Big\{\langle e|[Q_5^3,\mathcal{L}]|p\rangle + \sum_{\substack{\text{intermediate}\\ \text{spin}}} \langle e|\mathcal{L}|p\rangle\langle p|A_o^3(0)|p\rangle\Big\} \tag{3.3}$$

This procedure can be seen also as the expansion of eq. (2) to be

(1st term) + (2nd term) + (the rest)

$$= \text{commutator term} + q_o\,\frac{1}{q_o} + q_o \int \frac{\langle\rangle\langle\rangle}{q_o + \Delta p_o} + \cdots \tag{3.4}$$

where the third terms and the rest of the expansion should be of the order of q_o/m where m is a typical mass scale of the strong interaction ($\approx$ vector meson mass ~1 GeV). The first two terms of Eq. (4) are comparable and therefore neglecting the last term of Eq. (4) may yield an error of 10~15%.

The first term of Eq. (4) is computed using

$$Q_5^3 = \frac{1}{2} \int (u_j^+ \gamma_5 u_j - d_j^+ \gamma_5 d_j) d^3x \tag{3.5}$$

and

$$[Q_5^3, \mathcal{L}] = \left(\frac{4G\lambda}{\sqrt{2}}\right) \varepsilon_{ijk} \frac{1}{2} [\bar{e}^c_{kL} \gamma_\mu u_{jL} (r\bar{e}^+_L \gamma_\mu d_{iL} - \bar{e}^+_R \gamma_\mu d_{iR}) + \bar{u}^c_{kL} \gamma_\mu d_{jL} \bar{\nu}_{eR} \gamma_\mu d_{iR}] \tag{3.6}$$

Also use is made of

$$\langle p | A_o^3(0) | p \rangle = \frac{1}{2} g_A \frac{M}{E} \bar{u}(p) \gamma_4 \gamma_5 u(p) \tag{3.7}$$

Then Eq. (3) yields

$$f = \frac{-i}{F_\pi} \frac{4G\lambda}{\sqrt{2}} \varepsilon_{ijk} \left(\frac{m}{\varepsilon}\right)^{1/2} \bar{u}_e^{(\ell)}(p) \langle 0 | u_{ka} d_{ic} u_{jb} | p \rangle$$
$$\times \left(C^{-1} \gamma_\mu \frac{1+\gamma_5}{2}\right)_{ab} \left[\frac{1}{2} \left(\gamma_\mu \frac{1+3\gamma_5}{2}\right)_{ec} + \left(\gamma_\mu \frac{3+\gamma_5}{2}\right)_{ec} \times \right. \tag{3.8}$$
$$\left. \times \frac{M}{2E} g_A \bar{u}(p) \gamma_4 \gamma_5 u(p) \right]$$

where ε, $u^{(\ell)}(p)$ and E, u(p) are the energy and the Dirac spinor for the positron and the proton, respectively, g_A stands for the axial-vector coupling constant of neutron β decay (g_A $=1.260\pm0.007$). The amplitude in Eq. (8) is proportional to the three body Bethe-Salpeter (BS) amplitude for the proton. The general expression of the BS amplitude for the baryon octet B is given by[14)]

$$\langle 0 | T \psi^\alpha_{ia}(x_1) \psi^\beta_{jb}(x_2) \psi^\gamma_{kc}(x_3) | B \rangle = \left(\frac{M}{E}\right)^{1/2} \varepsilon_{ijk} \frac{1}{2}$$
$$\times (\chi^{(\xi)}_{abc} U^{(\xi)}_{\alpha\beta\gamma} + \chi^{(\eta)}_{abc} U^{(\eta)}_{\alpha\beta\gamma}) \phi(\xi,\eta,p) e^{ipX} \tag{3.9}$$

where

$$X = \frac{1}{3}(x_1+x_2+x_3), \quad \xi = x_1 - x_2, \quad \eta = \frac{1}{2}(x_1+x_2-2x_3),$$

$$p = p_1+p_2+p_3, \quad p_\xi = \frac{1}{2}(p_1-p_2), \quad p_\eta = \frac{1}{3}(p_1+p_2-2p_3), \tag{3.10}$$

$$\chi^{(\xi)}_{abc} = \left(\frac{-i\gamma p+M}{2M}\gamma_5 C\right)_{ab} u_c(p),$$

$$\chi^{(\eta)}_{abc} = \frac{1}{\sqrt{3}}(\chi^{(\xi)}_{bca} - \chi^{(\xi)}_{cab}), \tag{3.11}$$

and

$$U^{(\xi)}_{\alpha\beta\gamma} = \varepsilon_{\alpha\beta\delta} B^{\delta}_{\gamma}, \quad U^{(\eta)}_{\alpha\beta\gamma} = \frac{1}{\sqrt{3}}(U^{(\xi)}_{\beta\gamma\alpha} - U^{(\xi)}_{\gamma\alpha\beta}) \tag{3.12}$$

The normalization of the Bethe-Salpeter amplitude is determined by[12,13,15]

$$\int \frac{d^4p_\xi d^4p_\eta}{(2\pi)^8} |\bar{\phi}|^2 \left[\frac{(pp_\eta)^2}{M^2} + \frac{4}{3}\frac{(pp_\xi)^2}{M^2} - 4\left(\frac{1}{3}M - m_q\right)^2\right] = 1 \tag{3.13}$$

in the case where the quark masses are equal where $\bar{\phi}(p_\xi,p_\eta,p)$ is the Fourier transform of $\phi(\xi,\eta,p)$.

For the wave function of the relativistic harmonic oscillator potential[14],

$$\phi(\xi,\eta;p) = N \exp\left\{-\frac{\alpha}{6}\left[2\left(\frac{p\hat{\eta}}{M}\right)^2 + \hat{\eta}^2 + 2\left(\frac{p\hat{\xi}}{M}\right)^2 + \hat{\xi}^2\right]\right\} \tag{3.14}$$

with

$$\hat{\xi} = \frac{\xi}{\sqrt{2}}, \quad \hat{\eta} = (2/3)^{1/2}\eta, \tag{3.15}$$

the normalization condition gives

$$N = (\alpha/3\pi)^2 [\alpha - 2(M-3m_q)^2]^{-1/2} \times \frac{1}{\sqrt{72}} \tag{3.16}$$

Assuming that $M \approx 3m_q$, we obtain

$$\phi(0,0,p) \equiv N = \frac{1}{\sqrt{6\pi}}(\alpha/3\pi)^{3/2} \tag{3.17}$$

The decay amplitude of $p \to e^+ \pi^\circ$, Eq. (8), can be computed using Eq. (8)-(17),

$$f = \frac{-i}{F_\pi} \frac{4G\lambda}{\sqrt{2}} 6N \left(\frac{mM}{\varepsilon E}\right)^{1/2} [\bar{u}^{(\ell)}(p)\Gamma_1 u(p)$$

$$+ \sum_{\substack{spin \\ spin}} \bar{u}^{(\ell)}(p) \Gamma_2 u(p) \frac{M}{E} \rho \bar{u}(p)\gamma_4\gamma_5 u(p)] \tag{3.18}$$

where

$$\Gamma_i = c_i + d_i\gamma_5 \quad i = 1,2 \tag{3.19}$$

$$c_1 = -\frac{r+1}{2}, \; d_1 = \frac{r-1}{2}, \; c_2 = -(r-1), \quad d_2 = (r+1) \tag{3.20}$$

and

$$\rho = \frac{g_A}{2} \tag{3.21}$$

A further detailed calculation[12] gives

$$\frac{f}{\sqrt{2q_o}} = \frac{-i}{F_\pi} \frac{1}{(2q_o)^{1/2}} \frac{12G\lambda N}{\sqrt{2}} \left(\frac{mM}{\varepsilon E}\right)^{1/2} \bar{u}^{(\ell)}(p) \; (A+B\gamma_5)u(p) \tag{3.22}$$

where

$$A = c_1 - \rho d_2 = -\frac{(r+1)}{2}(1+g_A) = -3.39$$

$$B = d_1 - \rho c_2 = \frac{(r-1)}{2}(1+g_A) = 1.13 \tag{3.23}$$

The last expressions are for r=2 (SU(5) model). The decay amplitude is now

$$\Gamma(p \to e^+\pi^\circ) = \frac{M}{16\pi}\left(\frac{12G\lambda N}{\sqrt{2}F_\pi}\right)^2 (|A|^2+|B|^2) = \frac{M}{16\pi}\left(\frac{3g^2_{GUT}\lambda N}{2F_\pi m^2_X}\right)^2$$

$$\times \frac{(r^2+1)}{2}(1+g_A)^2 \tag{3.24}$$

b) $n \to \bar{\nu}\pi^{\circ}$

A similar calculation gives

$$f = \frac{-i}{F_\pi}\left[\langle\bar{\nu}_e|[Q_5^3,\mathcal{L}]|n\rangle + \lambda\langle\bar{\nu}_e|\mathcal{L}|n\rangle\langle n|A_o^3(0)|n\rangle\right]$$

$$= \frac{-i}{F_\pi}\frac{4G\lambda}{\sqrt{2}}\sqrt{\frac{m_\nu}{\varepsilon}}\;\bar{u}_e^{(\bar{\nu})}\;\varepsilon_{ijk}\langle 0|u_{ka}d_{ic}d_{jb}|n\rangle$$

$$\times\left(C^{-1}\gamma_\mu\frac{1+\gamma_5}{2}\right)_{ab}\left[\frac{1}{2}\left(\gamma_\mu\frac{1-\gamma_5}{2}\right)_{ec} - \lambda\left(\gamma_\mu\frac{1-\gamma_5}{2}\right)_{ec}\right.$$

$$\left.\times\frac{g_A M}{2E}\,\bar{u}(p)\;\gamma_4\gamma_5\;u(p)\right] \qquad (3.25)$$

As before, we rewrite Eq. (25) as

$$f = \frac{-i}{F_\pi}\frac{12G\lambda N}{\sqrt{2}}\sqrt{\frac{m_\nu M}{\varepsilon E}}\left(\bar{u}^{(\ell)}\Gamma_1 u + \rho\frac{M}{E}\bar{u}^{(\ell)}\Gamma_2 u\;\bar{u}\gamma_4\gamma_5 u\right) \qquad (3.26)$$

$$= -\frac{i}{F_\pi}\frac{12G\lambda N}{\sqrt{2}}\sqrt{\frac{m_\nu M}{\varepsilon E}}\;\bar{u}^{(\ell)}(A+B\gamma_5)\;u(p) \qquad (3.27)$$

where

$$\Gamma_i = c_i + d_i\gamma_5 \qquad (3.28)$$

and

$$A = c_1 - \rho d_2$$
$$B = d_1 - \rho c_2\;. \qquad (3.29)$$

The values of c_i, d_i and ρ are given by

$$c_1 = \frac{1}{2}\,,\quad d_1 = \frac{1}{2},\quad c_2 = 1,\quad d_2 = 1\quad \rho = -\frac{g_A}{2}\,, \qquad (3.30)$$

for which we get

$$A = \frac{1}{2}(1+g_A)$$

$$B = \frac{1}{2}(1+g_A) \qquad (3.31)$$

Comparison of Eqs. (23) and (31) leads to the prediction

$$\frac{\Gamma(n\to\bar{\nu}\pi^{\circ})}{\Gamma(p\to e^{+}\pi^{\circ})} = \frac{(\frac{1}{4}+\frac{1}{4})(1+g_A)^2}{(\frac{9}{4}+\frac{1}{4})(1+g_A)^2} = \frac{1}{5} \tag{3.32}$$

b) $p\to\bar{\nu}_e\pi^+$

The relevant formulae are

$$f = \sqrt{2q_o}\ \langle\bar{\nu}(p)\ \pi^{+}(0)|\mathcal{L}|p(p)\rangle$$

$$= -\frac{i}{\sqrt{2}F_\pi}\ \{\langle\bar{\nu}(p)|[Q_5^{1-i2},\mathcal{L}]|p\rangle + \langle\bar{\nu}(p)|\mathcal{L}|n\rangle\langle u|A_o^{1-i2}|p\rangle, \tag{3.33}$$

where

$$Q_5^{1-i2} = \int d_j^{+}\gamma_5\ u_j d^3x \tag{3.34}$$

$$Q_5^{1+i2} = \int u_j^{+}\gamma_5 d_j\ d^3x \tag{3.35}$$

Using

$$[Q_5^{1-i2},\ d_{j\,L/R}] = \mp u_{j\,L/R}, \tag{3.36}$$

we obtain

$$[Q_5^{1-i2},\mathcal{L}] = \frac{4G\lambda}{\sqrt{2}}\varepsilon_{ijk}\ [-(\bar{u}^c_{kL}\gamma_\mu u_{jL})(\bar{\nu}^c_{eR}\gamma_\mu d_{iR})$$
$$+ (\bar{u}^c_{kL}\gamma_\mu d_{jL})(\bar{\nu}^c_{eR}\ \gamma_\mu u_{iR}) \tag{3.37}$$

Note also that

$$\langle n|A_o^{1-i2}|p\rangle = g_A\ \frac{M}{E}\ \bar{u}_n\gamma_4\gamma_5 u_p \tag{3.38}$$

A similar calculation leads to the amplitude

$$A = B = -\frac{1}{\sqrt{2}}(1+g_A) \tag{3.39}$$

Hence, we obtain

$$\frac{\Gamma(p\to\nu_e\pi^+)}{\Gamma(p\to e^+\pi^\circ)} = \frac{2}{5} \tag{3.40}$$

d) $n\to e^+\pi^-$

$$f = -\frac{i}{\sqrt{2}F_\pi}\{\langle e^+|[Q_5^{1+i2},\mathcal{L}]|n\rangle + \langle e^+|\mathcal{L}|p\rangle\langle p|A_o^{1+i2}|n\rangle\} \tag{3.41}$$

$$[Q_5^{1+i2},\mathcal{L}] = \frac{4G\lambda}{\sqrt{2}}\varepsilon_{ijk}(\bar{d}^c_{kL}\gamma_\mu u_{jL}-\bar{u}^c_{kL}\gamma_\mu d_{jL}) \times (r\,\bar{e}^+_L\gamma_\mu d_{iL}+\bar{e}^+_R\gamma_\mu d_{iR}) \tag{3.42}$$

The final result is

$$A = -\frac{(r+1)}{\sqrt{2}}(1+g_A), \qquad B = \frac{(r-1)}{\sqrt{2}}(1+g_A) \tag{3.43}$$

and

$$\frac{\Gamma(n\to e^+\pi^-)}{\Gamma(p\to e^+\pi^\circ)} = 2, \tag{3.44}$$

a result which is independent of the value of r.

Table 1 summarizes the results for an arbitrary value of r (r=2 for SU(5)as was mentioned earlier). These results are consistent with other calculations using the SU(6) wave function of the quark model[6].

IV. The Kaonic and η Modes

a) $p \to \mu^+ K^\circ$, $\bar{\nu}_\mu K^+$, $n \to \bar{\nu}_\mu K^\circ$

The PCAC relation is expressed as

$$K^\circ = \frac{\phi^{6-i7}}{\sqrt{2}} = \frac{1}{\sqrt{2}\, F_K\, \mu_K^2} \frac{\partial A^{6-i7}}{\partial x_\mu}$$
$$K^+ = \frac{\phi^{4-i5}}{\sqrt{2}} = \frac{1}{\sqrt{2}\, F_K\, \mu_K^2} \frac{\partial A^{4-i5}}{\partial x_\mu} \tag{4.1}$$

and

$$A_\mu^\alpha = i\bar{q}\ \frac{\lambda_\alpha}{2}\, \gamma_\mu\, \gamma_5\, q \tag{4.2}$$

The soft kaon approximation gives

$$f(p \to \mu^+ K^\circ) = \frac{-i}{\sqrt{2}\, F_K} \{\langle \mu^+ | [Q_5^{6-i7}, \mathcal{L}] | p\rangle + \langle \mu^+ | \mathcal{L} | \Sigma^+\rangle\langle \Sigma^+ | A^{6-i7} | p\rangle\} \tag{4.3}$$

$$f(p \to \bar{\nu}_\mu K^+) = \frac{-i}{\sqrt{2} F_K} \{\langle \bar{\nu}_\mu | [Q_5^{4-i5}, \mathcal{L}] | p\rangle + \langle \bar{\nu}_\mu | \mathcal{L} | \Sigma^\circ\rangle\langle \Sigma^\circ | A^{4-i5} | p\rangle + \langle \bar{\nu}_\mu | \mathcal{L} | \Lambda\rangle\langle \Lambda | A^{4-i5} | p\rangle\} \tag{4.4}$$

and

$$f(n \to \bar{\nu}_\mu K^\circ) = \frac{-i}{\sqrt{2} F_K} \{\langle \bar{\nu}_\mu | [Q_5^{6-i7}, \mathcal{L}] | n\rangle + \langle \bar{\nu}_\mu | \mathcal{L} | \Sigma^\circ\rangle\langle \Sigma^\circ | A^{6-i7} | n\rangle + \langle \bar{\nu}_\mu | \mathcal{L} | \Lambda\rangle\langle \Lambda | A^{6-i7} | n\rangle\} \tag{4.5}$$

where

$$Q_5^{4-i5} = \int s_j^+ \gamma_5 u_j d^3x,$$

$$Q_5^{6-i7} = \int s_j^+ \gamma_5 d_j d^3x \tag{4.6}$$

The commutator terms are easily calculated,

$$[Q_5^{6-i7}, \mathcal{L}] = \frac{4G\lambda}{\sqrt{2}} \epsilon_{ijk} \{(\bar{u}^c_{kL}\gamma_\mu u_{jL})(-\bar{\mu}^+_L\gamma_\mu d_{iL}+\bar{\mu}_R\gamma_\mu d_{iR}) + (\bar{u}^c_{kL}\gamma_\mu d_{jL})(\bar{\nu}^c_{\mu R}\gamma_\mu d_{iR})\} \tag{4.7}$$

and

$$[Q_5^{4-i5}, \mathcal{L}] = \frac{4G\lambda}{\sqrt{2}} \epsilon_{ijk} \{(\bar{u}^c_{kL}\gamma_\mu u_{jL})(-\bar{\mu}^+_L\gamma_\mu u_{iL}+\bar{\mu}_R\gamma_\mu u_{iR}) + (\bar{u}^c_{kL}\gamma_\mu d_{jL})(\bar{\nu}^c_{\mu R}\gamma_\mu u_{iR})\} \tag{4.8}$$

The matrix elements $\langle B|A^\alpha|B\rangle$ can be read off from Table 2 where[11)]

$$g_A = D+F = 1.26\pm0.012$$

$$D/(D+F) = 0.658\pm0.007 \tag{4.9}$$

$$D-F = 0.398\pm0.021\ .$$

The amplitude for the kaonic mode can be written

$$f = \frac{-i}{\sqrt{2}\,F_K} \frac{12G\lambda}{\sqrt{2}} \sqrt{\frac{mM}{\epsilon E}}\, \bar{u}^{(\ell)}(A+B\gamma_5)u \tag{4.10}$$

where A and B are calculated in a similar way and are presented in Table 3. It is noticed that the amplitude for $p\to\mu^+K^\circ$ is purely s-wave. Table 3 also gives the decay rates relative to

$\Gamma(p\to e^+\pi^\circ)$.

b) $p\to e^+\eta$, $n\to\bar{\nu}_e\eta$

The amplitudes are given by

$$f(p\to e^+\eta) = -\frac{i}{F_\eta}\{\langle e^+|[Q_5^8,\mathcal{L}]|p\rangle + \langle e^+|\mathcal{L}|p\rangle\langle p|A_o^8|p\rangle\} \tag{4.11}$$

$$f(n\to\bar{\nu}_e\eta) = -\frac{i}{F_\eta}\{\langle\bar{\nu}_e|[Q_5^8,\mathcal{L}]|n\rangle + \langle\nu_e|\mathcal{L}|n\rangle\langle n|A_o^8|n\rangle\} \tag{4.12}$$

where

$$Q_5^8 = \frac{1}{2\sqrt{3}}\int(u_j^+\gamma_5 u_j + d_j^+\gamma_5 d_j - 2s_j^+\gamma_5 s_j)d^3x, \tag{4.13}$$

$$\begin{aligned}[Q_5^8,\mathcal{L}] = \frac{4G\lambda}{\sqrt{2}}\,\varepsilon_{ijk}\,&[(\bar{u}^c_{kL}\gamma_\mu u_{jL})(-r\bar{e}^+_L\gamma_\mu d_{iL}+\bar{e}_R^+\gamma_\mu d_{iR})\\ &+(\bar{u}^c_{kL}\gamma_\mu d_{jL})(\bar{\nu}^c_{eR}\gamma_\mu d_{iR})]\end{aligned} \tag{4.14}$$

Using Table 2 and a calculation similar to the previous sections, we obtain

$$f = \frac{-i}{F_\eta}\,\frac{12G\lambda N}{\sqrt{2}}\sqrt{\frac{mM}{\varepsilon E}}\,\bar{u}(A+B\gamma_5)u \tag{4.15}$$

where A and B are given in Table 4. We observe that these modes are completely negligible. (We have assumed that $F_\eta=F_\pi$).

V. <u>The Vector Meson Mode</u>

The vector meson modes for nucleon decay are

$$p\to e^+\omega\ ,\quad e^+\rho^\circ,\quad \bar{\nu}_e\rho^+ \tag{5.1}$$

and

$$n \to \bar{\nu}_e \omega, \quad \bar{\nu}_e \rho^\circ, \quad e^+\rho^- \tag{5.2}$$

We use vector meson dominance for the computation. This method is not as successful as PCAC and current algebra for pseudoscalar meson emission, but gives reasonable results in many cases for the vector meson phenomenology.

a) $p \to e^+\omega$

The basic formula for this process is

$$\begin{aligned} f &= \sqrt{2q_0}\, \langle e^+\omega|\mathcal{L}|p\rangle \\ &= -i\varepsilon_\mu \int \langle e^+|T(\mathcal{L}(0),\, j_\mu^{(\omega)}(x))|p\rangle\, e^{-iqx} d^4x \\ &= -i\varepsilon_\mu \Big[\sum_n \frac{\langle e^+|\mathcal{L}(0)|n\rangle\langle n|j_\mu^{(\omega)}(0)|p\rangle}{i(p_0+q_0-p_0^n+i\varepsilon)} (2\pi)^3\delta(p_n+q-p) \\ &\quad -\sum_n \frac{\langle e^+|j_\mu^{(\omega)}(0)|n\rangle\langle n|\mathcal{L}(0)|p\rangle}{i(k_0+q_0-p_0^n-i\varepsilon)} (2\pi)^3\delta(p_n+k-q)\Big] \end{aligned} \tag{5.3}$$

where the $j_\mu^{(\omega)}(x)$ is the source current coupled to the ω vector meson,

$$j_\mu^{(\omega)}(x) = (\mu_\omega^2 - \Box)\,\phi_\mu^{(\omega)}(x). \tag{5.4}$$

We estimate Eq. (5.3) by replacing the sum over intermediate states by the relevant lower resonance states. First, we may neglect the second terms of Eq. (3), since $|n\rangle = |e^+, x\rangle$ and $|x\rangle$ cannot be $|\omega\rangle$, $|\rho^\circ\rangle$. (The excited states of ω or ρ will not contribute to the sum either.) For the first term of Eq. (3), only p and the resonance N^+ with $I=1/2$ will contribute in the intermediate state sum. For simplicity, use only the p state in the sum and let us calculate the amplitude in the rest frame of the ω particle. Then

$$f = -i\varepsilon_\mu \frac{\langle e^+|\mathcal{L}(0)|p\rangle}{i\mu_\omega} \langle p|j_\mu^{(\omega)}(0)|p\rangle$$

$$= -i\varepsilon_\mu \frac{\langle e^+|\mathcal{L}(0)|p\rangle}{i\mu_\omega} \frac{if_\omega}{2} \bar{u}\, \gamma_\mu u \left(\frac{M}{E}\right). \tag{5.5}$$

This result follows directly from the vector meson dominance assumption. The coupling strength is given by[11)]

$$\frac{f_\omega^2}{4\pi} = 18.4 \pm 1.8 \tag{5.6}$$

and similarly

$$\frac{f_\rho^2}{4\pi} = 2.26 \pm 0.25 \tag{5.7}$$

fpr the ρ meson.

The calculation of the matrix element $\langle e^+|\mathcal{L}(0)|p\rangle$ is the same as before. Thus we obtain

$$f(p\to e^+\omega^\circ) = -i\varepsilon_\mu \frac{f_\omega}{2\mu_\omega} \frac{4G\lambda}{\sqrt{2}} \varepsilon_{ijk} \langle 0|u_{ka}u_{jb}d_{ic}|p\rangle$$

$$(-C^{-1}\gamma_\mu \frac{1+\gamma_5}{2})_{ab} [r(\gamma_\mu \frac{1+\gamma_5}{2})_{ec} + (\gamma_\mu \frac{1-\gamma_5}{2})_{ec}]$$

$$\cdot \frac{M}{E} (\bar{u}\gamma_\mu u) \tag{5.8}$$

$$= -i\varepsilon_\mu \frac{f_\omega}{\mu_\omega} \frac{12G\lambda N}{\sqrt{2}} \sqrt{\frac{mM}{\varepsilon E}}\, \bar{u}^{(\ell)} \left(-\frac{r-1}{2} + \frac{r+1}{2}\gamma_5\right) \frac{-i\gamma p+M}{2E} \gamma_\mu u$$

$$= -(\varepsilon p)\, f_\omega \frac{12G\lambda N}{\sqrt{2}} \sqrt{\frac{mM}{\varepsilon E}}\, \bar{u}^{(\ell)} \left(-\frac{r-1}{2} + \frac{r+1}{2}\gamma_5\right) u \frac{1}{\mu_\omega E} \tag{5.9}$$

The Lorentz covariance can be restored by replacing $\mu_\omega E$ by $-(qp) = 1/2\ (M^2+\mu_\omega^2)$. Comparing Eq. (9) with Eq. (2.19), (they differ by a factor $1/\sqrt{2q_0}$ in their definition), we have

$$g = i f_\omega \frac{12 G \lambda N}{\sqrt{2}} \tag{5.10}$$

$$C = -(r-1) \frac{M}{M^2+\mu_\omega^2} , \qquad D = (r+1) \frac{M}{M^2+\mu_\omega^2} \tag{5.11}$$

and

$$E=F=0 \tag{5.12}$$

Then we obtain

$$\Gamma(p\to e^+\omega) = \left(\frac{12G\lambda N}{\sqrt{2}}\right)^2 \frac{f_\omega^2}{4\pi} \frac{(r^2+1)}{8} \frac{\left(1-\frac{\mu_\omega^2}{M^2}\right)^4 M}{\mu_\omega^2 \left(1+\frac{\mu_\omega^2}{M^2}\right)} \tag{5.13}$$

Comparing Eq. (13) with Eq. (3.24), we obtain

$$R = \frac{\Gamma(p\to e^+\omega)}{\Gamma(p\to e^+\pi^\circ)} = \pi \frac{f_\omega^2}{4\pi} \frac{F_\pi^2}{\mu_\omega^2} \frac{4}{(1+g_A)^2} \frac{\left(1-\frac{\mu_\omega^2}{M^2}\right)^4}{\left(1+\frac{\mu_\omega^2}{M^2}\right)^2} \tag{5.14}$$

Even with the large value of the coupling constant $f_\omega^2/4\pi$, (Eq. (6)) we get

$$\frac{\Gamma(p\to e^+\omega)}{\Gamma(p\to e^+\pi^\circ)} = 0.002 \tag{5.15}$$

The ratio, Eq. (14) is a product of two small numbers $F_\pi^2/\mu_\omega^2 = 0.0143$, $(1 - \mu_\omega^2/M^2)^4 = 0.00860$ and a large number $f_\omega^2/4 = 57.8$.

We notice, however, that if we consider the SU(6) symmetric limit $\mu_\omega=\mu_\pi$ in Eq. (14), we obtain the value of the ratio R to be of the order of unity:

$$\begin{aligned} R &= 4.9 \quad \text{for } \mu_\omega = \mu_\pi \cong \sqrt{2}\, F_\pi = \frac{\mu_\omega+\mu_\pi}{2} = 0.46 \text{ GeV} \\ &= 2.1 \quad \text{for } \mu_\omega^2=\mu_\pi^2\cong 2F_\pi^2 = \frac{\mu_\omega^2+\mu_\pi^2}{2} = 0.316 \text{ GeV}^2 \end{aligned} \tag{5.16}$$

These results which may be viewed as the SU(6) symmetric results, are valid in a fictitious world where $\mu_\omega = \mu_\pi$, but are badly broken in the real world.

We must now determine what the effects of the higher excited states are in the computation of the amplitude, Eq. (3). In particular the nucleon state contributes only to the C and D amplitudes and gives a vanishing value for the E and F amplitudes (see Eqs. (11), (12) and Eqs. (2.19) and (2.20)). Notice that the C and D amplitudes are highly suppressed by a kinematical factor. We investigate this problem by using the experimental values of the photon-nucleon-resonance couplings[16]. In general, such couplings are essentially of the magnetic type and should be small as in the isoscalar magnetic moments of the nucleon. This is indicated in the opposite signs of the photon coupling to proton and neutron[16]. (Most of them are of this type). From what we discussed, we tentatively conclude that

$$\frac{\Gamma(p\to e^+\omega)}{\Gamma(p\to e^+\pi^\circ)} \lesssim 0.01 \tag{5.17}$$

b) $n \to \bar{\nu}_e \omega$

From the expression

$$f(n\to\bar{\nu}_e\omega) = -i\varepsilon_\mu \frac{\langle\bar{\nu}_e|\mathcal{L}(0)|n\rangle}{i\mu_\omega} \frac{if_\omega}{2} \bar{u}\gamma_\mu u \frac{M}{E} \tag{5.18}$$

it follows that

$$f(n\to\bar{\nu}_e\omega) = -(\varepsilon p)\, f_\omega \frac{12G\lambda}{\sqrt{2}} \sqrt{\frac{mM}{\varepsilon E}} \frac{1}{M^2+\mu_\omega^2} \bar{u}^{(\ell)}(1+\gamma_5)u$$

and therefore

$$\frac{\Gamma(n\to\bar{\nu}_e\omega)}{\Gamma(p\to e^+\omega)} = \frac{1}{r^2+1} = \frac{1}{5} \tag{5.19}$$

c) $p \to e^+\rho^\circ$, $\bar{\nu}_e\rho^+$, $n \to \bar{\nu}_e\rho^\circ$, $e^+\rho^-$

Similar considerations can be made for the ρ meson modes. In this case, the I=3/2 baryon resonance can contribute in the intermediate sums, but it has a smaller overall factor $f_\rho^2/4\pi \approx 1/9\ (f_\omega^2/4\pi)$. The net result is that it requires more involved calculations. However, we could use the SU(6) calculation for the ratios among the vector meson modes. In fact, our result for the ratio $\Gamma(n \to \bar{\nu}_e\omega)/\Gamma(p \to e^+\omega)$, Eq. (19), is quite consistent with the SU(6) calculation[6]. Therefore, we may conclude that

$$\Gamma(p \text{ or } n \to \text{lepton} + \rho)/\Gamma(p \to e^+\pi^\circ) \lesssim 0.01 \tag{5.20}$$

To summarize our results, we give a list of the two body branching ratios

	our result (%)	(SU(6) result)	
$p \to e^+\pi^\circ$	66	(30~40)	
$\bar{\nu}_e\pi^+$	26	(15)	
$e^+\omega$	< 1	(20 ~ 30)	(5.21)
$e^+\rho^\circ$	< 1	(~10)	
μ^+K°	7	(5~10)	

$n \to e^+\pi^-$	90	(70)	
$\bar{\nu}_e\pi^\circ$	9	(7)	
$e^+\rho^-$	0	(10~20)	(5.22)
$\bar{\nu}_\mu K^\circ$	1	(1)	

Tabel 5 gives more details of the comparison with other individual works.

VI. <u>The Total Two Body Decay Rates</u>

The total decay rates from the two body decay processes can be computed by adding all the partial rates computed and using the parameters determined from other experimental data. For proton decay, we have

$$\tau_p^{-1}\ (2\ \text{body}) = \frac{1}{0.66}\ \tau^{-1}\ (p\to e^+\pi^\circ) \tag{6.1}$$

$$= \frac{1}{0.66}\ \frac{M}{16\pi}\ \left(\frac{3g_{GUT}^2\ \lambda N}{2F_\pi\ m_x^2}\right)^2\ \frac{5}{2}\ (1+g_A)^2$$

where

$$N = \frac{1}{\sqrt{6\pi}}\ (\alpha/3\pi)^{3/2} \tag{6.2}$$

Using the values of the parameters mentioned earlier[7,14])

$$\frac{g_{GUT}^2}{4\pi} = 0.024, \quad \lambda = 3.5 \qquad \alpha = 0.45\ \text{GeV}^{-2} \tag{6.3}$$

we obtain

$$\tau_p\ (2\ \text{body}) = 4.5\times 10^{30}\ \text{years} \quad \text{for } m_X=6\times 10^{14}\ \text{GeV} \tag{6.4}$$

Or equivalently, for

$$\tau_p(2\text{body}) = \xi_p \times 10^{31}\ \text{years}\ \left(\frac{m_x(\text{GeV})}{6\times 10^{14}}\right)^4, \tag{6.5}$$

we have

$$\xi_p = 0.45 \tag{6.6}$$

For neutron decay

$$\tau_n^{-1}(2\ \text{body}) = 2\times\frac{0.66}{0.90}\ \tau_p\ (2\ \text{body})^{-1} \tag{6.7}$$

Hence

$$\tau_n(2\ \text{body}) = 3.1\times 10^{30}\ \text{years} \tag{6.8}$$

or

$$\xi_n/\xi_p = 0.68 \tag{6.9}$$

VII. Discussions

The field theoretical calculation presented in this lecture gives the relative branching ratios for the pionic modes which are quite consistent with the SU(6) symmetry calculation. However it seems to yield vastly different predictions for the vector meson modes. Because of the importance of the subject, we will continue the estimate of the vector meson decay rates including higher resonances in the intermediate states. It is also pertinent to estimate the effect of SU(6) symmetry breaking in the earlier SU(6) calculations.

Acknowledgment

The author is indebted to Jacques Leveille and Bob Levine for reading the manuscript. The work is partly supported by the U.S. Department of Energy.

References

1) J. Pati and A. Salam, Phys. Rev. D8, 1240 (1973); D10, 275 (1974).

2) H. Georgi and S.L. Glashow, Phys. Rev. Lett 32, 438 (1974).

3) H. Georgi and D.V. Nanopoulos, Phys. Lett. 82B, 392 (1979); Nucl. Phys. B155, 52 (1979).

4) e.g. Proc. of the Second Workshop on Grand Unified Gauge Theories held at the University of Michigan, April 24-26, 1981.

5) H. Georgi, H.R. Quinn and S. Weinberg, Phys. Rev. Lett. 33, 451 (1974); A.J. Buras, J. Ellis, M.K. Gaillard and D.V. Nanopoulos, Nucl. Phys. B135, 66 (1978); T.J. Goldman and D.A. Ross, Phys. Lett. 84B, 208 (1979); Nucl. Phys. B171, 273 (1980); S. Weinberg, Phys. Rev. Lett. 43, 1566 (1979); F. Wilczek and A. Zee, Phys. Rev. Lett. 43, 1571 (1979).

6) M. Machacek, Nucl. Phys. B159, 37 (1979); J.F. Donoghue, Phys. Lett. 92B, 99 (1980); G. Kane and G. Karl Phys. Rev.

D22, 2808 (1980); G. Karl and M.J. Lipkin, FNAL Report No. 80/54-THY (1980); M.B. Gavela et al. Phys. Lett 97B, 51 (1981); A.M. Din, G. Girardi and P. Sorba, Phys. Lett 91B, 77 (1980); E. Golowich, Phys. Rev. D22, 1148 (1980); H. Arisue, KUNS 583 HE(TH) 81/04.

7) J. Ellis, M.K. Gaillard and D.V. Nanopoulos, CERN Report No. CERN-TH-2833 (1980); P. Langacker, Phys. Reports C, to be published (1981).

8) M. Wise, R. Blankenbecler and L. Abbott SLAC PUB-2614 (1981).

9) Y. Tomozawa, Nuovo Cimento 46, 707 (1966).

10) S. Weinberg, Phys. Rev. Lett. 17, 616 (1966).

11) M.M. Nagel et al., Nucl. Phys. B109, 1 (1976).

12) Y. Tomozawa, Phys. Rev. Lett. 46, 463 (1981).

13) Y. Tomozawa, UM HE 80-20 (1980); Phys. Lett. 32B, 485 (1970); M. Machacek and Y. Tomozawa, Phys. Rev. D12, 3711 (1975).

14) R.P. Feynman, M. Kislinger and F. Ravndal, Phys. Rev. D3, 2706 (1971); R. Lipes, Phys. Rev. D5, 2849 (1972).

15) In Ref. 12, there was a mistake of a factor of 2 in Eq. (18) which results in the following change: In Ref. 12, Eq. (21), (22) should be multiplied by $1/\sqrt{2}$, and Eq. (32) should be multiplied by a factor of 2.

16) Particle Data Group, Rev. Mod. Phys. 52, No. 2 Part II (1980).

TABLE 1

Pionic decay amplitudes and relative rates
(r=2 for the SU(5) model)

	$p\to e^+\pi^\circ$	$p\to\bar{\nu}_e\pi^+$	$n\to e^+\pi^-$	$n\to\bar{\nu}_e\pi^\circ$
A	$-\frac{r+1}{2}(1+g_A)$	$-\frac{1}{\sqrt{2}}(1+g_A)$	$-\frac{r+1}{\sqrt{2}}(1+g_A)$	$\frac{1}{2}(1+g_A)$
B	$\frac{r-1}{2}(1+g_A)$	$-\frac{1}{\sqrt{2}}(1+g_A)$	$\frac{r-1}{\sqrt{2}}(1+g_A)$	$\frac{1}{2}(1+g_A)$
Γ	r^2+1	2	$2(r^2+1)$	1

TABLE 2

The matrix element $\langle B|A_0^\alpha|B'\rangle$

$$= h_\alpha \sqrt{\frac{MM'}{EE'}}\,\bar{u}_B\gamma_4\gamma_5 u_{B'}$$

	h_α
$\langle p\|A_0^3\|p\rangle$	$g_A/2$
$\langle n\|A_0^3\|n\rangle$	$-g_A/2$
$\langle\Sigma^+\|A_0^{6-i7}\|p\rangle$	D−F
$\langle\Sigma^\circ\|A_0^{4-i5}\|p\rangle$	$1/\sqrt{2}$ (D−F)
$\langle\Lambda\|A_0^{4-i5}\|p\rangle$	$-1/\sqrt{6}$ (D+3F)
$\langle\Sigma^\circ\|A_0^{6-i7}\|n\rangle$	$-1/\sqrt{2}$ (D−F)
$\langle\Lambda\|A_0^{6-i7}\|n\rangle$	$-1/\sqrt{6}$ (D+3F)
$\langle p\|A_0^8\|p\rangle$	$1/2\sqrt{3}$ (−D+3F)
$\langle n\|A_0^8\|p\rangle$	$1/2\sqrt{3}$ (−D+3F)

TABLE 3

Amplitudes and decay rates for the kaonic modes

	$p \to \mu^+ K^\circ$	$p \to \bar{\nu}_\mu K^+$	$n \to \bar{\nu}_\mu K^\circ$
A	2(1+D-F)	-2D/3	(1-2/3F)
B	0	-2D/3	(1-2F/3)
$\frac{\Gamma}{\Gamma(p \to e^+ \pi^\circ)}$	0.10	0.004	0.01

TABLE 4

Amplitudes and decay rates for the η modes

	$p \to e^+ \eta$	$n \to \bar{\nu}_e \eta$
A	$-\frac{r+1}{2}\,\frac{1-(3F-D)}{\sqrt{3}}$	$\frac{1}{2\sqrt{3}}(1-(3F-D))$
B	$-\frac{r-1}{2}\,\frac{1-(3F-D)}{\sqrt{3}}$	$\frac{1}{2\sqrt{3}}(1-(3F-D))$
$\frac{\Gamma}{\Gamma(p \to e^+ \pi^\circ)}$	0.01	0.001

TABLE 5

Two Body Branching Ratios for Nucleon Decay
(see Ref. 5,6 for the initials of the authors)

(a) Proton

	This work	M	GYOPR	D	G	DGS	KK	A
$e^+\pi^\circ$	66	33	37	9	13	31	38	36
$\bar{\nu}_e\pi^+$	26	9	15	3	5	11	15	13
μ^+K°	7	.3-.5	19	--	7	.5	5	9
$\bar{\nu}_\mu K^+$	0.4	--	0	--	.5	--	.6	.6
$e^+\eta$	0.8	12	7	3	.1	5	0	0
$e^+\omega$	< 1	22	18	56	46	19	26	27
$e^+\rho^\circ$	< 1	17	2	21	20	21	11	11
$\bar{\nu}_e\rho^+$	< 1	4	1	8	7	8	4	4

(b) Neutron

	This work	M	GYOPR	D	G	KK	A
$e^+\pi^-$	90	50	74	23	32	68	67
$\bar{\nu}_e\pi^\circ$	9	8	8	2	3	7	6
$\bar{\nu}_\mu K^\circ$	1.2	--	10	--	2	0.6	0.9
$\bar{\nu}_e\eta$	0.2	3	1.5	1	--	--	0
$\bar{\nu}_e\omega$	0	5	3.5	14	10	5	5
$\bar{\nu}_e\rho^\circ$	<1	4	.5	5	4	1.8	2
$e^+\rho^-$	<1	26	4	55	48	19	20

LATTICE GAUGE THEORY

Claude Itzykson

CEN Saclay, Division de la Physique Theorique,

B.P. N°2, 91190 Gif-sur-Yvette, France

in

Proceedings of the Fourth Kyoto Summer Institute on Grand Unified Theories and Related Topics, Kyoto, Japan, June 29 - July 3, 1981, ed. by M. Konuma and T. Maskawa (World Science Publishing Co., Singapore, 1981).

Contents

LATTICE GAUGE THEORY

C. Itzykson
DPh.T. CEN-Saclay
91191 Gif-sur-Yvette Cedex, France

1. Introduction

Beyond the original predictions of asymptotic freedom and its successful applications to deep inelastic high energy collisions and related processes, the theory of strong interactions based on a non abelian SU(3) color gauge invariance has had a hard time to produce numerical testable results. The difficulty lies in the confinement property which generates a spectrum very remote from the original dynamical degrees of freedom. On the other hand the appearance of new approaches like Wilson's Lattice quantization[1] have revealed to the theorist's delight a wealth of unexpected phenomena. Lattice gauge theory is still in an exploratory stage. However the general feeling is that the ground is safe eventhough practical developments are still scarce.

What has been achieved presently is a reasonable understanding of the confinement mechanism in the strong coupling regime. The Monte Carlo calculations initiated by Wilson[2] and Creutz Jacobs and Rebbi[3] and actively pursued, have demonstrated that the cutoff model approximates for small coupling the short distance asymptotically free continuous field theory. Their results had been partly anticipated by the developments of strong coupling series and the insight offered by mean field approximations[4].

Euclidean field theory using path integrals has provided numerous bridges with condensed matter physics. Local invariance seems in particular a common theme in a number of exciting problems of disordered media.

Very much as brownian motion is at the heart of the conventional field theory, the study of gauge invariant models suggests a theory of "Brownian surfaces" still in its infancy, in relation to the string model developped in the context of the dual resonance approximation to scattering amplitudes.

The role of topological defects, with its associate duality transformations in the abelian cases yield connections between strong and weak coupling situations but unfortunately cannot be extended except qualitatively to non abelian cases.

To extend this list of open problems let us mention the difficulties of introducing fermionic degrees of freedom, recently reanalized by Nielsen and Ninomiya[5] and others. The chiral anomaly reapears in the form of an explicit bearking due to regularization or in the disguise of the multiplication of the number of fermion species.

2. Lattice quantization

To cope with ultraviolet divergences numerous means have been developped ranging Pauli-Villars regularization to dimensional regularization, which had in common to be tailored for handling the perturbative Feynman integrals. Lattice quantization offers a mean of investigating the theory in the large. The task is then first to locate the possible transitions, those of interest being associated with very large correlation lengths on the scale of the lattice spacing a. The bare coupling constant and lattice spacing can be removed in favor of renormalized quantities characterizing a continuous field theory.

Fo a fixed bare theory the set of connected Green's functions satisfy a Callan-Symanzik equation

$$\left\{p.\frac{\partial}{\partial p} - \beta(g)\frac{\partial}{\partial g} - n\,\gamma(g)\right\} G_n(p,g) = \text{r.h.s.} \tag{1}$$

where the r.h.s. vanishes in the scaling limit and

$$\beta(g(\lambda)) = \lambda\frac{\partial}{\partial\lambda} g(\lambda) \tag{2}$$

dictates the variations of the "running" coupling constant with a scale of (euclidean) momenta λ . For non-abelian gauge fields and g small enough

$$\beta(g) = -\frac{b}{2} g^3 + \ldots \tag{3}$$

and as $\lambda \to \infty$

$$g^2(\lambda) \sim \frac{1}{b \ln \lambda} \tag{4}$$

i.e. at short distance the effective coupling constant goes to zero.

This motivates a lattice cutoff a to deal with momenta $p \ll \frac{1}{a}$ for which the discretization should play a negligible role as the coupling constant becomes vanishingly small, keeping some physical length fixed, string tension or mass of some excited state.

Having broken translational by construction, it is however important to retain as many symmetries of the sought for theory as possible. In particular gauge invariance plays a proeminent role. It expresses the fact that the physical content of the theory is unaffected when "matter" fields (Higgs or quark fields) are submitted to a local transformation

$$\phi(x) \to D(g_x)\ \phi(x) \tag{5}$$

where g_x is an x-varying element of the local group G(SU(3) for chromodynamics) and D the representation associated to ϕ . Since neighboring points on the lattice are separated by a finite distance, the vector potential of the continuous theory $A_\mu(x)$, has to be traded for an integral factor

$$g_{xy} = P \exp i \int_y^x A_\mu(x)\ dx^\mu \qquad g_{xy} \in G \tag{6}$$

a path ordered integral along a curve joining y to x, with the covariance property

$$g_{xy} \to g_x\ g_{xy}\ g_y^{-1} \tag{7}$$

under local transformations. Note that $\phi^+(y)\ D(g_{yx})\phi(x)$ is then a gauge invariant for a unitary representation D.

The Yang-Mills action is then replaced, in the Euclidean region, by a lattice sum, which up to a constant and a sign convention is

$$S = \beta \sum_P \chi(\Pi\ g_{xy}) \tag{8}$$

The sum runs over the elementary circuits of the lattice, called plaquettes (elementary squares of a cubic lattice), $\Pi\ g_{xy}$ is the ordered product of group variables along this circuit, and χ is a real class function (a weighted sum of irreducible characters) of a group G chosen to be maximum on the identity element.

In the simplest case it can be taken to be the real part of the trace in the fundamental representation. This is however not a unique choice and perhaps not suited for SU(N), N large. The reason is that with such a choice the plaquette action develops secondary maxima which might play an unwanted role. Perhaps the most natural choice is to use the heat kernel on the group, which reduces to the Villain form in the abelian case. Furthermore there is in principle no reason, except simplicity, to introduce only the shortest circuits. Indeed any real space renormalization will generate terms involving larger circuits.

If the action is expanded around its trivial maximum it takes the form

$$S = S_o - C\beta a^{4-d} \int d^d x\, (F^\alpha_{\mu\nu})^2 + \ldots \tag{9}$$

with $F^\alpha_{\mu\nu}$ the field strength. This shows that β is proportional to g_o^{-2}, with g_o the bare coupling constant. In the statistical analogy β is inversely proportional to temperature and the vacuum functional

$$Z = \int \prod_{xy} dg_{xy} \; \exp S \tag{10}$$

is interpeted as the partition function, with

$$F = \lim N^{-1} \ln Z \tag{11}$$

the free energy, N being the number of lattice points.

A coupling to Higgs particles can be added in the form of a sum over links and sites

$$S_m = \beta_m \sum_\ell \phi_x^+ D(g_{xy})\, \phi_y + \sum_x V(\phi_x^+ \phi_x) \tag{12}$$

where the first term would reduce in the naive continuum limit to a kinetic term with covariant derivatives instead of ordinary ones. One can also introduce fermions, using Grassmanian variables for the quark fields and appropriate γ-matrices

$$S_f = \beta_f \sum_\ell \bar{\psi}_x (\gamma_{xy} - 1)\, D(g_{xy})\, \psi_y + \sum_x \bar{\psi}_x \psi_x \tag{13}$$

Let us concentrate on the pure Yang-Mills theory. Gauge invariant observables are generated in the form of Wilson loops. These are ordered products along a connected closed loop C of variables g_{xy}. We then take the trace in an irreducible representation (a) and average over configurations

$$W_a(C) = \langle \chi_a [\prod_C g_{xy}] \rangle = Z^{-1} \int \prod_{xy} dg_{xy}\; e^S \chi_a [\prod_C g_{xy}] \tag{14}$$

For the quark confinement problem, χ_a corresponds to the fundamental representation. We take for C a large regular loop enclosing a minimal area A. The confinement criterion is then that, as C becomes large, W(C) behaves as

$$W(C) \sim \exp - KA \tag{15}$$

with K is the string tension - i.e. - the coefficient of the linearly rising part of the potential energy between two static colored sources. Recall that such a string in rotation generates a linear relation between angular momentum and square energy with the slope α' related to K through

$$K = \frac{1}{2\pi\alpha'} \qquad (16)$$

so that the order of magnitude of K is $\frac{1}{2\pi}$ GeV^2. Assuming that confinement persists in four dimensions down to vanishingly small bare coupling constant we should expect, according to asymptotic freedom that as $\beta \to \infty$ the dimensionless quantity Ka^2 behaves as

$$Ka^2 \underset{\beta \to \infty}{\sim} C_K \, e^{-b'\beta} \qquad (17)$$

where b' is related to the coefficient b in the Callan-Symanzik function. On the other hand in the strong coupling region

$$Ka \underset{\beta \to 0}{\sim} \ln \beta^{-1} \qquad (18)$$

Other observables can be studied, even within the unrealistic pure gauge sector. Of particular interest is the so called glue-ball mass, or inverse correlation length defined as the coefficient in the exponential decay law for the correlation between two widely separated Wilson loops

$$\langle \chi(\prod_{c_1} g_{xy}) \; \chi(\prod_{c_2} g_{xy}) \rangle \qquad \exp - L/\xi \qquad (19)$$

where L is an average distance between the loops . If the picture makes sense we might hope that in the scaling region, quantities as $K\xi^2$ reach a stable value and represent "physically" measurable quantities, as other dimensionless ratios in the more complete theory including quarks. Present values for $K\xi^2$ seem to be of order 0.1 within large incertainties[6]. This would put the glue ball mass above 1 GeV.

To develop a qualitative understanding of the phenomena occuring in lattice gauge fields it was useful to vary the space time dimension d, or the group, considering for instance the case when G was a finite group, either as an approximation to the realistic group,for instance finite subgroups of SU(2) or SU(3), or even the simplest case of a group Z_2 reduced to two elements ± 1. This toy model, a gauge generalization of the Ising model, allows to exhibit the techniques involved in handling lattice models while avoiding cumbersome algebra and may serve to test approximations or numerical simulations.

3. Strong coupling expansions and mean field approximation.

For large coupling, i.e. small β , an expansion in β or a related parameter is natural. Consider for simplicity the Z_2 gauge theory in arbitrary dimension d, with

$$Z = 2^{-Nd} \sum_{\sigma_\ell=\pm 1} \exp \beta \sum_p \sigma_p = \exp NF(\beta) \tag{20}$$

σ_p is the product of link invariables around a plaquette. Write

$$e^{\beta\sigma_p} = \cosh \beta(1 + t\,\sigma_p) \qquad t = \tanh \beta \tag{21}$$

(with a similar expansion in irreducible characters for the general case). Then

$$(\cosh \beta)^{-N^{d(d-1)}/2}\, Z = \sum_{\text{closed surfaces } S} t^{|S|} \tag{22}$$

$$= 1 + N\,\frac{d(d-1)(d-2)}{6}\, t^6 + N\,\frac{d(d-1)(d-2)}{2}\,(2d-5)\, t^{10} + \ldots$$

$$\frac{F}{d(d-1)} = \frac{1}{2}\ln \cosh\beta + \frac{1}{2}(d-2)\,[\tanh \beta^6 + (2d-5) \tanh \beta^{10} + \ldots]$$

In the expansion for Z the sum runs over "closed" surfaces - i.e. - sets of plaquettes such that any link of the lattice belongs to an even number of the selected plaquettes. The coefficient of a given power in t is an polynomial in N and the coefficient of the linear term, the reduced number of configurations, is the one appearing in the expansion for the free energy. The first term written above corresponds to a cube, the second to two adjacent cubes, the next would correspond to two disconnected cubes and so on. Enumerating the closed configurations, attaching in the general case the relevant group factors, becomes quickly cumbersome. More sophisticated techniques become necessary including the use of a computer[4][7]. Nevertheless such strong coupling series are or are getting available up to order 22 in certain cases, thanks to K. Wilson and others. They allow the study of the strong coupling region using extrapolation methods like Padéapproximants.

The factor (d-2) occuring in the expansion for the free energy in (22) is a reflection of the fact that pure gauge theory is trivial (and always confining) in two dimensions.

One can of course obtain expansions for the Wilson loop and therefore the string tension

$$-K\,a^2 \;=\; \ln(\tanh\beta) + 2(d-2)(\tanh\beta)^4 + \ldots \qquad (23)$$

Indeed the Wilson criterion is fulfilled in the strong coupling regime, of course assuming that the relevant selection rules on group representation are fulfilled in the realistic SU(3) case. The expansion for $\langle W(C)\rangle$ involves a summation over closed surfaces bounded by C and weighted by a factor decaying exponentially with their area.

It is interesting to compare the computer simulations for the string tension, or the internal energy (the derivative of the free energy) with the first few terms of the strong coupling series. The agreement is impressive over a large range of β , up to values of order unity (depending of course on the specific normalization adopted in each case) where a sudden turn over to the weak coupling regime takes place.

In the study of statistical models based on global symmetries a valuable tool of analysis has been the mean field approximation. At low temperature (small g_0) the systems tend to get ordered, the tendancy being favored by an increasing relative weight of energy (interpreted here as action) versus entropy. A way to achive this is to increase the dimension d. The effect of neighboring variables can then be accounted for by a mean ordering field, determines self-consistently, treating its fluctuations as perturbations. To illustrate the idea consider an Ising system with

$$Z \;=\; 2^{-N} \sum_{\sigma_x=\pm 1} \exp\beta \sum_{(xy)} \sigma_x\sigma_y \quad . \qquad (24)$$

Let us first derive an exact "equation of motion" using the invariance of the summation under the redefinition $\sigma_x \to \varepsilon\sigma_x$, $\varepsilon = \pm 1$. This leads to the relation

$$1 \;=\; \langle \exp - 2\beta\, \sigma_x \sum_{y(x)} \sigma_y \rangle \quad , \qquad (25)$$

where the sum in the exponential runs over the neighbors of the spin J_x. Set

$$H_x \;=\; \beta \,.\sum_{y(x)} \sigma_y \quad . \qquad (26)$$

Then we have the exact relation

$$\langle \cosh\, 2H_x - 1\rangle \;=\; \langle \sigma_x \sinh\, 2H_x \rangle \quad . \qquad (27)$$

For large coordination number -2d- we are tempted to approximate H_x by an average $\bar{H}$ (an infinitesimal symmetry breaking external field is in fact necessary to make this statement meaningfull) and (26) and (27) lead to the mean field approximation

$$\bar{H} = 2d\,\beta\,\langle\sigma\rangle$$

$$\langle\sigma\rangle = \tanh \bar{H} \quad , \qquad (28)$$

Above some critical value $\beta_c = \frac{1}{2d}$ these equations admit a non trivial solution indicating a magnetized phase while below β_c one has a disordered phase. The transition is second order with the magnetization vanishing as $(\beta - \beta_c)^{1/2}$ at the critical point. Of course, this is a crude approximation. In low enough dimensions (d < 4) critical fluctuations are important and sophisticated techniques based on the renormalization group are needed to obtain the correct critical behavior. In any case the mean field approximation can be turned into a rigorous inequality on the free energy and a systematic perturbation theory developed around it.

Let us blindly repeat the same steps for a Z_2 gauge theory. The exact relations are with obvious notations

$$\langle \cosh 2H_\ell - 1\rangle = \langle \sigma_\ell \sinh 2H_\ell\rangle$$

$$H_\ell = \beta \sum_{p(\ell)} \sigma_{\ell_1}\sigma_{\ell_2}\sigma_{\ell_3} \quad , \qquad (29)$$

and the corresponding meanfield approximation is

$$\bar{H} = 2\beta(d-1)\langle\sigma_\ell\rangle^3$$

$$\langle\sigma_\ell\rangle = \tanh \bar{H} \qquad (30)$$

$$F_{\text{mean field}} = F_o + \text{Sup}\left\{0,\ \frac{d(d-1)}{2}\ (\tanh \bar{H})^4 + d[\ln(\cosh \bar{H}) - \bar{H}\tanh \bar{H}]\right\} \ .$$

Analysis of these equations show a first order transition for a critical value β_c of order cst/2(d-1) and this behavior is generic no matter which gauge group is considered.

At first this looks totally unfounded. Indeed the approximation violates gauge invariance and a theorem of Elitzur [8] states that for all values of β we should find $\langle\sigma_\ell\rangle = 0$. However equations (30) can be shown to be saddle point equations which not only admit uniform solutions but gauge transformed thereof. Upon averaging over all saddle points, non invariant quantities vanish as they should, while gauge invariant ones are unaffected. For $\beta < \beta_c$ the Wilson loop vanishes in our approximation, while beyond β_c it obeys a perimeter law. It can be shown from the low temperature expansion that the predictions of mean field theory are valid term by term.

Of course the above techniques can be extended to more complex situations in particular to the case of Higgs couplings. The phase diagram admits two second order transitions connected by first order lines. The first one is the Higgs symmetry breaking in the absence of gauge fields. The second one occurs for non vanishing coupling. A recent analysis by Brézin and Drouffe[9] shows that it is of the ϕ^4 type.

For $\beta < \beta_c$ the mean field approximation predicts (up to $1/d$ corrections) a trivial confining phase. This may in fact not be correct as suggested by Drouffe, Parisi and Sourlas[10]. They observe that for β small and d large one can write an approximate equation for the internal energy $p = \langle\sigma_p\rangle$, expressing the fact that each plaquette can be decorated by a cube of plaquettes, the cubes being non overlapping in the large d limit. This is expressed by the equations

$$p = t + 2d\, p^5$$

$$F_{D.P.S.} = \frac{d(d-1)}{2} \ln \cosh\beta + \frac{d^3}{6} p^6 - d^4 p^{10} \qquad (31)$$

Or setting

$$u = 2d\, p^4 \qquad (32)$$

$$u(1-u)^4 = 2dt^4 \quad , \; F_{D.P.S.} = \frac{d(d-1)}{2} \ln \cosh\beta + \frac{d^{3/2} u^{3/2}}{12\sqrt{2}} (1-3u) \; . \qquad (33)$$

This shows that t scales like $d^{-1/4}$ and predicts a transition in a metastable phase beyond β_c.

4. Duality and topological excitations

Abelian models admit Kramers-Wannier duality transformations which interchange high and low temperatures. For simplicity let us again limit ourselves to Z_2 theories and consider first a three dimensional case where

$$Z_3^{gauge}(\beta) = (\cosh \beta)^{3N} \sum_{\text{closed surfaces } S} (\tanh \beta)^{|S|} . \qquad (34)$$

To describe a closed surface we can assign values $\tilde{\sigma}_i = 1$ at the center of cubes with the condition that neighboring cubes have opposite $\tilde{\sigma}$ if their common plaquette belongs to the surface. This yields $|S| = \sum_{(ij)} \frac{1 - \tilde{\sigma}_i\tilde{\sigma}_j}{2}$, a sum over nearest neighbors. We then recognize that the gauge model is dual to the 3-dimensional model at a dual temperature $\tilde{\beta}$ given by

$$\tanh\beta = e^{-2\tilde{\beta}} , \qquad (35)$$

the precise relation being

$$Z_3^{\text{gauge}}(\beta) = 2^{-N/2} (\sinh 2\beta)^{3N/2} \; Z_3^{\text{Ising}}(\tilde{\beta}) \quad . \qquad (36)$$

Similarly the 4-dimensional model is self dual. If it has a unique deconfining transition, this should occur at the self dual point

$$\tanh \beta_c = \sqrt{2} - 1 \quad . \qquad (37)$$

This has been confirmed by Monte Carlo simulations, the transition being found of first order.

Abelian models admit generalizations of this duality transformation. The Z_N models for $N > 4$ have three phases with dual transition points (presumably of second order), one of finite value the other increasing like N^2. The middle phase admits probably massless excitations of photon type and extending to infinity as $Z_N \to U(1)$.

Under duality a Wilson loop transforms into a 't Hooft loop[11]. The latter is a ratio $Z^F(\tilde{\beta})/Z(\tilde{\beta})$, where $Z^F(\tilde{\beta})$ is a frustrated partion function with couplings reversed for all plaquettes (in 4-dimensions) or links (in 3-dimensions) dual to those of a fiducial surface bounded by C. When the model is self dual we can consider both Wilson and 't Hooft loops at the same value of β. It turns out that when one of those has an area law decrease the other has a perimenter law. In the intermediate phase of Z_N models both have perimeter decrease. The concept of 't Hooft loop can be generalized to non abelian gauge fields provided the action admits transformations when the plaquette variables are modified by an element if the center of the group.

As mentioned above the U(1) case appears as a limit of Z_N models and exhibits a second order deconfining transition, to be contrasted with the Kosterlitz-Thouless transition of the two-dimensional XY model. Interactions among topolo-ical excitations, monopole strings for the U(1) model seem to be responsible for this transition.

It would be very interesting to be able to assess the role of instanton-configurations in the non-abelian case.

5. Roughening

Underlying gauge models is a theory of surfaces. It can be studied by continuous field theoretic methods[12] or approached from the lattice point of view. Several groups[13] have rediscovered recently a phenomenon studied several years ago by crystal growth physicists[14], i.e. surface roughening. This appears as a singularity (presumably an essential singularity) in the string tension computed on the lattice well before the deconfining transition. For instance any attempt at extrapolating the 3-dimensional Z_2 tension

$$-Ka^2 = \ln t + 2t^4 + 2t^6 + 10t^8 + 16t^{10} + 80\,\tfrac{2}{3}\,t^{12} + 150t^{14} + 734t^{16} + 1444\,\tfrac{2}{3}t^{18} + \ldots \qquad (38)$$

suggests erroneously that it vanishes well before the critical bulk transition at $t_c \simeq 0.6418$. Indeed roughening occurs at $t_R = 0.46$. The phenomenon can be understood through duality using the low temperature phase of the Ising model. We saw that the tension can then be identified with the interfacial free energy (per unit surface) of two coexisting phases. This can be modelled by a so called solid on solid (SOS) partition function which admits an XY type transition restoring translational symmetry in the direction perpendicular to the original interface. These deformations of surfaces can be tested using various indicators which generalize to other gauge groups and other dimensions.

It is interesting to note that in 4 dimension the roughening transition occurs in the region of rapid change between strong and weak coupling for SU(2) or SU(3) and almost at the self-dual point for the Z_2 gauge model. Beyond roughening the fluctuations of the surface generate a universal correction[15] to the $q\bar{q}$ static potential

$$\delta V = -\frac{(d-2)\pi}{24\,R} \qquad (39)$$

6. Conclusion

The beautiful data from Monte Carlo calculations have hardly been discussed above. A major advance is in the process of producing new results by including fermions into the picture. The phenomenology of lattice gauge theories is extremely rich and although we have not been able to address the questions of greatest interest to particle physicists, we have uncovered interesting transitions which still await a better explanation. Let us quote for instance the recent observation of first order transitions (presumably not deconfining) for SU(N), $N \geq 5$ in four dimensions using Wilson's action or similar phenomena for SO(3). Perhaps the greatest achievement so far has been the realization that on a rather small lattice, the simulations reveal that the ideas of asymptotic freedom are numerically verified.

REFERENCES

[1] K. Wilson, Phys. Rev. D10, 2445 (1974).

[2] K. Wilson in Proceedings of the Cargèse 1979 Summer School, 't Hooft et al. eds. Academic Press, New York (1980).

[3] M. Creutz, L. Jacobs, C. Rebbi, Phys. Rev. Lett. 42, 1390 (1979), Phys. Rev. D20, 1915 (1979).
M. Creutz, Phys. Rev. Lett. 43, 553 (1979).
C. Rebbi, Phys. Rev. D21, 3350 (1980).

[4] R. Balian, J.M. Drouffe, C. Itzykson, Phys. Rev. D10, 3376 (1974); D11, 2098, 2104 (1975).

[5] H.B. Nielsen and M. Ninomiya, Rutherford preprints ;
L.H. Karsten and J. Smit, Amsterdam preprint.

[6] B. Berg, Phys. Mett. 97B, 401 (1980).
G. Bhanot and C. Rebbi, to appear in Nuclear Phys. B.

[7] J.M. Drouffe, Nucl. Phys. B170, FS1 , 91 (1980).

[8] S. Elitzur, Phys. Rev. D12, 3978 (1975).

[9] E. Brézin, J.M. Drouffe, Saclay preprint.

[10] J.M. Drouffe, G. Parisi, N. Sourlas, Nucl. Phys. B161, 397 (1979).

[11] G. 't Hooft, Nucl. Phys. B153, 141 (1979).

[12] A.M. Polyakov, Landau Institute preprint, Moscou (1981).

[13] A, E. and P. Hasenfratz, Nucl. Phys. B (to appear), M. Luscher, G. Münster, P. Weisz, Nucl. Phys. B180, FS2 13 (1981), C. Itzikson, M. Peskin, J.B. Zuber, Phys. Lett. 95B, 259 (1980).

[14] For a review see J.D. Weeks, G.H. Gilmer, Adv. in Chem. Phys. 40, eds. Prigogine and Rice (J. Wiley and Sons, 1979).

[15] M. Lüscher, Nuclear Phys. B (to appear).

Summary of Contributions in the Discussion Session

SU(8) HYPERUNIFICATION AND HYBRID QUARKS AND LEPTONS

Jihn E. Kim

Department of Physics, Seoul National University,
Seoul 151, Korea

We have learned that the minimal grand unified model SU(5) is so attractive that any attempt to present a different GUT group requires a good motivation for the case. One of such motivations is to solve the family problem. In this seminar, I would like to present a grand unified theory based on SU(8) which I collaborated with H.S. Song[1]. The motivation for SU(8) is that it probably unified the hypercolour with other known interactions. As is well known, the problem of the hierarchy of 24 orders is not understood in the GUT scheme. The popular proposal for the gauge hierarchy is the existence of another non-abelian gauge interaction which can provide a strong force to dynamically break the electroweak gauge symmetry SU(2) × U(1). We will call this extra non-abelian gauge interaction as "hypercolour" and assume that the low energy gauge group is

$$SU(3)_c \times SU(2) \times U(1) \times G_h \tag{1}$$

instead of $SU(3)_c \times SU(2) \times U(1)$. Our objective is to unify (1) in a simple group.

I believe that the hypothetical simple hyperunification group is one of SU(N), O(N), or E_6, otherwise our existence is unnatural, due to the survival hypothesis[2]. Frampton[3] tried to find a simple group for unification of (1), but did not succeed. Therefore, I play the game by changing one of Frampton's assumptions to the following. Some of the quarks and leptons are composite. If all the quark and lepton components are fundamental, we do not have a solution, following Frampton. In this scenario, therefore, the simplicity we anticipate in the course of unification is <u>not</u> in the low energy building blocks, but in the dynamics at high energy above grand unification scale. Hence, we should form composite fermions out of hypercoloured fermions. For $SU(N)_h$ as a hypercolour group, this implies $SU(N)_h = SU(3)_h$ or $SU(5)_h$, etc. with N = odd integers. Even for $SU(5)_h$, there are too many fermions present, sufficient to destroy asymptotic freedom for $SU(5)_h$. For example, in SU(10) there is only one plausible complex anomaly-free non-repeating fermion representation

$$\psi^{\alpha\beta\gamma} + \psi_{\alpha\beta\gamma\delta} \text{ in } SU(10)$$
$$= (10^*,1) + (10,5) + (5,10) + (1,10^*)$$
$$+ (5,1) + (10,5^*) + (10^*,10^*)$$
$$+ (5^*,10) + (1,5)$$
$$\text{under } SU(5) \times SU(5)_h .$$

The β function for the hypercolour is

$$\beta_h = -\frac{g^3}{16\pi^2}\left\{ \frac{11}{3}\times 5 - \frac{1}{3}(10+5\cdot 3+3+10+10\cdot 3+5\cdot 3+1)\right\} \ln\frac{\tilde{M}}{\mu}$$
$$= -\frac{g^3}{16\pi^2}\left(\frac{55}{3} - \frac{84}{3}\right)\ln\frac{\tilde{M}}{\mu} \quad > 0$$

which shows asymptotically strong behaviour. Therefore, the only hope for the hypercolour group in SU(N) scenario is $SU(3)_h$. For O(4n+2) and E_6, the exact hypercolour group is $SU(3)_h$ also, which we will not discuss. For the Cremmer-Julia SU(8) gauge symmetry[4)], the possibility for the hypercolour is again $SU(3)_h$. Therefore, the case for $SU(3)_h$ seems to be quite plausible.

The simplest unification group we imagine for the unification of (1) is then SU(8). In SU(8), there exists only one set of interesting anomaly-free complex non-repeating representation which can be worked from the famous Georgi rule[2)]

$$\psi^{a} \oplus \psi^{ab} \oplus \psi_{abc} . \tag{2}$$

For confinement of the hypercolour to occur at a higher mass scale than the colour, differentiation of the coupling constants α_h and α_c is realized by the symmetry breaking at the grand unification mass scale. For simplicity, we assume the following symmetry breaking pattern:

$$SU(8) \xrightarrow[\text{at } \tilde{M}]{} SU(3)_c \times \tilde{U}(1) \times SU(5)$$
$$\xrightarrow[\text{at } M']{} SU(3)_c \times U(1) \times SU(2) \times SU(3)_h . \tag{3}$$

Since the mass scale for the hypercolour confinement is about 3000 times larger than that of QCD, we obtain the ratio

$$\frac{\tilde{M}}{M'} \approx (2700)^{\frac{1}{2}} \left(\frac{2700\,\mu_c}{\tilde{M}}\right)^{\frac{1}{11}} < 50 \tag{4}$$

which is an acceptable hierarchy in the Higgs scheme.

Below the scale M', Georgi and Glashow's SU(5) is not a good gauge symmetry. However, we will use $SU(5)_{GG}$ to compactly represent the fermions. The survived fermions below M' are

$$(10^*,3^*) + (5,3) + (5^*,3) + 2(10,1) + (5,1) \tag{5}$$

under $SU(5)_{GG} \times SU(3)_h$. Between M' and μ_h which is the hypercolour confining scale, nothing interesting happens except the respective coupling constants gradually change. At $\mu_h \approx$ 2-3 TeV, a dramatic thing happens. The hypercoloured fermions feel a strong force and they form hypercolour singlet bound states. The hypercolour singlets 2(10,1) + (5,1) do not feel the strong force, and they remain as spectator fermions (SF), while the process of confining occurs. For notational convenience, let us represent

$$\begin{aligned} (10^*,3^*) &= \psi_{\alpha\beta AL} \\ (5,3) &= \psi_{\alpha AR} \\ (5^*,3) &= \psi^{\alpha}_{AR} \quad , \end{aligned} \tag{6}$$

where α, β, etc. are $SU(5)_{GG}$ indices and A,B, etc. are $SU(3)_h$ indices. We call these hypercoloured fermions as most fundamental fermions (MFF).

The MFF's form the spin 1/2 fermion bound states

$$F_1 = \psi_{\alpha\beta AL}\, \psi_{\gamma\delta BL}\, C\, \psi_{\xi\eta CL}\, \varepsilon^{\alpha\beta\gamma\delta\xi}\, \varepsilon^{ABC} = 5^*_L \tag{7}$$

$$F_2 = \psi_{\alpha\beta AL}\, [\psi_{\gamma\delta BL}\, \psi^{\delta}_{CR}]\, \varepsilon^{\alpha\beta\gamma\mu\nu}\, \varepsilon^{ABC} = 10_L \tag{8}$$

$$F_3 = \psi_{\alpha\beta AL}\, \psi_{\gamma BR}\, C\, \psi_{\delta CR}\, \varepsilon^{\alpha\beta\gamma\delta\varepsilon}\, \varepsilon^{ABC} = 5_L \tag{9}$$

$$F_4 = [\psi_{\alpha\beta AL}\, \psi_{\gamma BR}]\, \psi_{\delta CR}\, \varepsilon^{\alpha\beta\gamma\delta\varepsilon}\, \varepsilon^{ABC} = 5_R \tag{10}$$

$$F_5 = \psi_{\alpha\beta AL}\, \psi_{\gamma BR}\, C\, \psi^{\gamma}_{CR}\, \varepsilon^{ABC} = 10^*_L \tag{11}$$

$$F_6 = [\psi_{\alpha\beta AL}\, \psi^{\alpha}_{BR}]\, \psi_{\gamma CR}\, \varepsilon^{\beta\gamma\mu\nu\rho}\, \varepsilon^{ABC} = 10^*_R \tag{12}$$

$$F_7 = \psi_{\alpha AR}\, C\, \psi_{\beta BR}\, \psi_{\gamma CR}\, \varepsilon^{\alpha\beta\gamma\mu\nu}\, \varepsilon^{ABC} = 10_R \tag{13}$$

$$F_8 = \psi_{\alpha AR}\, C\, \psi_{\beta BR}\, \psi^{\alpha}_{CR}\, \varepsilon^{ABC} = 5^*_R \tag{14}$$

$$F_9 = \psi_{\alpha AR}\, C\, \psi^{\alpha}_{BR}\, \psi^{\beta}_{CR}\, \varepsilon^{ABC} = 5_R \tag{15}$$

$$F_{10} = \psi^{\alpha}_{AR}\, C\, \psi^{\beta}_{BR}\, \psi^{\gamma}_{CR}\, \varepsilon_{\alpha\beta\gamma\mu\nu}\, \varepsilon^{ABC} = 10^*_R \tag{16}$$

where [ ] implies a spin zero combination and we have included only 5, 5*, 10 and 10* of $SU(5)_{GG}$.

We know that the theory is anomaly-free above the scale μ_h. The spectator fermions carry +3 units of anomaly with respect to 5 of SU(5). Following 't Hooft[5)], we require renormalizability also below μ_h. Therefore, the composite spin 1/2 fermions should carry +3 units of anomaly with respect to 5*

$$\ell_1 - \ell_2 - \ell_3 + \ell_4 + \ell_5 - \ell_6 + \ell_7 - \ell_8 + \ell_9 - \ell_{10} = 3 \tag{17}$$

where ℓ_i's are the number of composite fermions appearing in Eqs.(7)-(16).

We present a phenomenological argument to obtain composite fermions out of numerous possibilities satisfying (17). Our convention for 5 and 10 are the usual ones: 5^*_L contains the left-handed lepton doublet and 10_L contains the left-handed quark doublet. Though the spectator fermions have two 10_L's (d and s doublets), there is no spectator 5^*_L . Therefore, the left-handed lepton doublets should come from composite states. Namely, we have

$$\ell_1 + \ell_4 + \ell_9 \gtrsim 3 \quad . \tag{18}$$

The minimal solution (with the smallest number of SU(5) irreducible representations) is obtained if ℓ's with negative coefficients in (17) vanish

$$\ell_2 = \ell_3 = \ell_6 = \ell_8 = \ell_{10} = 0 \quad , \tag{19}$$

which implies

$$\ell_5 + \ell_7 = 0 \tag{20}$$

$$\ell_1 + \ell_4 + \ell_9 = 3 \quad . \tag{21}$$

Namely, the minimal solution constitutes the low energy fermions whose transformation properties under $SU(5)_{GG} \times SU(3)_h$ are

$$\left[(10,1)^0 + (5^*,1)^1\right]_e + \left[(10,1)^0 + (5^*,1)^1\right]_\mu + \left[(5,1)^0 + (5^*,1)^1\right]_\tau \; , \tag{22}$$

where the superscripts 0 and 1 refer to "fundamental" and "composite", respectively. We need (1,1) for τ to decay by "V-A" interaction

$$(1,1) = N_{\tau L} = \Psi_{\alpha\beta A L} \Psi^{\alpha}_{BR} C \Psi^{\beta}_{CR} \varepsilon^{ABC} \tag{23}$$

which is supposed to be heavier than τ.

The MFF's also form bosons, one of which is interpreted as a dynamical Higgs field whose condensate breaks the electroweak gauge group[6)]

$$H_{\alpha} \equiv \tilde{\Psi}_{\alpha\beta A L} C \Psi^{\beta A}_{L} \tag{24}$$

thus realizing the mechanism of Weinberg and Susskind.

Let us present two tests of our model, one of which we already passed. Brodsky and Drell[7)] observed that the composite model of leptons with dynamics present should not lead to too large anomalous magnetic moments of e and μ so that they do not contradict the observation. Let us suppose that the confining dynamics is strong at $m^* >> m_\ell$. If the cancellation or suppression does not occur for the $(g-2)_{e,\mu}$, we expect they are of a $\sim m_\ell/m^*$. From the good agreement of QED calculation on $(g-2)_{e,\mu}$ with the measurement, we find $m^* \gtrsim 10^3$ TeV, which suggests that the hypercolour may not be a right confining force of MFF's. In our model, however, the Brodsky-Drell syndrome is evaded. Any conceivable diagram contributing to (g-2) does not involve hypergluon exchange. The other tests of our model is that there is only one operator for the nucleon decay (neglecting generation variation)

$$O = e_R C u^{\alpha}_R q^{\beta i}_L C q^{\gamma j}_L \varepsilon_{\alpha\beta\gamma} \varepsilon_{ij} \tag{25}$$

predicting 100% polarized charged antileptons in the final state[8)].

The model presented here does not include the t quark but contains all the observed fermions. To give masses to fermions, we break the chiral symmetry of fermions (5) by four fermion condensates around GeV region.

REFERENCES

1) J.E. Kim and H.S. Song, Phys. Rev. D23RC, 2102 (1981) and Seoul National University, preprint, May (1981).

2) H. Georgi, Nucl. Phys. B156, 126 (1979).

3) P. Frampton, Phys. Rev. Letters 43, 1912 (1979).

4) E. Cremmer and B. Julia, Nucl. Phys. B159, 141 (1979).

5) G. 't Hooft, Lecture given at the Cargese Summer Institute (1979).

6) L. Susskind, Phys. Rev. D20, 2619 (1979);
S. Weinberg, Phys. Rev. D13, 974 (1976).

7) S.J. Brodsky and S.D. Drell, Phys. Rev. D22, 2236 (1980).

8) J.E. Kim, Seoul National University, preprint, May (1981).

AN SU(7) GRAND UNIFIED MODEL AND ITS INTERESTING FEATURES

Katsuji Yamamoto

Department of Nuclear Engineering

Kyoto University, Kyoto 606, Japan

There are two motivations to enlarge the grand unification group from the SU(5). One is to incorporate the horizontal interaction into the GUT scheme. The other is to expect extended colour and weak interactions such as $SU(4)_C \times SU(3)_W \times U(1)$ in the intermediate energy region 10^2GeV$\sim 10^{15}$GeV. Then, we might come across oases in the physical desert in extended grand unified theories such as SU(N)[1)~3)]. According to this motivation, we take an SU(7) model[3),4)] as a simple possibility and investigate its interesting features.

Our SU(7) model is as follows. The electric charge is defined for the fundamental representation 7 by

$$Q = (q ,-1/3,-1/3,-1/3, 1 , 0 ,-q).$$

Then, the fundamental representation 7 consists of SU(5) 5 and its two singlets with the charge $\pm q$. We suppose here that q can be nonzero. If q=0, we have usual SU(7) model. But we take q=1/2 for the reason mentioned below, though it is a rather radical possibility[3),4)]. In Kim's model, q is taken to be -1[5)].

As for the fermions, we take the following combination of totally antisymmetric representation and a singlet one, which can be embedded into the spinor representation of SO(14).

$$1 + [2] + [4] + [6] .$$

This combination is anomaly free and real for $U(1)_{em}$ so that there are no massless charged fermions, and contains 2 generation of "ordinary" fermions and the same number of "mirror" ones. "Ordinary" ("mirror") fermions have V-A (V+A) coupling for the $SU(2)_W$. In accordance with the charge definition, the charges of "ordinary" fermions $(u,d,\nu_e,e;c,s,\nu_\mu,\mu)$ are as usual and those of "mirror" ones $(U,D,N_E,E;C,S,N_M,M)$ are as follows:

$$Q(U,D,N_E,E) = (\ 7/6,\ 1/6,\ 1/2,-1/2),$$

$$Q(C,S,N_M,M) = (\ 1/6,-5/6,-1/2,-3/2).$$

Since the charge patterns of "ordinary" fermions and "mirror" ones are different from each other, the dangerous mixing of them is forbidden by the charge conservation. Thus the fermions in our model are liberated from the survival hypothesis by virtue of a bold choice $q\neq 0$.

The symmetry breaking pattern with $q=1/2$ which we recommend here is

$$\begin{aligned} &SU(7) \to SU(4)_C\times SU(3)_W\times U(1) \text{ at } M \\ &\to SU(3)_C\times SU(2)_W\times U(1)_{1/2} \text{ at } M_1 \to SU(3)_C\times U(1)^{em}_{1/2} \text{ at } M_W . \end{aligned}$$

The first stage at M is caused by the adjoint 48 $\phi^\alpha{}_\beta$ of Higgs and the second stage at M_1 by the 21 $h^{\alpha\beta}$ and the 224 $H^{\alpha\beta\gamma}{}_\delta$ of Higgs $(\alpha,\beta,\ldots=0\sim 6)$. It should be noticed that the 224 is inevitable to realize $q=1/2$ since $H^{456}{}_0$ of the 224 is electrically neutral only when $q=1/2$. The mass scales M and M_1 are estimated by considering the renormalization effects on relevant gauge coupling constants. We find that in the case of $q=1/2$ the intermediate mass scale M_1 lies in 10^4GeV$\sim 10^7$GeV consistently with $M > 10^{15}$GeV ans $\sin^2\theta_W(M_W) = 0.23 \pm 0.02$.

Finally we consider the neutrino masses. In addition to ν_{eL} and $\nu_{\mu L}$, there are two neutral Wyle fields ψ and ψ^{06} in 1 and [2] respectively. They form a Dirac mass through a Yukawa coupling with $h^{\alpha\beta}$:

$$f \cdot h_{\alpha\beta}\psi\psi^{\alpha\beta} \rightarrow f \cdot \langle h_{06} \rangle \psi\psi^{06}.$$

Since $\langle h_{06} \rangle$ causes the second stage of symmetry breaking, this mass is heavy enough ($\sim M_1$) to suppress the ν_{eL} and $\nu_{\mu L}$ masses sufficiently.

References

1) H. Georgi, Nucl. Phys. B156 (1979), 126.
2) S. Dawson and H. Georgi, Phys. Rev. Letters 43 (1979), 821.
3) I. Umemura and K. Yamamoto, Prog. Theor. Phys. 64 (1980), 278.
4) I. Umemura and K. Yamamoto, Phys. Letters 100B (1981), 34.
5) J.E. Kim, Phys. Rev. Letters 45 (1980), 1916.

GROUP TABLE FOR GRAND UNIFIED THEORIES

I. G. Koh

Fermi National Accelerator Laboratory, Batavia, Illinois 60510

Y. D. Kim, K. Y. Kim, Y. J. Park, and W. S. l'Yi

Sogang University, C.P.O.Box 1142, Seoul, Korea

Abstract

Group table is constructed with wide applications in grand unified theories. Table includes 1) complete set of weights for all classical groups, 2) projection operators, and 3) anomaly table.

1. Motivations for Group Table in Grand Unified Theories

Particle physicists working in grand unified theories are interested in the following questions:

a) what groups are good candidates?

b) what representations are available? Fermions should belong to the complex anomaly free representations.

c) how the above representations break into the representations of subgroup?

These group calculations are not generally simple. Better strategy would be to compile all these results in table and to refer to them.

There are several excellent tables and articles[1)-6)] answering the above problems. We have found the following important problems still not completely answered:

a) complete set of weights,

b) projection operators,

c) anomaly table.

2. Group Table for Grand Unified Theories

There are nine types of complex Lie algebras: four classical series A_n(=SU(n+1)), B_n(=SO(2n+1)), C_n(=Sp(2n)), D_n(=SO(2n)), and five types of isolated exceptional algebras E_6, E_7, E_8, F_4 and G_2.

2.1 Complete Set of Weights

All irreducible representations with dimension less than 1000 are tabulated for all classical groups.[7] One can easily identify complex representations in our table. A complex irreducible representation has the feature that weights at level k are not negatives of those at level $T(\Lambda)-k$, where $T(\Lambda)$ is the maximum level for the given irreducible representation Λ.

Also with the projection operators, particle contents of theories can simply be read from this table.

Table I. Examples of complex, real, pseudoreal representations

	complex		real		pseudoreal	
group	A 4		C 2		A 1	
	level	weight	level	weight	level	weight
	0	1 0 0 0	0	0 1	0	1
	1	-1 1 0 0	1	2-1	1	-1
	2	0-1 1 0	2	0 0		
	3	0 0-1 1	3	-2 1		
	4	0 0 0-1	4	0-1		

2.2 Projection Operators for Branching

Although the branching rules can be read from ref. 4, the projection operators are more powerful and convenient due to the independence of representation. All the projection operators for all classical groups into all their subgroups are included in our table[7]. U(1) subgroup is included in the projection operators, since U(1) plays a very important role in determining charge structure. Our convention for the projection operators is that the highest weight of an irreducible representation is projected onto the highest weight of subgroup's representations.

2.3 Anomaly Table

There are several simple methods to calculate the anomaly values for a limited class of irreducible representations. But it is apparently more convenient to tabulate and to look up. Anomaly values for all irreducible representations with dimension less than 1000 are included in our table[7].

Acknowledgment

We thank Prof. H. Georgi, Prof. D. V. Nanopoulos, Prof. P. Ramond, Prof. A. Zee for very useful comments. It is a pleasure to thank Prof. M. Konuma, Prof. Z. Maki, and Prof. T. Muta for hospitality in '81 Kyoto Summer Institute. This research was supported in part by KOSEF and Korea Trader's Foundation.

References

1. M. Konuma, K. Shima, and M. Wada, Suppl. Prog. Theor. Phys. 28 (1963).
2. C. Itzykson and M. Nauenberg, Rev. Mod. Phys. 38, 95 (1966).
3. B. Wybourne, "Symmetry Principles and Atomic Spectroscopy", Wiley-Interscience, 1970.
4. J. Patera and D. Sankoff, "Tables of Branching Rules for Represen-

tations of Simple Lie Algebras", Presses de l'Universite de Montreal, Montreal, 1974; W. G. McKay and J. Patera, "Tables of Dimensions, Indices, and Branching Rules for Representations of Simple Lie Algebras", Dekker, 1981.

5. M. Fischler, Fermilab-Pub-80/49-Thy, 1980.
6. R. Slansky, LA-UR-80-3495, 1980.
7. Y. D. Kim, K. Y. Kim, I. G. Koh, Y. I. Park, W. S. I'Yi, "Group Tables for Grand Unified Theories", Sogang University, Korea, 1981.

E_6 SUBCOLOR

Seiichiro Takeshita, Hiromasa Komatsu†, Akira Kakuto* and Kenzo Inoue

Department of Physics, Kyushu University 33, Fukuoka 812

*The Second Department of Engineering, Kinki University, Iizuka 820

Abstract

The subcolor theory based on the group E_6 is discussed. It is shown that the model with n left-handed preons assigned to $\underline{27}$ of E_6 has unique massless composite fermion spectra which are independent of n.

One of the important dynamical problems in subquark models is how to construct almost massless fermionic bound states. Paying special attention to this problem, 't Hooft stressed the importance of a chiral symmetry in subcolor theories.[1] He discussed the SU(N)-subcolor theory with n flavors of preons $\psi_i(\underline{N})_L + \psi_i(\underline{N})_R$ $(i=1,\cdots,n)$. The basic dynamical assumptions of subcolor theories are (i) the subcolor is unbroken and confined, and (ii) the chiral symmetry $G = SU(n)_L \times SU(n)_R \times U(1)$ associated with the preon fields is realized in the Wigner mode. The consistency of these two assumptions requires that the anomaly of three G currents[2] on the composite level is equal to that on the preon level:

$$\sum_{\text{composite } \underline{R}} \ell_R A_R = N,$$

where the preon contribution N on the right-hand-side comes from the subcolor degree of freedom of preons, the index ℓ_R, which specifies the number of massless composite fermionic states belonging to the irreducible representation $\underline{R}$ of G, is defined as the number of left-handed bound states minus that of right-handed ones which belong to $\underline{R}$ of G and A_R is its contribution to the anomaly.

't Hooft discussed the further condition which is concerned with the

† Fellow of the Japan Society for the Promotion of Science.

Appelquist-Carrazone (A-C) decoupling.[3] If we give a bare mass m to one of the preons, say ψ_1, and vary m to infinity, any effects of ψ_1 in low energy phenomena should disappear. By this mass term, the chiral symmetry G is reduced to $G' = SU(n-1)_L \times SU(n-1)_R \times U(1)$. On the composite level, this decoupling is established only through the G' invariant mass of composites which contain the massive preon ψ_1.

't Hooft found that there are no set of integers ℓ_R which satisfy these conditions except $n = 2$. The reason for the non-existence is that, the anomaly equation supplemented by the A-C decoupling condition requires that ℓ_R is independent of the flavor number n, so the anomaly equation must hold for any value of n. On the other hand, at $n = 0$, A_R takes the value 0 or $\pm N^2$ where N comes from the fact that the composites are made of N preons.[4] Therefore the left-hand-side of the anomaly equation becomes a multiple of N^2 and cannot be matched to the right-hand-side by any integer ℓ_R.

It is evident that the necessary condition for the theory to have integer-valued solution ℓ_R is that, if the composites are made of N preons, then the dimension of the preon ψ in the subcolor group must be a multiple of N^2.

We found[5] that, if a subcolor gauge group is restricted to simple groups, the only one which can have a nontrivial solution is the E_6 subcolor theory with preons being assigned to the fundamental representation 27. The product $27 \times 27 \times 27$ contains E_6 singlet, so composites are made of three preons and evidently 27 is a multiple of 3^2.

Here we introduce n flavors of left-handed preons which belong to 27 of E_6. The theory has a global symmetry $G = SU(n)$. The composite fermions transform under this SU(n) as follows:

$$\square\square\square : \ell_{1+}, \qquad \begin{array}{c}\square\\\square\end{array} : \ell_{1-}, \qquad \begin{array}{cc}\square&\square\\\square&\end{array} : \ell_2.$$

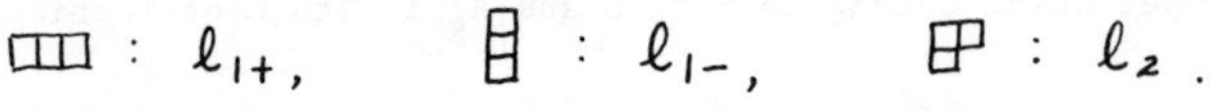

The anomaly equation for three SU(n) currents is given as

$$\ell_{1+} A(\square\square\square) + \ell_{1-} A\left(\begin{smallmatrix}\square\\\square\\\square\end{smallmatrix}\right) + \ell_2 A\left(\begin{smallmatrix}\square&\square\\\square&\end{smallmatrix}\right) = 27\,,$$

$$A\left(\begin{smallmatrix}\square\square\square\\ \begin{smallmatrix}\square\\\square\\\square\end{smallmatrix}\end{smallmatrix}\right) \equiv \frac{(n\pm 3)(n\pm 6)}{2}\,, \quad A\left(\begin{smallmatrix}\square&\square\\\square&\end{smallmatrix}\right) \equiv n^2-9\,.$$

In this model, the A-C decoupling condition is not applicable because the E_6 invariant preon mass is forbidden. If we impose an alternative condition that indices ℓ_R are independent of the flavor number n, we obtain a unique solution

$$\ell_{1+} = \ell_{1-} = -\ell_2 = 1.$$

That is, the composite fermions appear in the following representations of SU(n):

$$\Psi(\square\square\square)_L + \Psi\left(\begin{smallmatrix}\square\\\square\\\square\end{smallmatrix}\right)_L + \Psi\left(\begin{smallmatrix}\square&\square\\\square&\end{smallmatrix}\right)_R\,.$$

The asymptotic freedom of E_6 requires $n < 22$.

If we introduce both chiralities of preons belonging to 27 of E_6, the A-C decoupling condition becomes applicable. Also in this case we obtain integral valued solution for ℓ_R, but it contains an undetermined integral valued free parameter.

References

1) G. 't Hooft, Cargese Summer Institute Lecture (1979).

2) S.L.Adler, Phys. Rev. 177(1969), 2426.
   J.S.Bell and R.Jackiw, Nuovo Cimento 60A(1969), 47.

3) T.Appelquist and J.Carrazone, Phys. Rev. D11(1975), 2856.

4) Y.Frishman, A.Schwimmer, T.Banks and S.Yankielowicz, Nucl. Phys. B177 (1981), 157.
   G.R.Farrar, Phys. Letters 96B(1980), 273.

5) S.Takeshita, H.Komatsu, A.Kakuto and K.Inoue, preprint KYUSHU-81-HE-6 (1981).

COMPOSITE SYMMETRIES AND THE PREON MODEL

Liao-Fu Luo — Neimenggu University

Guo-Chen Yang — Hebei Institute of Technology

People's Republic of China

Abstract

The composite symmetries of leptons and quarks are derived from an SO(4) preon model.

Recently, a number of composite models of leptons and quarks have been proposed. In this short talk we shall suggest a simple preon model which can be summarized in the following way:

(1) Leptons and quarks possess three internal degrees of freedoms, denoted as A-spin, B-spin and C-spin, which may be called as preons;

(2) A-spin, B-spin and C-spin wave functions are (2,2) representations of $SO^A(4)$, $SO^B(4)$ and $SO^C(4)$ respectively. $SO(4) = SU(2) \times SU(2)'$. Suppose A-spin functions be denoted by $(A_2, A_1, -\bar{A}_1, \bar{A}_2)$. (A_2, A_1) and $(-\bar{A}_1, \bar{A}_2)$ are spinors of SU(2) and $(A_2, -\bar{A}_1)(A_1, \bar{A}_2)$ are that of SU(2)'. The quantum numbers are

	I_3		I'_3	
A_2	1/2		1/2	
A_1	-1/2			1/2
$-\bar{A}_1$		1/2	-1/2	
$\bar{A}_2$		-1/2		-1/2

The same is for B-spin and C-spin;

(3) The charge of particles (leptons and quarks)

$$Q = \frac{1}{3} I_3 + I_3'$$

in which

$$I_3 = I_{A3} + I_{B3} + I_{C3}$$

$$I_3' = I_{A3}' + I_{B3}' + I_{C3}' \quad .$$

From the three basic assumptions we obtain wave functions of particles as Table 1:

Table 1. Wave functions of fundamental fermions

$\nu_e=(A_1B_1\bar{C}_2)$	$u_R=(A_2B_2\bar{C}_2)$	$u_Y=(-A_1B_2\bar{C}_1)$	$u_G=(-A_2B_1\bar{C}_1)$
$e^+=(-A_2B_2\bar{C}_1)$	$\bar{d}_R=(-A_1B_1\bar{C}_1)$	$\bar{d}_Y=(A_2B_1\bar{C}_2)$	$\bar{d}_G=(A_1B_2\bar{C}_2)$
$e^-=(\bar{A}_2\bar{B}_2C_1)$	$d_R=(\bar{A}_1\bar{B}_1C_1)$	$d_Y=(-\bar{A}_2\bar{B}_1C_2)$	$d_G=(-\bar{A}_1\bar{B}_2C_2)$
$\bar{\nu}_e=(\bar{A}_1\bar{B}_1C_2)$	$\bar{u}_R=(\bar{A}_2\bar{B}_2C_2)$	$\bar{u}_Y=(-\bar{A}_1\bar{B}_2C_1)$	$\bar{u}_G=(-\bar{A}_2\bar{B}_1C_1)$
$\nu_\mu=(A_1\bar{B}_2C_1)$	$c_R=(-A_2\bar{B}_1C_1)$	$c_Y=(-A_1\bar{B}_1C_2)$	$c_G=(A_2\bar{B}_2C_2)$
$\mu^+=(-A_2\bar{B}_1C_2)$	$\bar{s}_R=(A_1\bar{B}_2C_2)$	$\bar{s}_Y=(A_2\bar{B}_2C_1)$	$\bar{s}_G=(-A_1\bar{B}_1C_1)$
$\mu^-=(\bar{A}_2B_1\bar{C}_2)$	$s_R=(-\bar{A}_1B_2\bar{C}_2)$	$s_Y=(-\bar{A}_2B_2\bar{C}_1)$	$s_G=(\bar{A}_1B_1\bar{C}_1)$
$\bar{\nu}_\mu=(\bar{A}_1B_2\bar{C}_1)$	$\bar{c}_R=(-\bar{A}_2B_1\bar{C}_1)$	$\bar{c}_Y=(-\bar{A}_1B_1\bar{C}_2)$	$\bar{c}_G=(\bar{A}_2B_2\bar{C}_2)$
$\nu_\tau=(\bar{A}_2B_1C_1)$	$t_R=(-\bar{A}_1B_2C_1)$	$t_Y=(\bar{A}_2B_2C_2)$	$t_G=(-\bar{A}_1B_1C_2)$
$\tau^+=(-\bar{A}_1B_2C_2)$	$\bar{b}_R=(\bar{A}_2B_1C_2)$	$\bar{b}_Y=(-\bar{A}_1B_1C_1)$	$\bar{b}_G=(\bar{A}_2B_2C_1)$
$\tau^-=(A_1\bar{B}_2\bar{C}_2)$	$b_R=(-A_2\bar{B}_1\bar{C}_2)$	$b_Y=(A_1\bar{B}_1\bar{C}_1)$	$b_G=(-A_2\bar{B}_2\bar{C}_1)$
$\bar{\nu}_\tau=(A_2\bar{B}_1\bar{C}_1)$	$\bar{t}_R=(-A_1\bar{B}_2\bar{C}_1)$	$\bar{t}_Y=(A_2\bar{B}_2\bar{C}_2)$	$\bar{t}_G=(-A_1\bar{B}_1\bar{C}_2)$
$\bar{E}(2)=(\bar{A}_2\bar{B}_2\bar{C}_2)$	$\bar{E}_R(\frac{4}{3})=(\bar{A}_1\bar{B}_1\bar{C}_2)$	$\bar{E}_Y(\frac{4}{3})=(\bar{A}_2\bar{B}_1\bar{C}_1)$	$\bar{E}_G(\frac{4}{3})=(\bar{A}_1\bar{B}_2\bar{C}_1)$
$\bar{E}(1)=(-\bar{A}_1\bar{B}_1\bar{C}_1)$	$\bar{E}_R(\frac{5}{3})=(-\bar{A}_2\bar{B}_2\bar{C}_1)$	$\bar{E}_Y(\frac{5}{3})=(-\bar{A}_1\bar{B}_2\bar{C}_2)$	$\bar{E}_G(\frac{5}{3})=(-\bar{A}_2\bar{B}_1\bar{C}_2)$
$E(1)=(A_1B_1C_1)$	$E_R(\frac{5}{3})=(A_2B_2C_1)$	$E_Y(\frac{5}{3})=(A_1B_2C_2)$	$E_G(\frac{5}{3})=(A_2B_1C_2)$
$E(2)=(A_2B_2C_2)$	$E_R(\frac{4}{3})=(A_1B_1C_2)$	$E_Y(\frac{4}{3})=(A_2B_1C_1)$	$E_G(\frac{4}{3})=(A_1B_2C_1)$

We see that the 64 fundamental fermions are divided into 4 groups: 1st group contains ν_e, e, u, d, corresponding to 1st generation, 2nd group contains ν_μ, μ, c, s, corresponding to 2nd generation, 3rd group contains ν_τ, τ, t, b, corresponding to 3rd generation and finally the 4th group contains the exotic states which correspond to the generation singlet.

We call the symmetries of preons as basic symmetries and the symmetries of leptons and quarks as composite symmetries. Suppose the generators of $SO^\alpha(4)$ (α=ABC) denoted by $\vec{\tau}_\alpha$ and $\vec{\tau}'_\alpha$. Due to the correlation between different degree of freedom one may assume the composite symmetries generated by the direct products of $\vec{\tau}_\alpha$ and $\vec{\tau}'_\alpha$. From Table 1 and this assumption one can deduce various symmetries, such as:

(1) Color SU(3) symmetry

The generators of color SU(3) are

$$T_+ = \tau_{A3}\tau_{B+}\tau_{C-} \quad T_- = \tau_{A3}\tau_{B-}\tau_{C+} \quad 2T_3 = \frac{1}{2}(\tau_{B3}-\tau_{C3})$$

$$V_+ = \tau_{A+}\tau_{B-}\tau_{C3} \quad V_- = \tau_{A-}\tau_{B+}\tau_{C3} \quad 2V_3 = \frac{1}{2}(\tau_{A3}-\tau_{B3})$$

$$U_+ = \tau_{A-}\tau_{B3}\tau_{C+} \quad U_- = \tau_{A+}\tau_{B3}\tau_{C-} \quad 2U_3 = \frac{1}{2}(\tau_{C3}-\tau_{A3})$$

The symmetry can be generalized to SU(4) which includes leptons as the fourth color. The U(1) generator in SU(4) is

$$2I_3 = \tau_{A3} + \tau_{B3} + \tau_{C3} = 3(B-L) \quad .$$

(2) Horizontal SU(3) symmetry

The generators are those in color SU(3) symmetry by replacing $\vec{\tau}_\alpha$ with $\vec{\tau}'_\alpha$. The horizontal symmetry can be generalized to SU(4) including exotic particles as generation singlet.

(3) Weak SU(2) symmetry

The generators of $SU(2)_W$ are

$$I_W^+ = \tau'_{A+}\tau'_{B+}\tau'_{C+} + \tau'_{A+}\tau'_{B-}\tau'_{C+} + \tau'_{A-}\tau'_{B+}\tau'_{C+}$$

$$I_W^- = \tau'_{A-}\tau'_{B-}\tau'_{C+} + \tau'_{A-}\tau'_{B+}\tau'_{C-} + \tau'_{A+}\tau'_{B-}\tau'_{C-}$$

$$2I_W^3 = \frac{1}{4}(\tau'_{A3} + \tau'_{B3} + \tau'_{C3} - 3\tau'_{A3}\tau'_{B3}\tau'_{C3}) \ .$$

More generally, by using the trident products of $\vec{\tau}_\alpha(\vec{\tau}'_\alpha)$ and $1_\alpha(1'_\alpha)$ one could obtain U(8)(U(8)') algebra. The grand unification can be established by gauging the semi-simple group SU(8)×SU(8)'.

Our preon model offers a new standpoint for discussing the breakdown of symmetries in addition to Higgs mechanism. For example one may assume a coupling between different degrees of freedom in the mass matrix

$$M = 3a + a\vec{\tau}'_B\cdot\vec{\tau}'_C + b\vec{\tau}'_C\cdot\vec{\tau}'_A + c\vec{\tau}'_A\cdot\vec{\tau}'_B \ .$$

If a~b~c, the mass gap between the 3rd generation and the other two is deduced. The Cabibbo angle – mass relation may turn out, too.

The authors are indebted to Prof. T. Lu for his encouragement.

ON A DIPOLE ELECTRIC MOMENT OF FERMION IN THE KLEIN-KAŁUŻA THEORY

M. W. Kalinowski
Institute of Philosophy, Polish Academy of Sciences,
00-330 Warsaw, Nowy Świat 72, Poland

Abstract

A generalization of minimal coupling is proposed and the usual Dirac equation is generalized within the Klein-Kałuża theory. The dipole electric moment of fermion of order 10^{-31} cm is obtained. The difficulties which appear in Thirring's approach to the problem are avoided.

Introduction

In the paper we consider Dirac equation in the Klein-Kałuża theory describes both gravitational and electromagnetic fields in a geometrical framework. However, in such a theory "interference" effects between both fields are absent. But one may obtain some gravitational-electromagnetical effects if one introduces spinor fields on 5-dimensional manifold and generalize minimal coupling scheme. In this way we may obtain new effect i.e. dipole electric moment of fermion. A similar problem was studied by W. Thirring[1]. However, Thirring's results were obtained at some price namely, there exist an unwanted minimal rest mass of fermion /of order 1 μg/ in his theory. In our approach we avoid this trouble.

Let us introduce the principal fibre bundle P over space-time E with the structure group U (1) and with the projection π . We define on P a connection α /electromagnetic connection/.

For a curvature Ω of α we have

$$\Omega = \pi^{*}\left(\frac{1}{2} F_{\mu\nu}\ \bar{\theta}^{\mu} \wedge \bar{\theta}^{\nu}\right) \qquad /1.1/$$

where π^{*} is a horizontal lift and $F_{\mu\nu}$ is a strength of electromagnetic field.

We introduce on P a frame

$$\theta^{A} = \left(\pi^{*}(\bar{\theta}^{\alpha}),\ \theta^{5} = \lambda\alpha\right),\ \lambda = 2\frac{\sqrt{G}}{c^{2}} > 0 \qquad /1.2/$$

where $\bar{\theta}^{\alpha}$ is a frame on E.

In the Klein-Kałuża theory P is metrized and has a Riemann connection ω_{AB}. Metric tensor on P is as follow[2]

$$\gamma = \pi^* g - \theta^5 \otimes \theta^5 \qquad /1.3/$$

where g is a metric tensor on E and has signature /- - - +/, γ has signature /- - - + -/.

The following notations is used all along the paper. Capital latin indeces A, B, C run 1, 2, 3, 4, 5, lower greec indeces α, β, γ run 1, 2, 3, 4; /E,g/ is a manifold with a metric tensor g and a Riemann connection

We also introduce a dual frame (ξ_A)

$$\gamma(\xi_A) = \gamma_{AB}\theta^B, \text{ where } \xi_A = (\xi_\alpha, \xi_5) \text{ and} \qquad /1.4/$$

$$\mathcal{L}_{\xi_5}\gamma = 0 \qquad /1.5/$$

Thus ξ_5 is Killing's vector of metric γ [2]. Let us now introduce a Riemann connection ω_{AB} on P and exterior covariant derivation D with respect to ω_{AB}. We suppose

$$D\gamma_{AB} = 0, \quad D\theta^A = 0 \qquad /1.6/$$

and we get

$$\omega_{AB} = \left(\begin{array}{c|c} \pi^*(\bar{\omega}_{\alpha\beta}) + \frac{\lambda}{2}\pi^*(F_{\alpha\beta})\theta^5 & \frac{\lambda}{2}\pi^*(F_{\alpha\gamma}\bar{\theta}^\gamma) \\ \hline -\frac{\lambda}{2}\pi^*(F_{\beta\gamma}\bar{\theta}^\gamma) & 0 \end{array} \right) \qquad /1.7/$$

On a manifold /E,g/ i.e. on a space-time we introduce a Levi--Civita symbol and dual Cartan's base

$$\bar{\eta}_{\alpha\beta\gamma\delta}, \quad \bar{\eta}_{1234} = \sqrt{-\det g} \qquad /1.8/$$

$$\bar{\eta}_\alpha = \frac{1}{2\cdot 3}\bar{\theta}^\delta \wedge \bar{\theta}^\gamma \wedge \bar{\theta}^\beta \bar{\eta}_{\alpha\beta\gamma\delta} \qquad \bar{\eta} = \frac{1}{4}\bar{\theta}^\alpha \wedge \bar{\eta}_\alpha \qquad /1.9/$$

For details concerning elements of geometry we deal with in this section see[3,4,5,6,7,8,9,10]

2. Dirac equation in the Klein-Kałuża theory

In this section we deal with a generalization of Dirac's equation on manifold P /a metricized electromagnetic bundle/. We introduce several kind of derivatives and using them we get a generalization of Dirac's equation.

We define spinor fields Ψ , $\bar{\Psi}$ on P, $\Psi : P \to \mathbb{C}^4$. For Ψ , $\bar{\Psi}$ we have we have

$$\Psi(pg_1) = \sigma(g_1^{-1})\Psi(p) , \quad \sigma \in \mathcal{L}(\mathbb{C}^4)$$

$$\bar{\Psi}(pg_1) = \bar{\Psi}(p)\sigma(g_1) \qquad /2.1/$$

where $p = (x,g) \in P$, g, $g_1 \in U(1)$.

Let us define a gauge derivative d_1 of field Ψ. One gets

$$d\Psi = \xi_\mu \Psi \theta^\mu + \xi_5 \Psi \theta^5 \text{ and } d_1\Psi = \text{hor } d\Psi = \xi_\mu \Psi \theta^\mu \qquad /2.2/$$

Let $\gamma_\mu \in \mathcal{L}(\mathbb{C}^4)$ be Dirac's matrices obeying the conventional relation

$$\{\gamma_\mu, \gamma_\nu\} = 2\,\eta_{\mu\nu} \qquad /2.3/$$

/where $\eta_{\mu\nu}$ is Minkowski's tensor of a sygnature $(- - - +)$/ and

$$\sigma_{\mu\nu} = \frac{1}{8}[\gamma_\mu, \gamma_\nu] \qquad /2.4/$$

It is easy to see that

$$\{\gamma_A, \gamma_B\} = 2\,\bar{g}_{AB} \qquad /2.5/$$

where $g_{AB} = \text{diog}(-1,-1,-1,-1,+1,-1)$, $\gamma^5 = \gamma^1\gamma^2\gamma^3\gamma^4 \in \mathcal{L}(\mathbb{C}^4)$ and $\gamma^A = (\gamma^\alpha, \gamma^5)$. We perform a infinitesimal change of frame θ^A

$$\theta^{A'} = \theta^A + \delta\theta^A = \theta^A - \varepsilon^A{}_B\theta^B , \; \varepsilon_{AB} + \varepsilon_{BA} = 0 \qquad /2.6/$$

If spinor field Ψ corresponds to θ^A, and Ψ' to $\theta^{A'}$ then we have:

$$\Psi' = \Psi + \delta\Psi = \Psi - \varepsilon^{AB}\hat{\sigma}_{AB}\Psi \qquad \bar{\Psi}' = \bar{\Psi} + \delta\bar{\Psi} = \bar{\Psi} + \bar{\Psi}\hat{\sigma}_{AB}\varepsilon^{AB} \qquad /2.7/$$

where $\hat{\sigma}_{AB} = \frac{1}{8}[\gamma_A, \gamma_B]$.

Notice that the dimension of spinor space for 2 n-dimensional space is 2^n and it is the same for a $(2n + 1)$- dimensional one. We take a spinor field for a 5-dimensional space P and assume that dependence on 5th dimension is trivial, i.e. Eq. /2.1/ holds. Taking a section we obtain spinor fields on E /of the same dimension of spinor space/.

Let us introduce a Langrange 4-form od Dirac field with a minimal coupling to an electromagnetic field

$$/2.8/ \quad \mathcal{L}_D(\Psi, \Psi, d_1) = i\frac{\hbar c}{2}(\bar{\Psi}\, l \wedge d_1\Psi + d_1\bar{\Psi} \wedge l\,\Psi) + m\eta\bar{\Psi}\Psi , \quad l = \gamma_\mu \bar{\eta}^\mu$$

And let us consider a covariant derivation of spinor fields Ψ and $\bar{\Psi}$ on P with respect to ω_{AB}

$$D\Psi = d\Psi + \omega^{AB}\hat{\sigma}_{AB}\Psi \quad , \quad D\bar{\Psi} = d\bar{\Psi} - \omega^{AB}\bar{\Psi}\hat{\sigma}_{AB} \qquad /2.9/$$

Now introduce derivatives $\mathfrak{D}$ i.e. "gauge" derivatives of new kind. These derivatives may be treated as a generalization of minimal coupling between spinor and gravitatio -electromagnetic field on P

$$\mathfrak{D}\Psi = \text{hor}\, D\Psi \quad , \quad \mathfrak{D}\bar{\Psi} = \text{hor}\, D\bar{\Psi} \qquad /2.10/$$

Using /1.7/ we obtain:

$$\mathfrak{D}\Psi = \bar{\mathfrak{D}}\Psi - \frac{\lambda}{8}\bar{F}^{\alpha}{}_{\mu}[\gamma_\alpha,\gamma_5]\Psi\theta^{\mu}, \quad \mathfrak{D}\bar{\Psi} = \bar{\mathfrak{D}}\bar{\Psi} + \frac{\lambda}{8}F^{\alpha}{}_{\mu}\bar{\Psi}[\gamma_\alpha,\gamma_5]\theta^{\mu} \qquad /2.11/$$

The derivative $\bar{\mathfrak{D}}$ is a covariant derivative with respect to both $\pi^*(\omega\alpha\beta)$ and "gauge" at once. It introduces an interaction between electromagnetic and gravitational fields with Dirac's spinor in a classical form.

Now let us turn to the Lagrange form /2.8/ and pass from "d_1" to „$\mathfrak{D}$"

Using formulae /2.12/ we obtain after some algebra

$$\mathcal{L}_D(\Psi,\bar{\Psi},\mathfrak{D}) = \mathcal{L}_D(\Psi,\bar{\Psi},\bar{\mathfrak{D}}) - i\,\frac{2\sqrt{G}}{c}\hbar F^{\mu\nu}\bar{\Psi}\gamma_5\sigma_{\mu\nu}\Psi\bar{\eta} \qquad /2.12/$$

Lagrangian $\mathcal{L}_D(\Psi,\bar{\Psi},\bar{\mathfrak{D}})$ describes the interaction between both electromagnetic and gravitational fields with spinor fields and is of well known classical form. The new term in /2.12/ is an interaction of electromagnetic field with a dipole electric moment of fermion of the value:

$$-\frac{2\sqrt{G}}{c}\hbar = -2\frac{l_{pl}}{\sqrt{\alpha}}q \simeq -0{,}95\cdot 10^{-31} q\,[\text{cm}] \qquad /2.13/$$

α - fine structure constant,

q - elementary charge, l_{pl} is a Planck length.

The additional term in lagrangian /2.12/ breaks Π C or T symmetries similarly as in Thirring's theory[1/], but Thirring defines operator Π C is a different way. Of course this breaking is very weak and it cannot be linked to nonconservation of Π C in decays of mesons K. Neverthelessnon-conservation of Π C in these decays has a good support in Kobayashi-Maskava model: the appearance of this dipole electrical moment should be rather related to quite different and more fundamental gravitational--electromagnetic effects than to week interactions of hadrons.

REFERENCES

1. Thirring W.: Act. Ph. Austr. Suppl. IX, 256-271 /1972/.
2. Trautman A.: Rep. of Math. Ph. 1, 29-62 /1970/.
3. Bergman P.G.: Introduction to the theory of relativity, Prentice Hall, New York 1942.
4. Kaluza T., Sitzungsber Preuss. Akad. Wiss., 966-1002 /1921/.
5. Lichnerowicz A.: Thèorie relativistes de la gravitation et de l'électromagnetisme, Masson Paris 1955.
6. Rayski J.: Act. Ph. Polonica 31, 87-97 /1965/.
7. Tonnelat M.A.: Les theories unitaires de l'électromagnetisme et de la gravitation, Gautier-Villars,Paris /1965/.
8. Kobayashi S., Nomizu K.: Foundations of differential geometry Vol. I and Vol. II, Interscience, New York /1963/.
9. Lichnerowicz A.: Thèorie globale des connexions et de groupe d holonomie, Ed. Cremonese Rome /1955/.
10. Rayski J.: Acta Ph. Austr. Suppl. XVIII, 463-469 /1977/.

Acknowledgement

I thank very much Prof. M. Konuma for an invitation to present these results during 4th Kyoto Summer Institute and financial support.

Closing Address

Ziro MAKI
Research Institute for Fundamental Physics
Kyoto University, Kyoto 606, Japan

Dear Participants!

The 4th Kyoto Summer Institute is now coming to close. My duty here is to address some words at the end of the schedule.

It seems extremely difficult for me to summarize the whole contents of lectures and seminars delivered in these five days. So, instead, let me try to show you, for entertainment, a few pictures which would describe my personal impression of this Summer Institute.

(Figure 1)

My "Fairy Tales of the GUT Age" begins with a well-known scene from the famous fables of Aesop's; a hare (GUT) (兎) is climbing rather quickly since it started in 1974. On the other hand another animal, a tortoise (亀), is also climbing along another pathway to win the race. Here, I suggest to modify the standard scenario, and to look for other possible solutions; Since we don't know what will happen because we have been told that there live many Devils (鬼) behind deep forests (of Higgs trees) near the top of Mt. Daimonji (大文字), which you have seen every day through windows of this lecture room. I would like to suggest two conceivable ends of the story, (a) and (b).

(Figure 2)

The case (b) looks a little bit "unnatural" to me, but I don't know which end is better. Anyway, so much for the fantasy!

On behalf of the Organizing Committee, I would like to express our deep gratitude to all the lecturers for giving us well-prepared and extremely stimulating lectures; we are also very grateful to all the seminar speakers and participants for their active contributions. Finally, we wish to thank all the secretarial staff of the Institute. Without their earnest help the Kyoto Summer Institute '81 could have never been successful.

Fairy Tales of the GUT Age

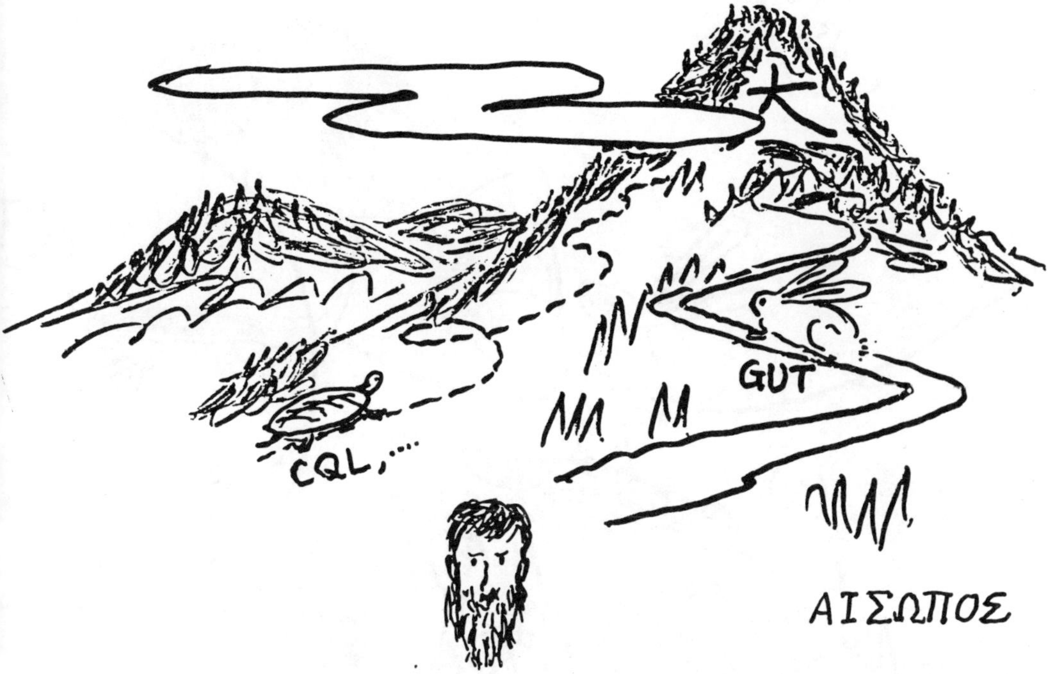

Picture 1

Picture 2

List of Participants

AIZAWA Nobuyuki	Dept. of Nuclear Engineering, Kyoto Univ., Kyoto 606, Japan
AOKI Ken-ichi	Dept. of Physics, Kyoto Univ., Kyoto 606, Japan
ARISUE Hiroaki	Dept. of Physics, Kyoto Univ., Kyoto 606, Japan
BANDO Masako	Dept. of Physics, Kyoto Univ., Kyoto 606, Japan
CHIKASHIGE Yūichi	Faculty of Engineering, Seikei Univ., Musashino, Tokyo 180, Japan
DOI Masaru	Osaka College of Pharmacy, Matsubara, Osaka 580, Japan
EBATA Takeshi	Dept. of Physics, College of General Education, Tohoku Univ., Sendai 980, Japan
FUJII Kanji	Dept. of Physics, Hokkaido Univ., Sapporo 060, Japan
FUJIKAWA Kazuo	Inst. for Nuclear Study, Univ. of Tokyo, Tanashi, Tokyo 188, Japan
FUJIWARA Toshiaki	Dept. of Physics, Kyoto Univ., Kyoto 606, Japan
FUKAI Tomoki	Physics Dept., Waseda Univ., Shinjuku, Tokyo 160, Japan
FUKUDA Reijiro	Research Inst. for Fundamental Physics, Kyoto Univ., Kyoto 606, Japan
FUKUI Takao	Dept. of Liberal Arts, Dokkyo Univ., Soka-city, Saitama 340, Japan
GEORGI Howard	Dept. of Physics, Harvard Univ., Cambridge, MA 02138 U.S.A.
HAMAMOTO Shinji	Dept. of Physics, Toyama Univ., Toyama 930, Japan
HARA Osamu	Atomic Energy Research Inst., College of Science and Technology, Nihon Univ., Tokyo 101, Japan
HATA Hiroyuki	Dept. of Physics, Kyoto Univ., Kyoto 606, Japan
HAYASHI Hirofumi	Physics Dept., Shizuoka Univ., Ohya, Shizuoka 422, Japan
HAYASHI Mitsuo	Dept. of Physics, Tokai Univ., Numazu 410, Japan
HIGUCHI Atsushi	Dept. of Physics, Kyoto Univ., Kyoto 606, Japan
HIKASA Ken-ichi	Dept. of Physics, Univ. of Tokyo, Hongo, Tokyo 113, Japan

HIOKI Zenrō	Research Inst. for Fundamental Physics, Kyoto Univ., Kyoto 606, Japan
HIRAI Shiro	Dept. of Nuclear Engineering, Kyoto Univ., Kyoto 606, Japan
HORIBE Minoru	Physics Dept., Fukui Univ., Fukui 910, Japan
ICHINOSE Shoichi	Inst. of Physics, Univ. of Tsukuba, Ibaraki 305, Japan
IDA Masakuni	Dept. of Physics, Kobe Univ., Kobe 657, Japan
IGARASHI Yuji	Univ. Dortmund, Theoretical Physics III, Postfach 500500, D-4600 Dortmund 50, West Germany
IGI Keiji	Dept. of Physics, Univ. of Tokyo, Hongo, Tokyo 113, Japan
IKEDA Minoru	Dept. of Physics, Osaka City Univ., Osaka 558, Japan
IKETANI Kazuhito	Dept. of Physics, Kyushu Univ., Fukuoka 812, Japan
INOUE Kenzo	Dept. of Physics, Kyushu Univ., Fukuoka 812, Japan
INOUE Takuo	Dept. of Physics, Kyoto Univ., Kyoto 606, Japan
ISHIKAWA Kiyoshi	Dept. of Physics, Osaka Univ., Toyonaka, Osaka 560, Japan
ITO Hiroaki	Dept. of Physics, Nagoya Univ., Nagoya 464, Japan
ITZYKSON Claude	CEN Saclay, Division de la Physique Theorique, B.P.N°2, 91190 Gif-sur-Yvette, FRANCE
KAI Naoyuki	Dept. of Physics, Univ. of Tokyo, Hongo, Tokyo 113, Japan
KAKAZU Kiyotaka	Dept. of Physics, Univ. of the Ryukyus, Nakajo-mura, Okinawa 901-24, Japan
KALINOWSKI Marek W.	Inst. of Philosophy, Polish Academy of Sciences, Warsaw 00-330 NOWYSWIAT 72, Poland
KASHIBAYASHI Shigeyuki	Dept. of Physics, Osaka City Univ., Osaka 558, Japan
KATO Mitsuhiro	Dept. of Physics, Kyoto Univ., Kyoto 606, Japan
KATUYA Mituaki	Lab. of Physics, Shizuoka Women's Univ., Shizuoka 422, Japan
KAWABE Rokuo	Dept. of Physics, Osaka Medical College, Takatsuki, Osaka 569, Japan

KAWAI Ei-ichiro	Physics Dept., Ehime Univ., Matsuyama, Ehime 790, Japan
KAWAI Hikaru	Dept. of Physics, Univ. of Tokyo, Hongo, Tokyo 113 Japan
KAWATI Syôzi	Dept. of Physics, Kwansei Gakuin Univ., Uegahara, Nishinomiya 662, Japan
KAYAMA Yoshihiko	Dept. of Physics, Kobe Univ., Kobe 657, Japan
KHANNA Mohinder P.	Dept. of Physics, Univ. of Alberta, Edmonton, Alberta, T6G 2J1, CANADA
KIANG David B.I.	Physics Dept., Dalhousie Univ., Halifax, N.S. CANADA B3H 3J5
KIM Jihn Eui	Dept. of Physics, Seoul National Univ., Seoul 151, Korea
KITANI Kohei	Physics Dept., Sagami Women's Univ., Sagamihara 228, Japan
KITAZOE Tetsuro	Dept. of Physics, Kobe Univ., Kobe 657, Japan
KIZUKURI Yoshiki	Physics Dept., Waseda Univ., Shinjuku, Tokyo 160, Japan
KOBAYASHI Akizo	Faculty of Education, Niigata Univ., Igarashi, Niigata 950-21, Japan
KOGO Kumiko	Dept. of Physics, Univ. of Tokyo, Hongo, Tokyo 113, Japan
KOH In-Gyu	Physics Dept., Sogang Univ., C.P.O. Box 1142, Seoul, Korea
KOIDE Yoshio	Lab. of Physics, Shizuoka Women's Univ., Shizuoka 422, Japan
KOIKAWA Takao	Physics Dept., Hiroshima University, Hiroshima 730 Japan
KOMATSU Hiromasa	Dept. of Physics, Kyushu Univ., Fukuoka 812, Japan
KONASHI Hiroshi	Dept. of Physics, Osaka Univ., Toyonaka, Osaka 560, Japan
KONDO Hiroki	Dept. of Physics, Faculty of Science and Engineering, Saga Univ., Saga 840, Japan
KONUMA Michiji	Research Inst. for Fundamental Physics, Kyoto Univ., Kyoto 606, Japan

KUBOTA Takahiro	Dept. of Physics, Univ. of Tokyo, Hongo, Tokyo 113, Japan
KUGO Taichiro	Dept. of Physics, Kyoto Univ., Kyoto 606, Japan
KURAMOTO Tetsuji	Dept. of Physics, Kyoto Univ., Kyoto 606, Japan
MAEHARA Toshinobu	Dept. of Physics, Hiroshima Univ. Hiroshima 730, Japan
MAKI Ziro	Research Inst. for Fundamental Physics, Kyoto Univ., Kyoto 606, Japan
MASKAWA Toshihide	Research Inst. for Fundamental Physics, Kyoto Univ., Kyoto 606, Japan
MATSUDA Masahisa	Dept. of Physics, Aichi Univ. of Education, Kariya 448, Japan
MATSUDA Satoshi	Dept. of Physics, Kyoto Univ., Kyoto 606, Japan
MATUMOTO Ken-iti	Dept. of Physics, Toyama Univ., Toyama 930, Japan
MIDORIKAWA Shoichi	Research Inst. for Fundamental Physics, Kyoto Univ., Kyoto 606, Japan
MINAMIKAWA Toshiyuki	Dept. of Physics, Tokyo Univ. of Mercantile Marine, Tokyo 135, Japan
MIYATA Hidenori	Dept. of Physics, Kwansei Gakuin Univ., Uegahara, Nishinomiya 662, Japan
MIYATA Hideo	Kanazawa Inst. of Technology, 7-1 Ogigaoka, P.O. Kanazawa-South, Ishikawa 921, Japan
MORITA Katsusada	Dept. of Physics, Nagoya Univ., Nagoya 464, Japan
MURAYAMA Akihiro	Physics Dept., Shizuoka Univ., Ohya, Shizuoka 422, Japan
MUTA Taizo	Research Inst. for Fundamental Physics, Kyoto Univ., Kyoto 606, Japan
NAKAHARA Haruhiko	Inst. of Physics, University of Tokyo, Komaba, Tokyo 153, Japan
NAKAJIMA Hideo	Dept. of Engineering Math., Faculty of Engineering, Utsunomiya Univ., Utsunomiya 321, Japan
NAKANISHI Noboru	Research Inst. for Mathematical Sciences, Kyoto Univ., Kyoto 606, Japan
NAKAMURA Seitaro	Dept. of Physics, Tokai Univ., Hiratsuka, Kanagawa 259-12, Japan

NAKAYAMA Ryuichi	Dept. of Physics, Univ. of Tokyo, Hongo, Tokyo 113, Japan
NANOPOULOS Demetrios V.	CERN, CH-1211 Geneve 23, Switzerland
NELSON Charles A.	Research Inst. for Fundamental Physics, Kyoto Univ., Kyoto 606, Japan and Physics Dept., SUNY, Binghamton, New York 13901, U.S.A.
NINOMIYA Masao	Rutherford Lab., Chilton, Didcot, Oxfordshire OX11 0QX, England
NISHIURA Hiroyuki	Research Inst. for Fundamental Physics, Kyoto Univ. Kyoto 606, Japan
NIUYA Takayuki	Dept. of Nuclear Engineering, Kyoto Univ., Kyoto 606, Japan
NOJIRI Shin-ichi	Dept. of Physics, Kyoto Univ., Kyoto 606, Japan
OGAWA Kaku	Dept. of Physics, Kyoto Univ., Kyoto 606, Japan
OHKUWA Yoshiaki	Dept. of Physics, Osaka Univ., Toyonaka, Osaka 560, Japan
OHNUKI Yoshio	Dept. of Physics, Nagoya Univ., Nagoya 464, Japan
OKABAYASHI Masao	Faculty of Education, Shiga Univ., Ōtsu, Shiga 520, Japan
OHTA Nobuyoshi	Inst. of Physics, Univ. of Tokyo, Komaba, Tokyo 153, Japan
OHTANI Yasuaki	Dept. of Physics, Kyushu Univ., Fukuoka 812, Japan
OTSUKI Shoichiro	Dept. of Physics, Kyushu Univ., Fukuoka 812, Japan
OHYA Katsuhiko	Fukuyama City Junior Col. for Women, Fukuyama, Hiroshima 720, Japan
OKA Takamitsu	Research Inst. for Theoretical Physics, Hiroshima Univ., Takehara, Hiroshima 725, Japan
OKADA Yasuhiro	Dept. of Physics, Univ. of Tokyo, Hongo, Tokyo 113, Japan
OKAZAKI Takashi	Dept. of Physics, Kanazawa Univ., Kanazawa 920, Japan
OSHIMO Noriyuki	Physics Dept., Waseda Univ., Shinjuku, Tokyo 160, Japan

OTOKOZAWA Jun	Dept. of Science & Engineering, Nihon Univ., Kanda, Tokyo 101, Japan
OZAKI Kazuhiko	Osaka Inst. of Technology, Osaka 535, Japan
PAKVASA Sandip	Physics Dept., Univ. of Hawaii, Honolulu, HI 96822, U.S.A.
PECCEI Roberto	Max Planck Inst. für Physik und Astrophysik, Fohringer Ring 6, Postfach 40 12 12, D-8000 Munich 40, Germany
PHUA Kok-Khoo	Research Inst. for Fundamental Physics, Kyoto Univ., Kyoto 606, Japan and Dept. of Physics, Univ. of Singapore, Bukit Timah Road, Singapore 10, SINGAPORE
RAJASEKARAN Guruswany	National Laboratory for High Energy Physics (KEK), Tsukuba, Ibaraki 305, Japan and Theoretical Physics Dept., Univ. of Madras, Guindy Campus, Madras 600025, India
RAMOND Pierre	Dept. of Physics, Univ. of Florida, 215 Williamson Hall, Gainesville, Florida 32611, U.S.A.
SAKAGAMI Masaaki	Dept. of Physics, Osaka Univ., Toyonaka, Osaka 560, Japan
SAKAMOTO Makoto	Dept. of Physics, Kyushu Univ., Fukuoka 812, Japan
SATO Katsuhiko	Dept. of Physics, Kyoto Univ., Kyoto 606, Japan
SATO Yuichi	Dept. of Physics, Kyoto Univ., Kyoto 606, Japan
SAWAYANAGI Hirofumi	Dept. of Physics, Hokkaido Univ. Sapporo 060, Japan
SHIN-MURA Mamoru	Dept. of Physics, Nagoya Univ., Nagoya 464, Japan
SO Hiroto	Dept. of Physics, Kyushu Univ., Fukuoka 812, Japan
SOGAMI Ikuo	Dept. of Physics, Kyoto Sangyo Univ., Kyoto 603, Japan
SONG Hi Sung	Dept. of Physics, Seoul National Univ., Seoul 151, Korea
SUGIYAMA Yûki	Dept. of Physics, Nagoya Univ., Nagoya 464, Japan
SUZUKI Chitao	Dept. of Physics, Nagoya Univ., Nagoya 464, Japan
SUZUKI Tsuneo	Dept. of Physics, Kanazawa Univ., Kanazawa 920, Japan

TAKAGI Fujio	Dept. of Physics, Tohoku Univ., Sendai 980, Japan
TAKASUGI Eiichi	Inst. of Physics, Coll. of General Education, Osaka Univ., Toyonaka, Osaka 560, Japan
TAKESHITA Seiichiro	Dept. of Physics, Kyushu Univ., Fukuoka 812, Japan
TAKEUCHI Yousuke	Physics Dept., Ehime Univ., Matsuyama, Ehime 790, Japan
TANAKA Sho	Dept. of Physics, Kyoto Univ., Kyoto 606, Japan
TANIMOTO Morimitsu	Physics Dept., Ehime Univ., Matsuyama, Ehime 790, Japan
TOMOZAWA Yukio	The Harrison M. Randall Lab. of Physics, Univ. of Michigan, Ann Arbor, Michigan 48109, U.S.A.
TORIU Takashi	Dept. of Physics, Kyoto Univ., Kyoto 606, Japan
TSUCHIDA Tetsuya	Dept. of Physics, Ibaraki Univ., Mito, Ibaraki 310, Japan
UEHARA Shozo	Dept. of Physics, Kyoto Univ., Kyoto 606, Japan
URANISHI Yoshie	Dept. of Physics, Osaka City Univ., Osaka 558, Japan
USHIO Kenichi	Dept. of Physics, Osaka Univ., Toyonaka, Osaka 560, Japan
WATAMURA Satoshi	Inst. of Physics, Coll. of General Education, Univ. of Tokyo, Komaba, Tokyo 153, Japan
WATANABE Tadashi	Dept. of Physics, Kobe Univ., Kobe 657, Japan
YAMAMOTO Katsuji	Dept. of Nuclear Engineering, Kyoto Univ., Kyoto 606, Japan
YAMAMOTO Noboru	Dept. of Physics, Osaka Univ., Toyonaka, Osaka 560, Japan
YAMANAKA Yoshiya	Physics Dept., Waseda Univ., Shinjuku, Tokyo 160, Japan
YAMAWAKI Koichi	Dept. of Physics, Nagoya Univ., Nagoya 464, Japan
YANAGIDA Tsutomu	Physics Dept., Coll. of General Education, Tohoku Univ., Sendai 980, Japan
YANG Guo-Chen	Hebei Inst. of Technology, Tianjin, China
YOSHIMURA Motohiko	National Lab. for High Energy Physics (KEK), Tsukuba, Ibaraki 305, Japan
ZEE Anthony	Dept. of Physics, FM-15, Univ. of Washington, Seattle, Washington 98195, U.S.A.

Program

Speakers and chairmen in parentheses

Time	June 29 (Mon.)	June 30 (Tues.)	July 1 (Wed.)	July 2 (Thurs.)	July 3 (Fri.)
9:00–10:15	Registration				
9:10–10:20		Zee (K. Igi)	Georgi (K. Matumoto)	Ramond (K. Fujikawa)	Itzykson (R. Fukuda)
10:15–10:30	Coffee break				
10:20–10:50		Coffee break	Coffee break	Coffee break	Coffee break
10:30–10:50	Opening (S. Tanaka)				
10:50–12:00	Nanopoulos (S. Tanaka)	Yoshimura (K. Igi)	Zee (K. Matumoto)	Yoshimura (K. Fujikawa)	Georgi (R. Fukuda)
12:00–13:30	Lunch	Lunch	Lunch	Lunch	Lunch
13:30–14:40	Ramond (S. Matsuda)	Nanopoulos (S. Otsuki)		Nanopoulos (T. Suzuki)	Sato (N. Nakanishi)
14:40–15:10	Coffee break	Coffee break			
15:10–16:20	Georgi (T. Muta)	Ramond (K. Fujii)		Tomozawa (M. Ida)	Zee (Y. Ohnuki)
16:20–16:30	Coffee break	Coffee break			concluding session (Y. Ohnuki) (16:20–16:40)
16:30–17:40	Peccei (T. Muta)	Pakvasa (K. Fujii)		discussion sessions (M. Ida, T. Kugo) (16:30–18:15)	
18:00–20:00	Welcome party				

List of Topics of the Previous Kyoto Summer Institutes and Proceedings

1. Particle Physics and Accelerator Projects, September 1-5, 1978

   Proceedings ed. by M. Ida, T. Kamae and M. Kobayashi
   Published by the Organizing Committee of Kyoto Summer Institute,
   Research Institute for Fundamental Physics, Kyoto University,
   Kyoto 606, Japan, 1978.

2. Physics of Low Dimensional Systems, September 8-12, 1979

   Proceedings ed. by Y. Nagaoka and S. Hikami
   Published by Publication Office, Progress of Theoretical Physics,
   c/o Yukawa Hall, Kyoto University, Kyoto 606, Japan, 1979.

3. Fundamental Physics of Amorphous Semiconductors, September 8-11, 1980.

   Proceedings ed. by F. Yonezawa
   Published by Springer-Verlag, Berlin, Heidelberg, New York, 1981.